Introduction to Matrix Analytic Methods in Stochastic Modeling

ASA-SIAM Series on Statistics and Applied Probability

The ASA-SIAM Series on Statistics and Applied Probability is published jointly by the American Statistical Association and the Society for Industrial and Applied Mathematics. The series consists of a broad spectrum of books on topics in statistics and applied probability. The purpose of the series is to provide inexpensive, quality publications of interest to the intersecting membership of the two societies.

Latouche, G. and Ramaswami, V., *Introduction to Matrix Analytic Methods in Stochastic Modeling*

Peck, R., Haugh, L. D., and Goodman, A., *Statistical Case Studies: A Collaboration Between Academe and Industry, Student Edition*

Peck, R., Haugh, L. D., and Goodman, A., *Statistical Case Studies: A Collaboration Between Academe and Industry*

Barlow, R. E., *Engineering Reliability*

Czitrom, V. and Spagon, P. D., *Statistical Case Studies for Industrial Process Improvement*

Introduction to Matrix Analytic Methods in Stochastic Modeling

G. Latouche
Université Libre de Bruxelles
Brussels, Belgium

V. Ramaswami
AT&T Labs
Holmdel, New Jersey

siam.
Society for Industrial and Applied Mathematics
Philadelphia, Pennsylvania

ASA
American Statistical Association
Alexandria, Virginia

10 9 8 7 6 5 4 3 2 1

Library of Congress Cataloging-in-Publication Data

Latouche, G. (Guy)
Introduction to matrix analytic methods in stochastic modeling / G. Latouche, V. Ramaswami.
p. cm. -- (ASA-SIAM series on statistics and applied probability)
Includes bibliographical references and index.
ISBN 0-89871-425-7 (pbk.)
1. Markov processes. 2. Queuing theory. 3. Matrices.
I. Ramaswami, V. II. Title. III. Series
QA274.7.L38 1999
519.2'3.--dc21 98-48647

Contents

Preface

Matrix analytic methods constitute a success story, illustrating the enrichment of a science, applied probability, by a technology, that of digital computers. Marcel Neuts has played a seminal role in these exciting developments, promoting numerical investigation as an essential part of the solution of probability models. He wrote in 1973,

> "To do work in computational mathematics is ... a commitment to a more demanding definition of what constitutes the solution to a mathematical problem. When done properly, it conforms to the highest standard of scientific research." (see [73, Page 20]).

This had been long accepted among numerous scientific communities but was not at the time the prevalent view among applied probabilists. Neuts's program of research led in a few years to the celebrated "matrix-geometric distribution" and "phase-type processes," brought nowadays under the umbrella title of "matrix analytic methods."

The excitement one feels when dealing with that subject stems from the synergy resulting from keeping algorithmic considerations in the forefront when solving stochastic problems. The connection was already noted in 1959 by Kemeny and Snell who wrote,

> "One of the practical advantages of this new treatment of the subject is that these elementary matrix operations can easily be programmed for a high-speed computer. The authors have developed a pair of programs ... which will find a number of interesting quantities for a given problem directly from the transition matrix. These programs were invaluable

> in the computation of examples and in the checking of conjectures for theorems.
>
> A significant feature of the new approach is that it makes no use of the theory of eigen-values. The authors found, in each case, that the expressions in matrix form are simpler than the corresponding expressions usually given in terms of eigen-values. This is presumably due to the fact that the fundamental matrices have direct probabilistic interpretations, while the eigen-values do not." (see [39, Page v]).

Matrix analytic methods are popular as modeling tools because they give one the ability to construct and analyze, in a unified way and in an algorithmically tractable manner, a wide class of stochastic models. The methods are applied in several areas, of which the performance analysis of telecommunication systems is one of the most notable at the present time. The methods also offer to mathematicians the delight of discovering the stochastic process at work in the computational procedure, as when one finds that the successive steps in an iterative algorithm have a probabilistic significance.

This book presents the basic mathematical ideas and algorithms of the matrix analytic theory. Our approach uses probabilistic arguments to the fullest extent and allows us to show clearly the unity of argumentation in the whole theory. With this in mind, we have had to develop new proofs for some of the results so as to shed the dependence on the Perron–Frobenius theory of finite dimensional matrices, fixed point theorems of functional analysis, etc., which are prevalent in some of the early texts on the subject. While most of these probabilistic proofs have been published over the years, a few appear here for the first time.

The methods themselves are presented within the simpler framework of quasi-birth-and-death processes (QBDs). These are Markov processes in two dimensions, the *level* and the *phase*, such that the process does not jump across several levels in one transition. The advantage of working with QBDs, instead of the more general GI/M/1- and M/G/1-type Markov chains, is that we may present the basic features of matrix analytic methods without being encumbered by side issues which arise as soon as the processes are put into some application framework. The restriction to QBDs, however, is not unduly limiting; we show in Chapter 13 that the theory for the more general class of models can be deduced from the QBD analysis.

We separate as much as possible the elucidation of the *structure* of some quantity from the determination of its *numerical value.* One advantage in presenting the probabilistic results separate from the algorithms is that we make it clearly appear that the structural properties do not depend on whether there are finitely or infinitely many values for the phase dimension: only when doing actual matrix computations does it become necessary to deal with a finite state space for the phase.

A major hurdle to overcome, in order to delve into matrix analytic theory, is to start thinking in terms of blocks of states and transition submatrices, instead of keeping track of individual states and scalar transition probabilities. We gently lead the reader in that direction by introducing one aspect of the theory at a time. To begin with, we present several simple examples from queueing theory in the first chapter. They are intended to help the reader become familiar with the matrix notations which we use throughout the text, as well as to get some feeling for the variety of models which are hidden by the general block notations. Chapter 1 may safely be skipped by readers who wish to jump immediately into the mathematics.

We next give two chapters on phase-type distributions and phase-type processes. Chapter 2 is the place where individual states lose their identity and merely become members of a vague "phase set"; by drawing analogies with the exponential distribution, we show that most of the results may be seen as matrix versions of familiar scalar properties. In Chapter 3 we associate a counter to the phases, thereby creating a rudimentary two-dimensional process which is the matrix generalization of the Poisson process.

The structure of the stationary distribution of QBDs is determined in Chapters 4 and 6. Chapter 4 deals with birth-and-death processes. To present that material, we heavily make use of renewal theory; this is an unusual approach since we do not rely as much as one usually does on specific features of the birth-and-death models which lead to an easier analysis. It is, however, that approach which we extend to the general case, and we think it is advantageous to give it first in Chapter 4, without the added complexity of having to deal, as in later chapters, with matrices instead of scalar quantities.

The properties we analyze in the intermediary Chapter 5 are fundamental to our arguments and are repeatedly used in the remainder of the text. In addition, together with Chapter 4, this chapter gives us the opportunity to present in detail both discrete and continuous time processes. We show the strong similarity between the arguments which we use in both cases, as well as the equations which are obtained. In later chapters, this allows us to develop the theory for discrete time QBDs and to skip the details of the derivation of the results for the continuous time processes. Finally, we show how the application of theorems from Markov chain theory may be used to interpret the Gaussian elimination method for the solution of linear systems.

Chapter 8 is very important as it covers material where algorithmic and probabilistic reasoning are most intimately connected. In three steps, we take the reader from one of the simplest iterative procedures to the fastest, relating the successive approximations to the dynamic behavior of the stochastic process itself. The quadratic convergence of the fastest algorithm is proved by a probabilistic argument. At present, work is progressing on developing computational procedures based on the spectral analysis of various component matrices. In Chapter 9 we briefly give the argument showing why the spectral approach works in principle. We do not elaborate on this material as probabilistic reasoning is not very prominent and reported numerical experience is scanty. Chapters 10 and 11 on QBDs with a finite number of levels give us the opportunity to bring together material from several different authors, about different problems, and to show how these results may be interpreted in the light of the general properties in Chapter 5.

Lest a reader should leave this book feeling that the study of QBDs is a closed subject, we conclude with five extremely short chapters on various major extensions, ranging from the mostly algorithmic to the essentially structural: QBDs with level-dependent transitions, processes skip-free in one direction, processes on a tree-like state space, product form networks, and processes with a general state space for the phase dimension.

We barely mention M/G/1- and GI/M/1-type Markov chains. This glaring omission of an important subject might be explained away by the fact that these may be viewed as special cases of QBDs, as we prove in Chapter 13. More to the point, we find that these processes are better discussed in the context of the queueing or other applications from which they stem; in them arise a number of questions, such as the study of waiting time distributions, etc., which clearly are beyond our present goal to describe the fundamental aspects of matrix analytic methods.

We have presented the material in different short courses to various audiences. It has been received well, both by newcomers to the area and by those who have some familiarity with the subject matter. Our nearly fanatical devotion to probabilistic reasoning makes it possible for us to rely on basic properties for the most part, about Markov processes and renewal theory. The prerequisites are advanced calculus and linear algebra, at the level usually taught in undergraduate curricula, and a course in stochastic processes at the level of Çinlar [12], Feller [20, 21], Karlin and Taylor [36, 37], or Resnick [106]. In the few places where more advanced material is required, in particular in Chapter 7, we point the reader to specific references. Our bibliography does not form an exhaustive survey of the field but only contains references cited in the text.

Chapters 1–6, 8, 10, and 12 constitute the core of the methodology and may form the basis for a one-semester course at the senior undergraduate or graduate level. Chapter 7 on ergodicity conditions is essential but not easy, and an instructor might choose to walk the students through the main ideas. Chapter 9 on spectral analysis is peripheral, except for its first two sections. The reader who has absorbed the core material may work her way through Chapter 11 on passage times without the need of an instructor. The last chapters are on extensions and areas of work that have opened up recently; they are more appropriate for seminar topics after a basic course.

It is a pleasant duty to acknowledge the influence of our teachers, who instilled in us the desire to pursue mathematical research and who shaped our interest for probability theory and algorithmic thinking: Guy Louchard and Jean Teghem of the Université Libre de Bruxelles and K. Balasubramanian and K. N. Venkataraman of the University of Madras. Marcel Neuts has been a worthy mentor and role model to both of us; this book grew out of the sapling he planted and nurtured over many years.

Many colleagues and students have read the manuscript at several stages of its development and have pointed out errors and possible improvements. Our heartfelt thanks go to Allan T. Andersen, Nigel Bean, Nadjet Benseba, Brigitte Bertrand, Olivier De Deckere, Ed Kao, Bernard Larue, Herlinde Leemans, David Lucantoni, Marcel F. Neuts, Bo Friis Nielsen, Colm O'Cinneide, Marie-Ange Remiche, Volker Schmidt, and Peter Taylor.

A special mention must be made of Don Gaver, the series editor, who encouraged us and acted decisively at the right time and of Pascaline Browaeys, who made this project her own and tirelessly prodded us back to work whenever she felt we were straying away.

The first author has found much stimulation in the scholarly atmosphere of the Université Libre de Bruxelles and benefited much from visits to Bellcore, Purdue University, University of Delaware, University of Adelaide, and the Naval Post Graduate School in Monterey.

The second author owes the benefit of higher education to the willingness of his mother and her parents to endure abject poverty for a longer period. Instrumental to his progress have also been Charles D. Pack and Patricia Wirth, two perceptive managers who understand the value and power of innovation. He also thanks Bellcore and AT&T for providing an environment where he could indulge in scholarly pursuits and enjoy their fruition in meaningful practical uses.

Our families have provided an enviable refuge from worldly pressures and constant understanding at those not infrequent times when we are there but in body. To Monique and Soundaram, to Michael, Cindy, Priya, and Prem, this book is dedicated.

G. LATOUCHE
V. RAMASWAMI

Part I

Quasi-Birth-and-Death Processes

Chapter 1

Examples

We begin by describing several examples of quasi-birth-and-death processes (QBDs). These examples serve various purposes. First, they are intended to help the reader acquire some feeling for the range of applications of QBDs. Our second purpose is to motivate some of the terminology and notations which will be used throughout the text. Finally, these examples will be used as illustrations in later chapters.

Our presentation here will not be very formal. The reader who already has some familiarity with the subject may safely skip this chapter; all the definitions introduced here will be restated later with precision.

1.1 The M/M/1 Queue

A birth-and-death process on the nonnegative integers is a Markov process in which the only allowed transitions are from the state n to the next higher state $n+1$ for all $n \geq 0$ and from n to $n-1$ for $n \geq 1$.

The canonical example of homogeneous birth-and-death processes is the M/M/1 queue. This is a single server queueing system with infinite waiting room and is customarily described by the diagram of Figure 1.1. Customers join the system at the renewal epochs of a Poisson process with parameter λ. They enter a waiting room if there are other customers already present, or immediately begin service if no other customer is present. Customers are served in the order of arrival. The durations of services are independent and identically distributed (i.i.d.)

Figure 1.1: Diagrammatic representation of the M/M/1 queue. The arrival rate of new customers is λ; the service rate is μ.

random variables, independent of the arrival process. The service time distribution is exponential with parameter μ, i.e., with probability density $f(x) = \mu e^{-\mu x}$ for $x \geq 0$.

We denote by $N(t)$ the number of customers present in the system at time t, either waiting or being served. The process $\{N(t), t \geq 0\}$ is a continuous time Markov chain on the state space $\{0, 1, 2, \ldots\}$. The possible transitions, and the corresponding instantaneous rates, are given in the table below.

From	To	Rate	
n	$n-1$	μ	for $n \geq 1$
n	$n+1$	λ	for $n \geq 0$

Suppose time t is such that $N(t) = n$. Then the probability that $N(t+h) = n+1$, due to an arrival in $(t, t+h)$, is $\lambda h + o(h)$. If $n \geq 1$, then the server is busy at time t and the probability that a service is completed is $\mu h + o(h)$.

Another characterization sometimes used for Markov processes is the state transition diagram. We present that of the M/M/1 queue in Figure 1.2.

A third characterization is in terms of the infinitesimal generator Q, the matrix such that q_{ij} is the instantaneous transition rate from i to j for $i \neq j$ and such that $q_{ii} = -\sum_{\forall j \neq i} q_{ij}$. It is well known that $|q_{ii}|$ is the parameter of the exponentially distributed sojourn time in i. For

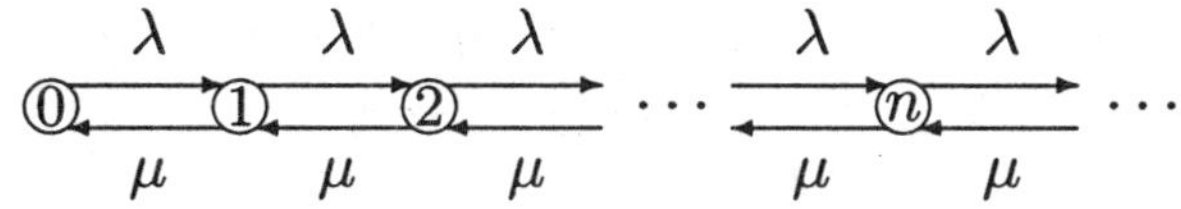

Figure 1.2: State transition diagram for the M/M/1 queue. The arrows represent the possible transitions and are labeled with the corresponding instantaneous rates.

the M/M/1 queue, we have that

$$Q = \begin{bmatrix} -\lambda & \lambda & \cdot & \cdot & \cdots \\ \mu & -(\lambda+\mu) & \lambda & \cdot & \cdots \\ \cdot & \mu & -(\lambda+\mu) & \lambda & \cdots \\ \cdot & \cdot & \mu & -(\lambda+\mu) & \cdots \\ \vdots & \vdots & \vdots & \vdots & \ddots \end{bmatrix}. \tag{1.1}$$

The most important feature here is that Q is a tridiagonal matrix: the elements of the upper diagonal are all equal; so are the elements of the lower diagonal; as for the elements on the main diagonal, with the exception of the upper left corner, the rest are equal.

The singularity of the upper left element reflects the fact that state 0 is a *boundary state*: we see in Figure 1.2 that 0 is the only state from which it is impossible to move to the left.

Remark 1.1.1 Unless otherwise stated, sample paths are right continuous, i.e., "state at time t" means "state at time $t+$," so that if t is an epoch of transition, then the state at t is the new state entered into.

1.2 The M/M/1 Queue in a Random Environment

A simple example of a QBD is the M/M/1 queue in a Markovian environment (Neuts [79, Chapter 6]). In short, this system behaves like an M/M/1 queue, but the arrival and service rates vary over time. In addition to the arrival and service processes, one defines a so-called environmental process $\{E(t), t \geq 0\}$ on a finite state space $\{1, 2, \dots, m\}$ with instantaneous transition rates $s_{ij}, 1 \leq i \neq j \leq m$.

The environment controls the arrival and service processes as follows. Suppose time t is such that $E(t) = j$. Then the arrival rate is λ_j and the service rate is μ_j, provided that the server is busy at time t.

Thus, the whole system is a two-dimensional continuous time Markov chain $\{(N(t), E(t)), t \geq 0\}$ on the state space $\{(n, i); n \geq 0, 1 \leq i \leq m\}$, where $N(t)$ is the number of customers present and $E(t)$ is the state of the environment at time t.

Changes of state occur when the environment changes, when a new customer arrives, or when a service is completed. The possible transitions, and the corresponding instantaneous rates, are given in the table below.

From	To	Rate	
(n, i)	$(n-1, i)$	μ_i	for $n \geq 1$
(n, i)	(n, j)	s_{ij}	for $n \geq 0, i \neq j$
(n, i)	$(n+1, i)$	λ_i	for $n \geq 0$

It would be unnecessarily confusing to depict the state transition diagram in full generality, and therefore we show in Figure 1.3 a particular example only. In that example, there are $m = 3$ states in the environment and the instantaneous rates s_{21} and s_{13} are equal to zero.

In order to display the infinitesimal generator Q of the system, it is necessary to define a linear ordering of the states. Of particular interest is the lexicographical ordering $\{(0, 1), (0, 2), \dots, (0, m), (1, 1), (1, 2), \dots, (1, m), (2, 1), (2, 2), \dots, (2, m), \dots, (n, 1), (n, 2), \dots, (n, m), \dots\}$: we first enumerate all the states with 0 customers, then all the states with 1 customer, then all the states with 2 customers, etc.

We shall call by *level* the subset of all states corresponding to a fixed number of customers in the system. The levels appear as columns in the diagram of Figure 1.3 while the rows in that diagram correspond to the various environmental states. The lexicographical ordering corresponds to the enumeration of the states level by level.

The infinitesimal generator Q corresponding to the diagram of Figure 1.3 is given below. We have labeled each row and column by the

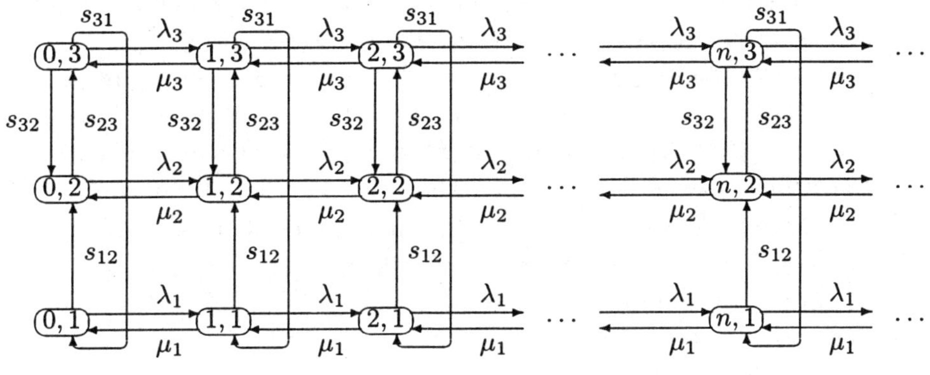

Figure 1.3: State transition diagram for the M/M/1 queue in a random environment. There are 3 environmental states. The arrows represent the possible transitions; for this example, it is assumed that s_{21} and s_{13} are equal to 0.

---

corresponding state for easy reference:

$$
Q = \begin{array}{c} \\ (0,1) \\ (0,2) \\ (0,3) \\ (1,1) \\ (1,2) \\ (1,3) \\ (2,1) \\ (2,2) \\ (2,3) \\ \vdots \end{array}
\begin{array}{c}
\begin{array}{ccccccccccc} (0,1) & (0,2) & (0,3) & (1,1) & (1,2) & (1,3) & (2,1) & (2,2) & (2,3) & \cdots \end{array} \\
\left[\begin{array}{ccc|ccc|ccc|c}
q_1^\circ & s_{12} & \cdot & \lambda_1 & \cdot & \cdot & \cdot & \cdot & \cdot & \cdots \\
\cdot & q_2^\circ & s_{23} & \cdot & \lambda_2 & \cdot & \cdot & \cdot & \cdot & \cdots \\
s_{31} & s_{32} & q_3^\circ & \cdot & \cdot & \lambda_3 & \cdot & \cdot & \cdot & \cdots \\
\hline
\mu_1 & \cdot & \cdot & q_1^* & s_{12} & \cdot & \lambda_1 & \cdot & \cdot & \cdots \\
\cdot & \mu_2 & \cdot & \cdot & q_2^* & s_{23} & \cdot & \lambda_2 & \cdot & \cdots \\
\cdot & \cdot & \mu_3 & s_{31} & s_{32} & q_3^* & \cdot & \cdot & \lambda_3 & \cdots \\
\hline
\cdot & \cdot & \cdot & \mu_1 & \cdot & \cdot & q_1^* & s_{12} & \cdot & \cdots \\
\cdot & \cdot & \cdot & \cdot & \mu_2 & \cdot & \cdot & q_2^* & s_{23} & \cdots \\
\cdot & \cdot & \cdot & \cdot & \cdot & \mu_3 & s_{31} & s_{32} & q_3^* & \cdots \\
\hline
\vdots & \vdots & \vdots & \vdots & \vdots & \vdots & \vdots & \vdots & \vdots & \ddots
\end{array}\right]
\end{array},
$$

where $q_1^\circ = -s_{12} - \lambda_1, q_2^\circ = -s_{23} - \lambda_2, q_3^\circ = -s_{31} - s_{32} - \lambda_3, q_1^* = -\mu_1 - s_{12} - \lambda_1, q_2^* = -\mu_2 - s_{23} - \lambda_2$, and $q_3^* = -\mu_3 - s_{31} - s_{32} - \lambda_3$. Note that $-s_{12}, -s_{23}$, and $-s_{31} - s_{32}$ are the diagonal elements of the infinitesimal generator S of the environmental Markov process $\{E(t), t \geq 0\}$.

The important feature here is that the matrix Q is block-tridiagonal. For general values of the size m and the generator S of $\{E(t), t \geq 0\}$, the generator Q of the M/M/1 queue in a random environment is given by

$$Q = \begin{bmatrix} S-\Lambda & \Lambda & 0 & 0 & \cdots \\ M & S-\Lambda-M & \Lambda & 0 & \cdots \\ 0 & M & S-\Lambda-M & \Lambda & \cdots \\ 0 & 0 & M & S-\Lambda-M & \cdots \\ \vdots & \vdots & \vdots & \vdots & \ddots \end{bmatrix}, \tag{1.2}$$

where S, Λ, and M are matrices of order m (the number of environmental states), S is the generator of the environmental process, Λ and M are diagonal matrices with $\Lambda_{ii} = \lambda_i$, and $M_{ii} = \mu_i$ for $1 \leq i \leq m$.

Notice how similar the two structures (1.1) and (1.2) are: the matrix (1.2) is block-tridiagonal and the blocks on the upper diagonal are all equal, as are the blocks on the lower diagonal and the blocks on the main diagonal, with the exception of the block in the upper left corner.

This is the characteristic of the Markov processes which we shall study in later chapters.

1.3 Phase-Type Queues

These queues constitute a generalization of the M/M/1 queue, where the service time and interarrival time distributions are not restricted to the exponential; instead, they are assumed to be of phase-type (Neuts [79, Chapter 3]). Phase-type distributions (abbreviated as PH distributions) will be defined with precision in Chapter 2; we now give an informal description only and two specific examples.

Assume that services are comprised of a number of operations, numbered 1 to m. Each operation i has a random duration, exponentially distributed, with parameter ν_i; when that operation is completed, the service proceeds to another operation j with probability p_{ij}, until eventually the whole service is completed. Thus, two services may differ by the operations which are performed, as well as by the durations of the operations. The total service duration, from start to finish, is said to be of *phase-type*; the individual operations are called phases.

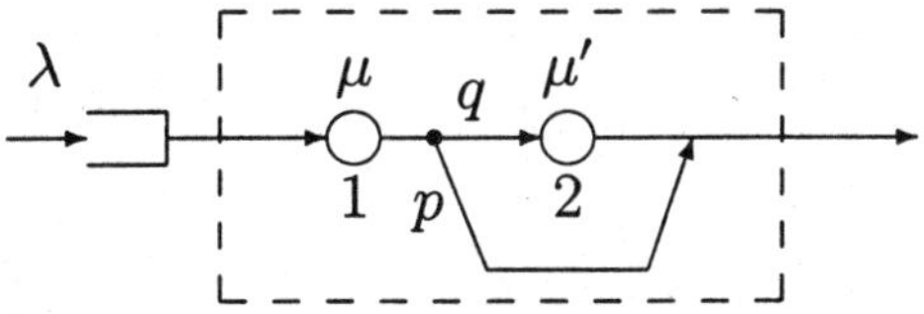

Figure 1.4: Diagrammatic representation of an M/PH/1 queue. The arrival rate is λ. The service time distribution has two phases: the first phase has parameter μ; the second phase has parameter μ' and is performed with probability q.

The M/PH/1 Queue

In our first example, we assume that there are two service phases. The first is exponentially distributed with parameter μ; when it is completed, the work performed is inspected. With probability p, the work is found to be satisfactory and the customer departs. With probability $q = 1 - p$, the customer needs additional work, which is exponentially distributed with parameter μ'; at the end of this second operation, the customer automatically departs. If we further assume that customers arrive according to a Poisson process with rate λ, then we have a particular example of the M/PH/1 queue, which may be described by the diagram of Figure 1.4.

The diagram is to be read as follows: at any given time, there may be at most one customer in the dashed box. When a customer's service begins, the customer enters node 1; upon leaving node 1, he leaves the system with probability p, and with probability q he enters node 2, from which he leaves the system. The times spent in the nodes 1 and 2 are exponentially distributed with respective parameters μ and μ'.

With this description, the queue may be represented as a continuous time Markov process on the state space $\ell(0) \cup \ell(1) \cup \ell(2) \cup \ldots$, where $\ell(0) = \{0\}$ and $\ell(n) = \{(n,1),(n,2)\}$ for $n \geq 1$; state 0 corresponds to an empty system, and for $n \geq 1$, the state (n,j) signifies that there are n customers in the queue, $n-1$ of whom are in the waiting room, and the customer in service is in the node j.

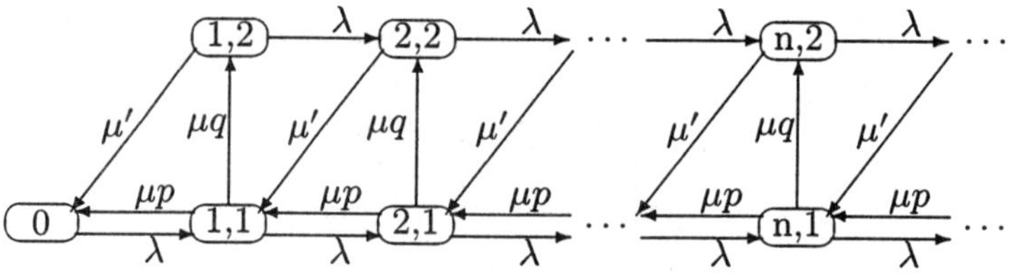

Figure 1.5: State transition diagram for an M/PH/1 queue. The arrival rate is λ. The service time distribution has two phases: the first phase has parameter μ; the second phase has parameter μ' and is performed with probability q.

Changes of state occur when there is a new arrival, when a service is completed, or when a customer in service at node 1 moves to node 2. These transitions are given in the table below. The state transition diagram is shown in Figure 1.5.

From	To	Rate	
0	$(1,1)$	λ	
$(1,1)$	0	μp	
$(n,1)$	$(n,2)$	μq	for $n \geq 1$
$(n,1)$	$(n-1,1)$	μp	for $n \geq 2$
$(1,2)$	0	μ'	
$(n,2)$	$(n-1,1)$	μ'	for $n \geq 2$
(n,j)	$(n+1,j)$	λ	for $n \geq 1$, $j = 1,2$

A few words of comment are in order here. The Markov process leaves the state 0 with instantaneous rate λ when a new customer arrives; this new customer immediately begins its service in node 1. When the Markov process is in the state $(n,2)$ for some $n \geq 1$, an end of service occurs at the instantaneous rate μ', and if $n \geq 2$, a new customer begins its service in node 1. When the Markov process is in the state $(n,1)$ for some $n \geq 1$, the customer being served enters node 2 with probability q; with probability p it leaves the system and a new service may begin.

If we order the states by level (that is, by the number of customers) and, within a level, by the label of the service node, we find that the infinitesimal generator Q is given as follows. We have again labeled each row and column by the corresponding state:

$$
Q = \begin{array}{c@{\;}c}
 & \begin{array}{cccccccc} 0 & (1,1) & (1,2) & (2,1) & (2,2) & (3,1) & (3,2) & \cdots \end{array} \\
\begin{array}{c} 0 \\ (1,1) \\ (1,2) \\ (2,1) \\ (2,2) \\ (3,1) \\ (3,2) \\ \vdots \end{array} &
\left[\begin{array}{c|cc|cc|cc|c}
-\lambda & \lambda & \cdot & \cdot & \cdot & \cdot & \cdot & \cdots \\ \hline
\mu p & q_1^* & \mu q & \lambda & \cdot & \cdot & \cdot & \cdots \\
\mu' & \cdot & q_2^* & \cdot & \lambda & \cdot & \cdot & \cdots \\ \hline
\cdot & \mu p & \cdot & q_1^* & \mu q & \lambda & \cdot & \cdots \\
\cdot & \mu' & \cdot & \cdot & q_2^* & \cdot & \lambda & \cdots \\ \hline
\cdot & \cdot & \cdot & \mu p & \cdot & q_1^* & \mu q & \cdots \\
\cdot & \cdot & \cdot & \mu' & \cdot & \cdot & q_2^* & \cdots \\ \hline
\vdots & \vdots & \vdots & \vdots & \vdots & \vdots & \vdots & \ddots
\end{array}\right]
\end{array},
$$

where $q_1^* = -\lambda - \mu$ and $q_2^* = -\lambda - \mu'$. The generator may also be written as

$$
Q = \begin{bmatrix}
-\lambda & \lambda\boldsymbol{\tau} & \mathbf{0} & \mathbf{0} & \cdots \\
\boldsymbol{t} & T - \lambda I & \lambda I & 0 & \cdots \\
\mathbf{0} & \boldsymbol{t}\cdot\boldsymbol{\tau} & T - \lambda I & \lambda I & \cdots \\
\mathbf{0} & 0 & \boldsymbol{t}\cdot\boldsymbol{\tau} & T - \lambda I & \cdots \\
\vdots & \vdots & \vdots & \vdots & \ddots
\end{bmatrix}, \tag{1.3}
$$

where

$$
\boldsymbol{\tau} = \begin{bmatrix} 1 & 0 \end{bmatrix}, \tag{1.4}
$$

$$
\boldsymbol{t} = \begin{bmatrix} \mu p \\ \mu' \end{bmatrix}, \tag{1.5}
$$

$$
T = \begin{bmatrix} -\mu & \mu q \\ 0 & -\mu' \end{bmatrix}, \tag{1.6}
$$

and $\boldsymbol{t}\cdot\boldsymbol{\tau}$ is the direct product of $\boldsymbol{t}$ and $\boldsymbol{\tau}$: $(\boldsymbol{t}\cdot\boldsymbol{\tau})_{ij} = t_i\tau_j$ for $1 \le i, j \le m$. Here, as elsewhere, we denote by I the identity matrix of required order.

Again, we notice the similarity among the structures (1.1), (1.2), and (1.3). There is a structural difference between the matrices (1.2) and (1.3): while in the former all the blocks have the same size, we see here that the three blocks $-\lambda$, $\lambda\boldsymbol{\tau}$, and $\boldsymbol{t}$ have dimensions different from the other blocks. This is because the boundary level $\ell(0)$ contains fewer states than the interior levels $\ell(n)$ with $n \ge 1$.

We have now exemplified all the ingredients of a QBD. Before proceeding with the examples, we shall give their general definition in the continuous time case.

Definition 1.3.1 *A continuous time, homogeneous QBD is a Markov process with the following properties:*

(a) *It has a two-dimensional state space* $\cup_{n\geq 0}\ell(n)$*, where* $\ell(0) = \{(0,1), (0,2), \ldots, (0,m')\}$ *and* $\ell(n) = \{(n,1),(n,2),\ldots,(n,m)\}$ *for all* $n \geq 1$*;* m *and* m' *may be infinite. The subset of states* $\ell(n)$ *is called* level n.

(b) *One-step transitions from a state are restricted to states in the same or in the two adjacent levels. In other words, a transition from* (n,i) *to* (n',j) *is not possible if* $|n'-n| \geq 2$.

(c) *For* $n \geq 1$*, the instantaneous transition rate between two states in the same level* $\ell(n)$ *or between two states in the levels* $\ell(n)$ *and* $\ell(n+1)$ *do not depend on* n*. More formally, we have that for* n *and* $n' \geq 1$*, the transition rate from* (n,i) *to* (n',j) *may not depend on* n *and* n' *individually, but only on their difference.*

The infinitesimal generator Q *has the following structure:*

$$Q = \begin{bmatrix} B_1 & B_0 & 0 & 0 & \cdots \\ B_2 & A_1 & A_0 & 0 & \cdots \\ 0 & A_2 & A_1 & A_0 & \cdots \\ 0 & 0 & A_2 & A_1 & \cdots \\ \vdots & \vdots & \vdots & \vdots & \ddots \end{bmatrix}, \tag{1.7}$$

where $A_0, A_1,$ *and* A_2 *are square matrices of order* m*;* B_1 *is a square matrix of order* m'*;* B_0 *is an* $m' \times m$ *matrix; and* B_2 *is an* $m \times m'$ *matrix.*

The PH/M/1 Queue

In our second example of a PH queue, we use the Erlang distribution. We denote by $F_{m,\nu}(\cdot)$ the distribution function of an Erlang random

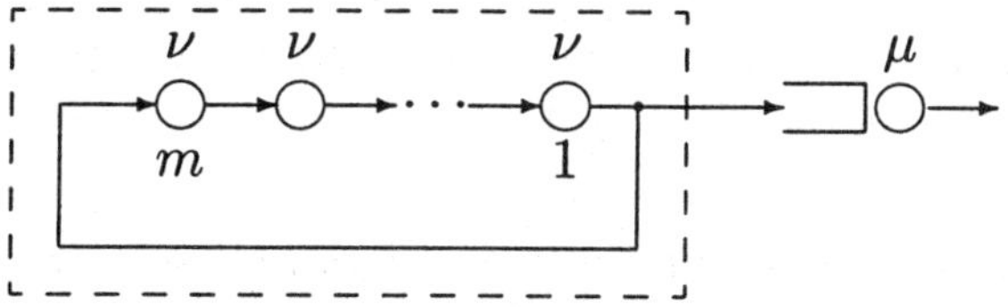

Figure 1.6: Diagrammatic representation of a PH/M/1 queue. The service rate is μ; the interarrival times have the Erlang distributions $F_{m\nu}(\cdot)$.

---

variable with parameters m and ν; the density function is given by

$$f_{m,\nu}(x) = \begin{cases} 0, & x < 0, \\ \frac{1}{(m-1)!}\nu^m x^{m-1} e^{-\nu x}, & x \geq 0. \end{cases}$$

We now consider a single server queue for which arrivals occur according to a renewal process and service times are exponentially distributed. The intervals between arrivals have the distribution $F_{m,\nu}(\cdot)$. This is an example of a PH/M/1 queue for which interarrival times have an Erlang distribution.

It is well known that the sum of m independent, exponentially distributed random variables with parameter ν has an Erlang (m, ν) distribution. If we use the method of phases, we find that this particular PH/M/1 queue may be described by the diagram of Figure 1.6, which is to be read as follows: a token circulates among the nodes in the dashed box, remaining in each node for an exponentially distributed interval. Whenever the token leaves node 1 and returns to node m, a new customer joins the queue.

With this description, the queue may be represented as a two-dimensional Markov process, $\{(N(t), \varphi(t)), t \geq 0\}$ on the state space $\cup_{n\geq 0}\ell(n)$, with $\ell(n) = \{(n, 1), \ldots, (n, m)\}$ for all $n \geq 0$, where $N(t)$ and $\varphi(t)$, respectively, represent the number of customers in the system and the position of the token at time t. The possible transitions are enumerated in the table below.

From	To	Rate	
(n,j)	$(n-1,j)$	μ	for $n \geq 1$
(n,j)	$(n,j-1)$	ν	for $n \geq 0, j \geq 2$
$(n,1)$	$(n+1,m)$	ν	for $n \geq 0$

The infinitesimal generator is given below in the special case where $m = 3$:

$$Q = \left[\begin{array}{ccc|ccc|ccc|c} q^\circ & \cdot & \cdot & \cdot & \cdot & \nu & \cdot & \cdot & \cdot & \cdots \\ \nu & q^\circ & \cdot & \cdot & \cdot & \cdot & \cdot & \cdot & \cdot & \cdots \\ \cdot & \nu & q^\circ & \cdot & \cdot & \cdot & \cdot & \cdot & \cdot & \cdots \\ \hline \mu & \cdot & \cdot & q^* & \cdot & \cdot & \cdot & \cdot & \nu & \cdots \\ \cdot & \mu & \cdot & \nu & q^* & \cdot & \cdot & \cdot & \cdot & \cdots \\ \cdot & \cdot & \mu & \cdot & \nu & q^* & \cdot & \cdot & \cdot & \cdots \\ \hline \cdot & \cdot & \cdot & \mu & \cdot & \cdot & q^* & \cdot & \cdot & \cdots \\ \cdot & \cdot & \cdot & \cdot & \mu & \cdot & \nu & q^* & \cdot & \cdots \\ \cdot & \cdot & \cdot & \cdot & \cdot & \mu & \cdot & \nu & q^* & \cdots \\ \hline \vdots & \vdots & \vdots & \vdots & \vdots & \vdots & \vdots & \vdots & \vdots & \ddots \end{array}\right],$$

where $q^\circ = -\nu$ and $q^* = -\nu - \mu$. In the general case, we have that

$$Q = \begin{bmatrix} S & \boldsymbol{s}\cdot\boldsymbol{\sigma} & 0 & 0 & \cdots \\ \mu I & S - \mu I & \boldsymbol{s}\cdot\boldsymbol{\sigma} & 0 & \cdots \\ 0 & \mu I & S - \mu I & \boldsymbol{s}\cdot\boldsymbol{\sigma} & \cdots \\ 0 & 0 & \mu I & S - \mu I & \cdots \\ \vdots & \vdots & \vdots & \vdots & \ddots \end{bmatrix},$$

where the vectors $\boldsymbol{s}$ and $\boldsymbol{\sigma}$ and the matrix S of order m are defined as follows:

$$\boldsymbol{\sigma} = \begin{bmatrix} 0 & 0 & \cdots & 1 \end{bmatrix}, \tag{1.8}$$

$$\boldsymbol{s} = \begin{bmatrix} \nu \\ 0 \\ \vdots \\ 0 \end{bmatrix}, \tag{1.9}$$

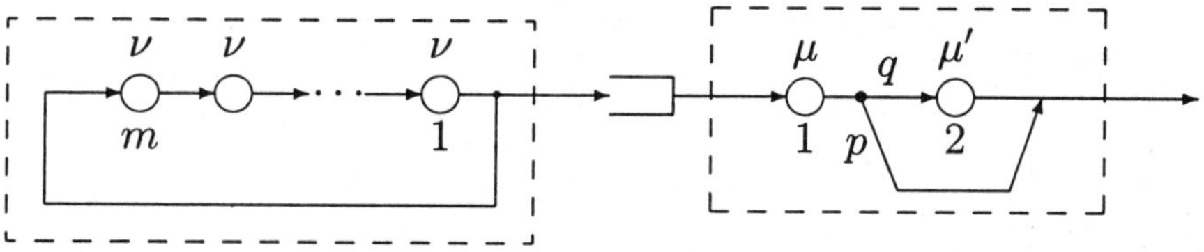

Figure 1.7: Diagrammatic representation of a PH/PH/1 queue. The interarrival times have the Erlang distributions $F_{m\nu}(\cdot)$. The service time distribution has two phases: the first phase has parameter μ; the second phase has parameter μ' and is performed with probability q.

and

$$S = \begin{bmatrix} -\nu & \cdot & \cdot & \cdots & \cdot & \cdot \\ \nu & -\nu & \cdot & \cdots & \cdot & \cdot \\ \cdot & \nu & -\nu & \cdots & \cdot & \cdot \\ \vdots & \vdots & \vdots & \ddots & \vdots & \vdots \\ \cdot & \cdot & \cdot & \cdots & -\nu & \cdot \\ \cdot & \cdot & \cdot & \cdots & \nu & -\nu \end{bmatrix}. \tag{1.10}$$

The PH/PH/1 queue

The PH/PH/1 queue is a single server system with renewal arrivals, where both the interarrival times and the service times have a PH distribution. In the example of Figure 1.7, the interarrival times have the Erlang distribution $F_{m,\nu}(\cdot)$ and the service time distribution is the same as for the M/PH/1 queue described earlier (see Figure 1.4).

This may still be represented as a Markov process $\{(N(t), \varphi(t)), t \geq 0\}$ on the state space $\cup_{n\geq 0}\ell(n)$, but the levels themselves become two dimensional here since we need to record $\varphi(t) = (\varphi_s(t), \varphi_a(t))$, where $\varphi_s(t)$ is the index of the node occupied by the customer in service, when the server is busy, while $\varphi_a(t)$ is the position of the token in the arrival process. Thus, we have that $\ell(0) = \{(0, j), 1 \leq j \leq m\}$ and $\ell(n) = \{(n, i, j) : i = 1 \text{ or } 2, 1 \leq j \leq m\}$ for $n \geq 1$. The possible transitions are enumerated in Table 1.1.

From	To	Rate	
$(0,j)$	$(0,j-1)$	ν	for $2 \le j \le m$
$(0,1)$	$(1,1,m)$	ν	
$(1,1,j)$	$(0,j)$	μp	for $1 \le j \le m$
$(n,1,j)$	$(n-1,1,j)$	μp	for $n \ge 2,\ 1 \le j \le m$
$(n,1,j)$	$(n,2,j)$	μq	for $n \ge 1,\ 1 \le j \le m$
$(1,2,j)$	$(0,j)$	μ'	for $1 \le j \le m$
$(n,2,j)$	$(n-1,1,j)$	μ'	for $n \ge 2,\ 1 \le j \le m$
(n,i,j)	$(n,i,j-1)$	ν	for $n \ge 1,\ i = 1,2,\ 2 \le j \le m$
$(n,i,1)$	$(n+1,i,m)$	ν	for $n \ge 1,\ i = 1,2$

Table 1.1: Transitions for the PH/PH/1 queue of Figure 1.7.

The states are grouped by levels, so that (n,i,j) precedes (n',i',j') if $n < n'$. Within each level, the states are grouped according to the service phase, so that $(n,1,j)$ precedes $(n,2,j')$ for all j,j'. Finally, for a fixed level and a fixed service phase, the states are ordered by arrival phase, so that (n,i,j) precedes (n,i,j') if $j < j'$. One may verify that the infinitesimal generator Q has the form (1.7).

With the increased number of states in each level, it becomes difficult to display in detail the infinitesimal generator. This is why we separately give the individual blocks below:

- The matrix B_1 is of order m and is equal to the matrix S of (1.10).
- The matrix B_0 has m rows and $2m$ columns and is equal to $[\boldsymbol{s} \cdot \boldsymbol{\sigma} \;\; 0]$, where $\boldsymbol{s}$ and $\boldsymbol{\sigma}$ are given in (1.8), (1.9).
- The matrix B_2 has $2m$ rows and m columns and is given by

$$B_2 = \begin{bmatrix} \mu p I \\ \mu' I \end{bmatrix}.$$

- The matrices A_0, A_1, and A_2 have order $2m$ and are given by

$$A_0 = \begin{bmatrix} \boldsymbol{s} \cdot \boldsymbol{\sigma} & 0 \\ 0 & \boldsymbol{s} \cdot \boldsymbol{\sigma} \end{bmatrix},$$

$$A_1 = \begin{bmatrix} -\mu I + S & \mu q I \\ 0 & -\mu' I + S \end{bmatrix},$$

$$A_2 = \begin{bmatrix} \mu p I & 0 \\ \mu' I & 0 \end{bmatrix}.$$

In a more compact form, using Kronecker products[1] we may write that

$$\begin{aligned} B_1 &= S, \\ B_0 &= \boldsymbol{\tau} \otimes \boldsymbol{s} \cdot \boldsymbol{\sigma}, \\ B_2 &= \boldsymbol{t} \otimes I, \\ A_0 &= I \otimes \boldsymbol{s} \cdot \boldsymbol{\sigma}, \\ A_1 &= T \otimes I + I \otimes S, \\ A_2 &= \boldsymbol{t} \cdot \boldsymbol{\tau} \otimes I. \end{aligned}$$

It is worth noting that Latouche and Ramaswami [55] describe an approach which leads to blocks of size $m_a + m_s$, where m_a and m_s are the number of phases in the arrival and service processes, respectively, instead of blocks of size $m_a m_s$ as here.

Transition diagrams are useful for simple processes only. Even for moderately complex systems, they are very hard to display and examine (observe that the diagram of Figure 1.5 corresponds to a very elementary M/PH/1 queue). For that reason, we shall no longer show transition diagrams; we shall enumerate the transition rates and directly display the infinitesimal generators.

[1]If A is an $n \times m$ matrix and B is an $n' \times m'$ matrix, the Kronecker product $A \otimes B$ is an $nn' \times mm'$ matrix defined by

$$A \otimes B = \begin{bmatrix} a_{11}B & \cdots & a_{1m}B \\ \vdots & & \vdots \\ a_{n1}B & \cdots & a_{nm}B \end{bmatrix}.$$

We refer to Graham [28] for further details.

1.4 A Queue with Two Priority Classes

In this system, there are two classes of customers who arrive according to independent Poisson processes, respectively, with rates λ_1 and λ_2. The service times are exponentially distributed, with rates μ_1 and μ_2. Customers of class 1 have preemptive priority over customers of class 2, so that the service of a class 2 customer may begin only if there is no class 1 customer and will be interrupted by a new arrival of class 1 (Miller [68]).

Clearly, this may be described as a two-dimensional Markov process $\{(N_1(t), N_2(t)), t \geq 0\}$, where $N_i(t)$ is the number of customers of class i in the system at time t. The state space is $\cup_{i\geq 0}\ell(i)$, where $\ell(n) = \{(n,0),(n,1),\ldots\}$, and the possible transitions are given below.

From	To	Rate	
(n_1, n_2)	$(n_1 - 1, n_2)$	μ_1	for $n_1 \geq 1, n_2 \geq 0$
$(0, n_2)$	$(0, n_2 - 1)$	μ_2	for $n_2 \geq 1$
(n_1, n_2)	$(n_1, n_2 + 1)$	λ_2	for $n_1, n_2 \geq 0$
(n_1, n_2)	$(n_1 + 1, n_2)$	λ_1	for $n_1, n_2 \geq 0$

A particular feature of this model is that the state space is infinite in both dimensions, so that it is not obvious which should play the role of the levels.

The natural choice might be to let $\ell(n)$ be the set of states where the number of *class* 1 customers is equal to n. The infinitesimal generator is then given by

$$Q = \begin{bmatrix} B_1 & \lambda_1 I & 0 & \cdots \\ \mu_1 I & A_1 & \lambda_1 I & \cdots \\ 0 & \mu_1 I & A_1 & \cdots \\ \vdots & \vdots & \vdots & \ddots \end{bmatrix}, \tag{1.11}$$

where

$$A_1 = \begin{bmatrix} -\mu_1 - \lambda_1 - \lambda_2 & \lambda_2 & \cdot & \cdots \\ \cdot & -\mu_1 - \lambda_1 - \lambda_2 & \lambda_2 & \cdots \\ \cdot & \cdot & -\mu_1 - \lambda_1 - \lambda_2 & \cdots \\ \vdots & \vdots & \vdots & \ddots \end{bmatrix}$$

and

$$\mathrm{B}_1 = \begin{bmatrix} -\lambda_1-\lambda_2 & \lambda_2 & \cdot & \cdots \\ \mu_2 & -\mu_2-\lambda_1-\lambda_2 & \lambda_2 & \cdots \\ \cdot & \mu_2 & -\mu_2-\lambda_1-\lambda_2 & \cdots \\ \vdots & \vdots & \vdots & \ddots \end{bmatrix}.$$

Alternately, we may choose to represent by $\ell(n)$ the set of states where the number of *class* 2 customers is equal to n. This corresponds to the ordering (n_2, n_1), and the infinitesimal generator is now given by

$$Q = \begin{bmatrix} B_1 & \lambda_2 I & 0 & \cdots \\ A_2 & A_1 & \lambda_2 I & \cdots \\ 0 & A_2 & A_1 & \cdots \\ \vdots & \vdots & \vdots & \ddots \end{bmatrix} \qquad (1.12)$$

with

$$\begin{aligned} A_1 &= \begin{bmatrix} -\mu_2-\lambda_1-\lambda_2 & \lambda_1 & \cdot & \cdots \\ \mu_1 & -\mu_1-\lambda_1-\lambda_2 & \lambda_1 & \cdots \\ \cdot & \mu_1 & -\mu_1-\lambda_1-\lambda_2 & \cdots \\ \vdots & \vdots & \vdots & \ddots \end{bmatrix}, \\ A_2 &= \begin{bmatrix} \mu_2 & 0 & 0 & \cdots \\ 0 & 0 & 0 & \cdots \\ 0 & 0 & 0 & \cdots \\ \vdots & \vdots & \vdots & \ddots \end{bmatrix}, \\ B_1 &= \begin{bmatrix} -\lambda_1-\lambda_2 & \lambda_1 & \cdot & \cdots \\ \mu_1 & -\mu_1-\lambda_1-\lambda_2 & \lambda_1 & \cdots \\ \cdot & \mu_1 & -\mu_1-\lambda_1-\lambda_2 & \cdots \\ \vdots & \vdots & \vdots & \ddots \end{bmatrix}. \end{aligned}$$

The second ordering is better because $B_1 = A_1 + A_2$ and A_2 has only one column which is not identically zero, two properties which are very useful, as we shall see in Chapters 6 and 8.

1.5 Tandem Queues with Blocking

QBDs are not restricted to single station queueing systems. Here, we consider a system of two queues in tandem, described in Figure 1.8.

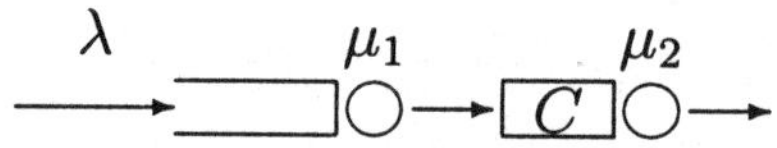

Figure 1.8: Diagrammatic representation of two queues in tandem. The arrival rate of new customers is λ; the service rates at the first and second servers are μ_1 and μ_2; the intermediate buffer has capacity C.

The assumptions are as follows. New customers arrive according to a Poisson process with rate λ and join an infinite waiting room in front of server 1. They receive a first service, exponentially distributed with rate μ_1, then they move to a second waiting room and eventually receive a second service, exponentially distributed with rate μ_2. The waiting room between the two servers has finite capacity, and there can be at most C customers (either waiting or being served) in the second half of the system. If at some time server 1 completes a service when there are already C customers in the second half, then the newly served customer may not leave the first half of the system, and server 1 becomes blocked. When a service completion occurs at server 2, the blocking customer may proceed, and server 1 may resume (Latouche and Neuts [50]).

If we denote by $N_1(t)$ the number of customers present at time t who have not yet completed their first service and by $N_2(t)$ the number of customers who have received the first but not the second service, we find that the process $\{(N_1(t), N_2(t)), t \geq 0\}$ is a Markov process on the state space $\cup_{i \geq 0} \ell(i)$ with $\ell(n) = \{(n, 0), \ldots, (n, C+1)\}$. When $N_2(t) = C + 1$, it means that the server 1 is blocked. The possible transitions are given below.

From	To	Rate	
(n_1, n_2)	$(n_1 - 1, n_2 + 1)$	μ_1	for $n_1 \geq 1, n_2 \leq C$
(n_1, n_2)	$(n_1, n_2 - 1)$	μ_2	for $n_1 \geq 0, n_2 \geq 1$
(n_1, n_2)	$(n_1 + 1, n_2)$	λ	for $n_1 \geq 0$

The infinitesimal generator has the form (1.7), where the blocks have order $C + 2$. The matrix A_0 is equal to λI, and the matrices A_1, A_2,

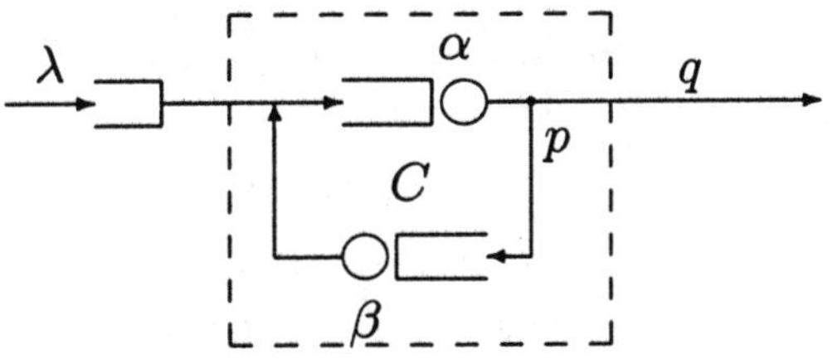

Figure 1.9: Diagrammatic representation of a multiprogramming system. The arrival rate of new customers is λ; the service rates at the first and second servers are α and β; the probability of feedback is p; the probability of departure is $q = 1 - p$; the total capacity of the loop is C.

and B_1 are displayed below in the representative case where $C = 3$; furthermore, $B_0 = A_0$ and $B_2 = A_2$:

$$A_1 = \begin{bmatrix} -\lambda-\mu_1 & \cdot & \cdot & \cdot & \cdot \\ \mu_2 & -\lambda-\mu_1-\mu_2 & \cdot & \cdot & \cdot \\ \cdot & \mu_2 & -\lambda-\mu_1-\mu_2 & \cdot & \cdot \\ \cdot & \cdot & \mu_2 & -\lambda-\mu_1-\mu_2 & \cdot \\ \cdot & \cdot & \cdot & \mu_2 & -\lambda-\mu_2 \end{bmatrix},$$

$$A_2 = \begin{bmatrix} \cdot & \mu_1 & \cdot & \cdot & \cdot \\ \cdot & \cdot & \mu_1 & \cdot & \cdot \\ \cdot & \cdot & \cdot & \mu_1 & \cdot \\ \cdot & \cdot & \cdot & \cdot & \mu_1 \\ \cdot & \cdot & \cdot & \cdot & \cdot \end{bmatrix},$$

and

$$B_1 = \begin{bmatrix} -\lambda & \cdot & \cdot & \cdot & \cdot \\ \mu_2 & -\lambda-\mu_2 & \cdot & \cdot & \cdot \\ \cdot & \mu_2 & -\lambda-\mu_2 & \cdot & \cdot \\ \cdot & \cdot & \mu_2 & -\lambda-\mu_2 & \cdot \\ \cdot & \cdot & \cdot & \mu_2 & -\lambda-\mu_2 \end{bmatrix}.$$

1.6 Multiprogramming Queues

The system depicted in Figure 1.9 is another example with multiple queues and blocking; it is a simple model for a multiprogramming computer system (Latouche [42]). The assumptions are as follows: the customers are programs which alternately require CPU and I/O operations. At the end of each CPU operation, a program may either require an I/O operation with probability p or leave the system with probability q. Each I/O operation is followed by a request for CPU. The total number of programs allowed in the loop is bounded by C, the multiprogramming level; the customers in excess must wait outside, in an infinite buffer. When a customer leaves the system, it is replaced by a customer from the outside buffer, if the latter is not empty. Arrivals occur according to a Poisson process with parameter λ, and CPU and I/O operations are exponentially distributed, respectively, with parameters α and β.

If we respectively denote by $N_0(t), N_1(t)$, and $N_2(t)$ the number of customers in the outside queue, the number requiring service at the CPU, and the number requiring service at the I/O server at time t, we see that $\{(N_0(t), N_1(t), N_2(t)),\ t \geq 0\}$ is a Markov process with the following constraints on the state space: first, we have that $N_1(t) + N_2(t) \leq C$; second, we have that if $N_0(t) \geq 1$, then necessarily $N_1(t) + N_2(t) = C$.

Because of the last constraint, the three-dimensional Markov process behaves very much as a two-dimensional process. More precisely, we shall assume that the state space is $\cup_{i \geq 0} s(i)$, where

$$s(n) = \{(n-C, 0), \ldots, (n-C, C)\} \qquad \text{for } n \geq C+1$$

and

$$s(n) = \{(0,0,n), (0,1,n-1), \ldots, (0, n-1, 1), (0, n, 0)\}$$

for $0 \leq n \leq C$.

The interpretation is as follows: if the state at time t is $(0, i, j)$, this means that $N_0(t) = 0, N_1(t) = i$, and $N_2(t) = j$; if the state at time t is (i, j), this means that $N_0(t) = i, N_1(t) = j$, and $N_2(t) = C - j$; the states are ordered by increasing values of the total number of customers in the system and for a fixed total, by increasing values of N_1. The transitions are enumerated in Table 1.2.

The structure of the infinitesimal generator is more complex than in the previous examples due to the more involved behavior at the boundary, as we see below. That matrix Q is given by

From	To	Rate	
$(0,i,j)$	$(0,i-1,j)$	αq	for $i \geq 1, i+j \leq C$
$(0,i,j)$	$(0,i-1,j+1)$	αp	for $i \geq 1, i+j \leq C$
$(0,i,j)$	$(0,i+1,j-1)$	β	for $j \geq 1, i+j \leq C$
$(0,i,j)$	$(0,i+1,j)$	λ	for $i+j \leq C-1$
$(0,i,j)$	$(1,i)$	λ	for $i+j = C$
$(1,i)$	$(0,i,C-i)$	αq	for $i \geq 1$
(n,i)	$(n-1,i)$	αq	for $n \geq 2, i \geq 1$
(n,i)	$(n,i-1)$	αp	for $n \geq 1, i \geq 1$
(n,i)	$(n,i+1)$	β	for $n \geq 1, i \leq C-1$
(n,i)	$(n+1,i)$	λ	for $n \geq 1$

Table 1.2: Transitions for the multiprogramming system of Figure 1.9.

$$Q = \begin{bmatrix} A_{00} & A_{01} & 0 & \cdots & 0 & 0 & 0 & 0 & \cdots \\ A_{10} & A_{11} & A_{12} & \cdots & 0 & 0 & 0 & 0 & \cdots \\ 0 & A_{21} & A_{22} & \cdots & 0 & 0 & 0 & 0 & \cdots \\ \vdots & \vdots & \vdots & \ddots & \vdots & \vdots & \vdots & \vdots & \ddots \\ 0 & 0 & 0 & \cdots & A_{C-1,C-1} & A_{C-1,C} & 0 & 0 & \cdots \\ 0 & 0 & 0 & \cdots & A_{C,C-1} & A_1 & \lambda I & 0 & \cdots \\ 0 & 0 & 0 & \cdots & 0 & A_2 & A_1 & \lambda I & \cdots \\ 0 & 0 & 0 & \cdots & 0 & 0 & A_2 & A_1 & \cdots \\ \vdots & \vdots & \vdots & \ddots & \vdots & \vdots & \vdots & \vdots & \ddots \end{bmatrix}, \tag{1.13}$$

where A_2 and A_1 are square matrices of order $C+1$, A_{ii} is a square matrix of order $i+1$ for $0 \leq i \leq C-1$, $A_{i,i+1}$ is a rectangular $(i+1) \times (i+2)$ matrix for $0 \leq i \leq C-1$, and $A_{i,i-1}$ is a rectangular $(i+1) \times i$ matrix for $1 \leq i \leq C$. These matrices are given by

$$A_2 = \begin{bmatrix} 0 & \cdot & \cdots & \cdot & \cdot \\ \cdot & \alpha q & \cdots & \cdot & \cdot \\ \vdots & \vdots & \ddots & \vdots & \vdots \\ \cdot & \cdot & \vdots & \alpha q & \cdot \\ \cdot & \cdot & \vdots & \cdot & \alpha q \end{bmatrix},$$

$$A_{i,i-1} = \begin{bmatrix} 0 & \cdot & \cdot & \cdots & \cdot & \cdot \\ \alpha q & 0 & \cdot & \cdots & \cdot & \cdot \\ \cdot & \alpha q & 0 & \cdots & \cdot & \cdot \\ \cdot & \cdot & \alpha q & \cdots & \cdot & \cdot \\ \vdots & \vdots & \vdots & \ddots & \vdots & \vdots \\ \cdot & \cdot & \cdot & \cdots & \alpha q & 0 \\ \cdot & \cdot & \cdot & \cdots & \cdot & \alpha q \end{bmatrix} \quad \text{for } 1 \le i \le C,$$

$$A_1 = \begin{bmatrix} -\lambda-\beta & \beta & \cdot & \cdots & \cdot & \cdot \\ \alpha p & -\lambda-\alpha-\beta & \beta & \cdots & \cdot & \cdot \\ \cdot & \alpha p & -\lambda-\alpha-\beta & \cdots & \cdot & \cdot \\ \vdots & \vdots & \vdots & \ddots & \vdots & \vdots \\ \cdot & \cdot & \cdot & \cdots & -\lambda-\alpha-\beta & \beta \\ \cdot & \cdot & \cdot & \cdots & \alpha p & -\lambda-\alpha \end{bmatrix},$$

$A_{00} = -\lambda$, A_{ii} has the same structure as A_1 for $1 \le i \le C-1$, and

$$A_{i,i+1} = \begin{bmatrix} 0 & \lambda & \cdot & \cdot & \cdots & \cdot & \cdot \\ \cdot & 0 & \lambda & \cdot & \cdots & \cdot & \cdot \\ \cdot & \cdot & 0 & \lambda & \cdots & \cdot & \cdot \\ \vdots & \vdots & \vdots & \vdots & \ddots & \cdot & \cdot \\ \cdot & \cdot & \cdot & \cdot & \cdots & \lambda & \cdot \\ \cdot & \cdot & \cdot & \cdot & \cdots & 0 & \lambda \end{bmatrix} \quad \text{for } 0 \le i \le C-1.$$

Despite the complex boundary behavior, the generator (1.13) does conform to the general form (1.7) if we set $m = C+1, m' = C(C+1)/2$,

$$B_1 = \begin{bmatrix} A_{00} & A_{01} & 0 & \cdots & 0 \\ A_{10} & A_{11} & A_{12} & \cdots & 0 \\ 0 & A_{21} & A_{22} & \cdots & 0 \\ \vdots & \vdots & \vdots & \ddots & \vdots \\ 0 & 0 & 0 & \cdots & A_{C-1,C-1} \end{bmatrix},$$

$$B_0 = \begin{bmatrix} 0 \\ 0 \\ 0 \\ \vdots \\ A_{C-1,C} \end{bmatrix},$$

and

$$B_2 = [0 \; 0 \; 0 \; \cdots \; A_{C,C-1}].$$

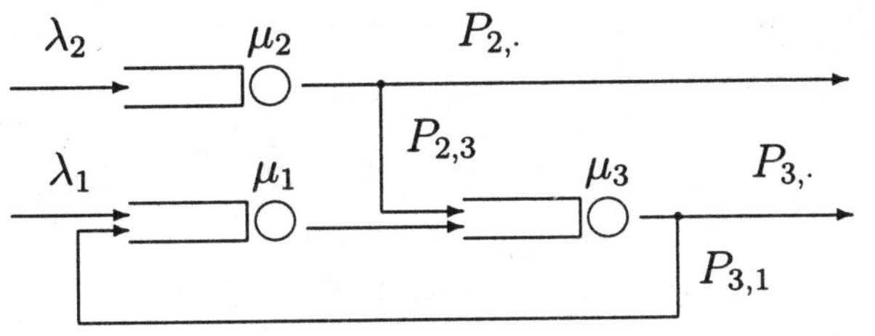

Figure 1.10: Diagrammatic representation of a Jackson network with three nodes. Arrivals may occur at the nodes 1 and 2 with rates λ_1 and λ_2, respectively; the service rate at node i is μ_i; departures occur from nodes 2 and 3 with probability $P_{2,\cdot}$ and $P_{3,\cdot}$, respectively.

1.7 Open Jackson Networks

The traditional treatment of Jackson networks (and product form networks in general) uses the ideas of partial and detailed balance, quasi-reversible queues, etc. (Kelly [38], Lavenberg [57]) and is very different from the way we shall analyze QBDs in this book. Nevertheless, Jackson networks may be viewed as QBDs as we show here.

An example of an open Jackson network is given in Figure 1.10. These are networks of exponential servers with infinite buffers, fed from the outside by Poisson processes. At the end of a service at station i, a customer may decide to leave the system with probability $P_{i,\cdot}$ or to enter the queue in front of server j with the probability $P_{i,j}$. In this particular example, there are three stations, two Poisson arrival processes to the stations 1 and 2 but no direct arrival to station 3; upon completion of their service at station 1, the customers move to station 3 with probability $P_{1,3} = 1$; upon completion of their service at station 2, the customers may either leave the system with probability $P_{2,\cdot}$, or move to station 3 with probability $P_{2,3}$; upon completion of their service at station 3, customers leave the system with probability $P_{3,\cdot}$ or feed back to station 1 with probability $P_{3,1}$.

This is a Markov process on a three-dimensional state space, infinite in each dimension: $\{(n_1, n_2, n_3); n_1 \geq 0, n_2 \geq 0, n_3 \geq 0\}$. The

From	To	Rate	
(n_1, n_2, n_3)	(n_1+1, n_2, n_3)	λ_1	for $n_1, n_2, n_3 \geq 0$
(n_1, n_2, n_3)	(n_1+1, n_2, n_3-1)	$\mu_3 P_{3,1}$	for $n_1, n_2 \geq 0; n_3 \geq 1$
(n_1, n_2, n_3)	(n_1-1, n_2, n_3+1)	μ_1	for $n_1 \geq 1; n_2, n_3 \geq 0$
(n_1, n_2, n_3)	(n_1, n_2+1, n_3)	λ_2	for $n_1, n_2, n_3 \geq 0$
(n_1, n_2, n_3)	(n_1, n_2-1, n_3)	$\mu_2 P_{2,\cdot}$	for $n_1, n_3 \geq 0; n_2 \geq 1$
(n_1, n_2, n_3)	(n_1, n_2-1, n_3+1)	$\mu_2 P_{2,3}$	for $n_1, n_3 \geq 0; n_2 \geq 1$
(n_1, n_2, n_3)	(n_1, n_2, n_3-1)	$\mu_3 P_{3,\cdot}$	for $n_1, n_2 \geq 0; n_3 \geq 1$

Table 1.3: Transitions for the Jackson network of Figure 1.10.

---

transitions are given in Table 1.3.

If we define the level $\ell(n)$ as the subset of states where there are n customers in the station 1, we find that $\ell(n)$ is two dimensional[2] since the states in $\ell(n)$ are indexed by (n_2, n_3). Thus, the infinitesimal generator has the structure (1.7), where the blocks B_i and A_i, $i = 0, 1, 2$, are doubly infinite. One may verify that

$$A_0 = \begin{bmatrix} \tilde{A}_0 & 0 & 0 & \cdots \\ 0 & \tilde{A}_0 & 0 & \cdots \\ 0 & 0 & \tilde{A}_0 & \cdots \\ \vdots & \vdots & \vdots & \ddots \end{bmatrix} = B_0,$$

with

$$\tilde{A}_0 = \begin{bmatrix} \lambda_1 & \cdot & \cdot & \cdots \\ \mu_3 P_{3,1} & \lambda_1 & 0 & \cdots \\ \cdot & \mu_3 P_{3,1} & \lambda_1 & \cdots \\ \vdots & \vdots & \vdots & \ddots \end{bmatrix};$$

[2]We have already observed one such example in section 1.3 on the PH/PH/1 queue.

that

$$A_2 = \begin{bmatrix} \tilde{A}_2 & 0 & 0 & \cdots \\ 0 & \tilde{A}_2 & 0 & \cdots \\ 0 & 0 & \tilde{A}_2 & \cdots \\ \vdots & \vdots & \vdots & \ddots \end{bmatrix} = B_2,$$

with

$$\tilde{A}_2 = \begin{bmatrix} \cdot & \mu_1 & \cdot & \cdots \\ \cdot & \cdot & \mu_1 & \cdots \\ \cdot & \cdot & \cdot & \cdots \\ \vdots & \vdots & \vdots & \ddots \end{bmatrix};$$

that

$$A_1 = \begin{bmatrix} \tilde{A}_{11} & \tilde{A}_{10} & 0 & \cdots \\ \tilde{A}_{12} & \tilde{A}_{11} & \tilde{A}_{10} & \cdots \\ 0 & \tilde{A}_{12} & \tilde{A}_{11} & \cdots \\ \vdots & \vdots & \vdots & \ddots \end{bmatrix},$$

with

$$\tilde{A}_{10} = \begin{bmatrix} \lambda_2 & \cdot & \cdot & \cdots \\ \cdot & \lambda_2 & \cdot & \cdots \\ \cdot & \cdot & \lambda_2 & \cdots \\ \vdots & \vdots & \vdots & \ddots \end{bmatrix},$$

$$\tilde{A}_{12} = \begin{bmatrix} \mu_2 P_{2,\cdot} & \mu_2 P_{2,3} & \cdot & \cdots \\ \cdot & \mu_2 P_{2,\cdot} & \mu_2 P_{2,3} & \cdots \\ \cdot & \cdot & \mu_2 P_{2,\cdot} & \cdots \\ \vdots & \vdots & \vdots & \ddots \end{bmatrix},$$

$$\tilde{A}_{11} = \begin{bmatrix} \cdot & \cdot & \cdot & \cdots \\ \mu_3 P_{3,\cdot} & -\mu_3 & \cdot & \cdots \\ \cdot & \mu_3 P_{3,\cdot} & -\mu_3 & \cdots \\ \vdots & \vdots & \vdots & \ddots \end{bmatrix} - (\lambda_1 + \lambda_2 + \mu_1 + \mu_2) I;$$

and that

$$B_1 = A_1 + \mu_1 I.$$

With this example, and those in the preceding sections, we see that the notion of *levels* covers situations which are very different. In section 1.2,

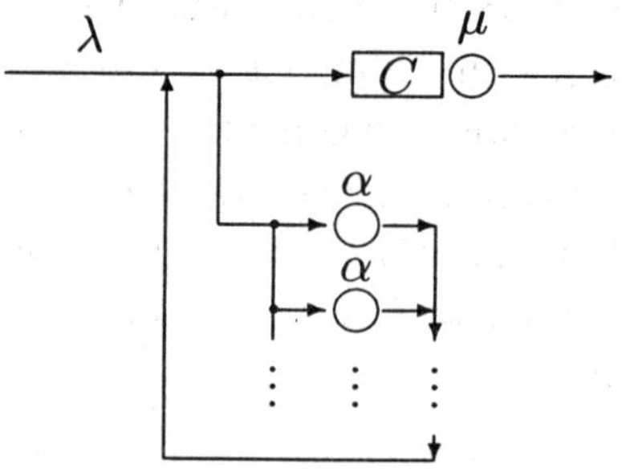

Figure 1.11: Diagrammatic representation of an M/M/1/C queue with retrial. The arrival rate is equal to λ; the system capacity is C; the service rate is μ; the retrial rate is α. The retrial delay is modeled as an infinite server queue.

the phase within the levels is merely an artifact, which allows one to model random variations in the arrival and service mechanisms. Here, the phase covers the whole remainder of the queueing network. We shall discuss in Chapter 15 the connections between Jackson networks and QBDs in a simple setting.

1.8 A Retrial Queue

We conclude this chapter by mentioning that there exist processes for which the infinitesimal generator is block-tridiagonal but does not conform to the model of (1.7) because the transition rates are level dependent. One such example is the M/M/1/C queue with retrial depicted in Figure 1.11 and described below (Bright and Taylor [10]).

The system comprises one server with a finite buffer and total capacity C; customers are served in the order in which they enter the buffer, the service times are exponentially distributed with parameter μ. New customers arrive according to a Poisson process with parameter λ. If the buffer is full, a newly arriving customer waits for an exponentially distributed interval with parameter α; at the end of the waiting period, the customer attempts again to enter the buffer; if the buffer is still full, the customer waits for a new interval, etc.

If we denote by $N_a(t)$ the number of customers who have arrived but have not yet entered the system and by $N_i(t)$ the number of customers who are in the system, either waiting or being served, we find that the process $\{(N_a(t), N_i(t)), t \geq 0\}$ is a Markov process on the levels $\cup_{n\geq 0}\ell(n)$, where $\ell(n) = \{(n,0),(n,1),\ldots,(n,C)\}$. The possible transitions are as follows.

From	To	Rate	
(n_a, C)	(n_a+1, C)	λ	for $n_a \geq 0$
(n_a, n_i)	(n_a, n_i+1)	λ	for $n_a \geq 0, 0 \leq n_i \leq C-1$
(n_a, n_i)	(n_a, n_i-1)	μ	for $n_a \geq 0, 1 \leq n_i \leq C$
(n_a, n_i)	(n_a-1, n_i+1)	$n_a\alpha$	for $n_a \geq 1, 0 \leq n_i \leq C-1$

The infinitesimal generator is given by

$$Q = \begin{bmatrix} B_1 & A_0 & 0 & 0 & \cdots \\ A_{2,1} & A_{1,1} & A_0 & 0 & \cdots \\ 0 & A_{2,2} & A_{1,2} & A_0 & \cdots \\ 0 & 0 & A_{2,3} & A_{1,3} & \cdots \\ \vdots & \vdots & \vdots & \vdots & \ddots \end{bmatrix},$$

where the blocks are of order $C+1$; they are displayed below in the case where $C = 4$:

$$B_1 = \begin{bmatrix} -\lambda & \lambda & \cdot & \cdot & \cdot \\ \cdot & -\lambda & \lambda & \cdot & \cdot \\ \cdot & \cdot & -\lambda & \lambda & \cdot \\ \cdot & \cdot & \cdot & -\lambda & \lambda \\ \cdot & \cdot & \cdot & \cdot & -\lambda \end{bmatrix},$$

$$A_0 = \begin{bmatrix} \cdot & \cdot & \cdot & \cdot & \cdot \\ \cdot & \cdot & \cdot & \cdot & \cdot \\ \cdot & \cdot & \cdot & \cdot & \cdot \\ \cdot & \cdot & \cdot & \cdot & \cdot \\ \cdot & \cdot & \cdot & \cdot & \lambda \end{bmatrix},$$

$$A_{2,n} = \begin{bmatrix} \cdot & n\alpha & \cdot & \cdot & \cdot \\ \cdot & \cdot & n\alpha & \cdot & \cdot \\ \cdot & \cdot & \cdot & n\alpha & \cdot \\ \cdot & \cdot & \cdot & \cdot & n\alpha \\ \cdot & \cdot & \cdot & \cdot & \cdot \end{bmatrix},$$

$$A_{1,n} = \begin{bmatrix} -\lambda - n\alpha & \lambda & \cdot & \cdot & \cdot \\ \mu & -\lambda - n\alpha - \mu & \lambda & \cdot & \cdot \\ \cdot & \mu & -\lambda - n\alpha - \mu & \lambda & \cdot \\ \cdot & \cdot & \mu & -\lambda - n\alpha - \mu & \lambda \\ \cdot & \cdot & \cdot & \mu & -\lambda - \mu \end{bmatrix}.$$

QBDs with level-dependent transitions will be discussed briefly in Chapters 4 and 12.

Part II

The Method of Phases

Chapter 2

PH Distributions

This part of the book deals with PH distributions and related point processes, which provide a versatile set of tractable models for applied probability. They are based on the method of stages, a technique introduced by A. K. Erlang and generalized by M. F. Neuts. The key idea is to model random time intervals as being made up of a (possibly random) number of exponentially distributed segments and to exploit the resulting Markovian structure to simplify the analysis.

We start with a chapter on PH distributions for several reasons. First and foremost, they provide us with the simplest introduction to matrix analytic methods: the distribution of a random variable is defined through a matrix; its density function, moments, etc. are expressed with this matrix. Furthermore, matrix analytic methods involve a constant interplay between formal algebraic manipulations of mathematical expressions and reasoning about the probabilistic interpretation of these expressions; this is amply illustrated in this chapter.

In addition, PH distributions provide a simple framework to demonstrate how one can extend many simple results on exponential distributions to more complex models without losing computational tractability. Finally, phase models provide convenient examples for the paradigms for which matrix analytic methods apply.

We begin by introducing continuous PH distributions as a natural generalization of the exponential and Erlang distributions. We next examine the properties that make PH distributions a versatile class for modeling purposes. We then define discrete PH distributions and show

how much the discrete and the continuous cases are similar. We shall finally discuss some algorithms for computing the distribution function and the probability density function.

2.1 The Exponential Distribution

The exponential distribution has density

$$f(x) = \lambda e^{-\lambda x} \qquad \text{for } x \geq 0$$

and enjoys the status of a mainstay in applied probability for several reasons. First and foremost, it has the memoryless property

$$\mathrm{P}[X > t + s | X > t] = \mathrm{P}[X > s] \qquad \text{for all } t, s \geq 0.$$

This provides tremendous ease in conditioning arguments and results in a Markovian structure of models which involve the exponential distribution. For instance, it is because of the memoryless property that we obtain a Markov process for queues like the M/M/C, that there is an embedded Markov chain at departure epochs for the M/G/1 queue, etc.

Example 2.1.1 Consider a Poisson process of rate λ; let $\{a_n : n \geq 0\}$ be the probability distribution of the number of events in a random interval of length X exponentially distributed with parameter μ. We have that

$$a_0 = \int_0^\infty e^{-\lambda x} \mu e^{-\mu x} dx = \frac{\mu}{\lambda + \mu}$$

because, given X, the conditional probability of having no events in the interval is $e^{-\lambda X}$.

For $n \geq 1$, we condition on the epoch U of the first event and write

$$a_n = \int_0^\infty \lambda e^{-\lambda u} e^{-\mu u} du \; a_{n-1} = \frac{\lambda}{\lambda + \mu} a_{n-1};$$

this is so because we have $n \geq 1$ events in an interval of length X if and only if

a) the random interval X is greater than U; the probability of this given U is of course $e^{-\mu U}$, and in addition,

b) $n-1$ events occur in the remaining part of the interval. The probability of this is a_{n-1}, due to the memoryless property which entails that given $X > u$, the conditional distribution of the remaining interval $X-u$ is exponential with parameter μ.

Thus, we see that $\{a_n\}$ is a geometric distribution.

A second important property of the exponential distribution is that the exponential kernel leads to significant simplifications in various integral and differential equations; we give an illustration of this in the example below.

Example 2.1.2 Consider the M/G/∞ queue with Poisson arrivals at rate λ and general service time distribution $H(\cdot)$. We denote by $N(t)$ the number of customers present at time t and by $P(z,t)$ the probability generating function of $N(t)$ given that $N(0)=0$:

$$P(z,t) = \mathrm{E}[z^{N(t)}|N(0)=0].$$

To compute $P(z,t)$, we make a simplifying assumption which does not change $N(t)$: the first arrival is served by server 1 and all other arrivals are served by servers 2, 3, ... With this, we can evaluate $P(z,t)$ through a simple regenerative argument. There are three cases to consider.

Case 1. There are no arrivals in $(0,t]$: The probability of this event is $e^{-\lambda t}$. Under this event, $N(t)=0$, and therefore the conditional probability generating function of $N(t)$ in this case is $z^0=1$.

Case 2. The first arrival occurs in $(t-u, t-u+du)$ and is served before time t: The elementary probability of this event is $\lambda e^{-\lambda(t-u)} du H(u)$. The first customer does not contribute to $N(t)$ since he leaves before t. In the time interval $[t-u,t]$, the servers 2, 3, etc. behave as an independent M/G/∞ system starting empty, and $N(t)$ is also the number of customers in that system at time t. Therefore, under the conditions of this case, $N(t)$ is distributed like $N(u)$ and its probability generating function is $P(z,u)$.

Case 3. The first arrival occurs in $(t-u, t-u+du)$ and is still in the system at time t: The elementary probability of this event is $\lambda e^{-\lambda(t-u)}du(1-H(u))$. Then, $N(t)$ is distributed like $1+N(u)$, and its conditional probability generating function is $zP(z,u)$.

Using the law of total probability, we can now write

$$P(z,t) = e^{-\lambda t} + \int_0^t \lambda e^{-\lambda(t-u)}[H(u) + z(1-H(u))]P(z,u)du. \qquad (2.1)$$

Multiplying both sides by $e^{\lambda t}$ and differentiating with respect to t, we obtain the following differential equation:

$$P'(z,t) = \lambda(z-1)[1-H(t)]P(z,t)$$

with

$$P(z,0) = 1,$$

which gives

$$P(z,t) = \exp\left\{\lambda(z-1)\int_0^t [1-H(u)]du\right\}.$$

Thus, $N(t)$ has a Poisson distribution with parameter $\lambda \int_0^t [1-H(u)]du$; furthermore, the limiting distribution as $t \to \infty$ is Poisson with parameter $\lambda \mathrm{E}[S]$, where $\mathrm{E}[S]$ is the mean service time.

For the GI/G/∞ queue, which has a general renewal arrival process instead of a Poisson process, we may repeat the argument and obtain an integral equation similar to (2.1). However, it will not, in general, be easy to solve. It is the exponential kernel which allows us to solve (2.1).

2.2 Simple Generalizations

Noting the simplicity of the exponential distribution, applied probabilists use it as a building block to construct more complex distributions. This is accomplished by considering simple operations like convolutions and finite mixtures of exponential distributions. For instance, the Erlang distribution $E(n,\lambda)$, which is a member of the Gamma family and has density function

$$f(x) = \frac{\lambda^n}{(n-1)!}x^{n-1}e^{-\lambda x} \qquad \text{for } x \geq 0, \qquad (2.2)$$

is the distribution of a sum of n independent exponential random variables with parameter λ.

For the exponential distribution, the Laplace–Stieltjes transform is given by $\phi(s) = \lambda/(\lambda + s)$, a rational function. From this it follows that distributions obtained as convolutions and finite mixtures of exponential distributions will have rational Laplace–Stieltjes transforms. A rational transform lends itself conveniently to analysis using partial fraction expansions, and one way of generalizing the construct is to consider a general distribution with rational Laplace–Stieltjes transform. That approach may be found in standard books on queueing theory; see, for example, Cohen [16].

Here, however, we generalize the exponential distribution on the basis of a different consideration: we shall exploit the fact that many distributions derived from the exponential can be viewed as the distribution of the time till absorption in suitably defined Markov processes. As many of our results will demonstrate, this allows one to deal with them without resorting to transform methods and to simplify the analysis by appealing to the simple dependency structure underlying Markov processes.

For instance, let us consider a Markov process with state space $\{0, 1, \ldots, n\}$, initial probability vector $(0, 1, 0, \ldots, 0)$, and infinitesimal generator

$$Q = \begin{bmatrix} 0 & 0 & 0 & \cdots & 0 & 0 \\ 0 & -\lambda & \lambda & \cdots & 0 & 0 \\ 0 & 0 & -\lambda & \cdots & 0 & 0 \\ \vdots & \vdots & \vdots & \ddots & \vdots & \vdots \\ 0 & 0 & 0 & \cdots & -\lambda & \lambda \\ \lambda & 0 & 0 & \cdots & 0 & -\lambda \end{bmatrix}.$$

The transition diagram is given in Figure 2.1. The Markov process starts in state 1 and traverses through the successive states until it is absorbed into state 0. The time to reach the absorbing state is the sum of the exponential holding times in each of the states $1, 2, \ldots, n$; thus, the first passage time to absorption has an Erlang $E(n, \lambda)$ distribution.

Thus, when service times have an Erlang distribution, we may view the services as consisting of n successive phases (or stages), each taking an independent, identically, and exponentially distributed amount

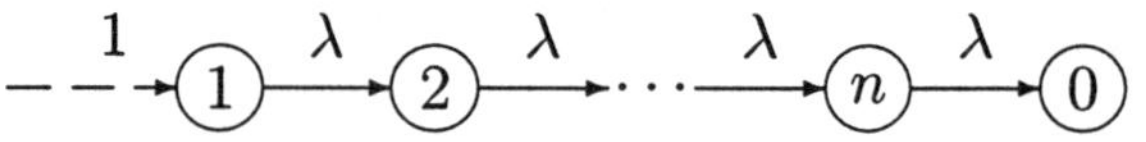

Figure 2.1: Phase-type representation of the E(n,λ) Erlang distribution.

of time. This idea of introducing an auxiliary variable called *phase* is a useful artifice in creating Markovian structures that enable easy computations.

As a second example, we consider the hyperexponential distribution

$$f(x) = \sum_{1 \leq k \leq n} \beta_k \lambda_k e^{-\lambda_k x}$$

($\beta_k > 0$ for all k, $\sum_{1 \leq k \leq n} \beta_k = 1$), which is the convex mixture of n exponential distributions. It is the absorption time of the Markov process shown in Figure 2.2.

The exponential distribution itself may be so represented as the distribution of the time till absorption in the trivial Markov process described below:

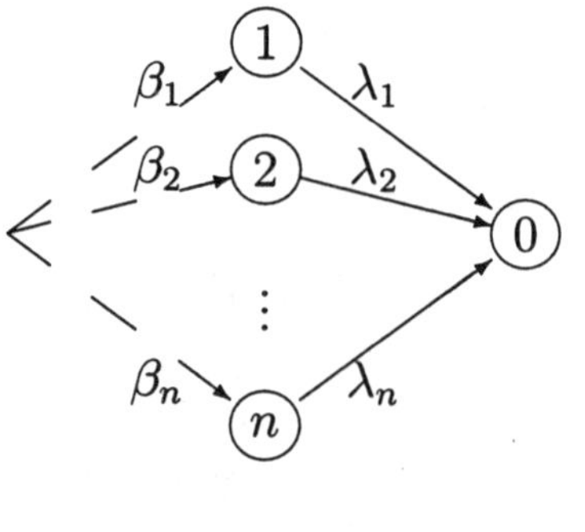

Figure 2.2: Phase-type representation of the hyperexponential distribution.

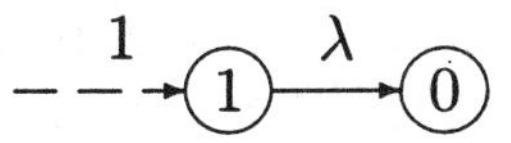

We are now ready to consider the general framework of PH distributions along the lines of Neuts [74, 79].

2.3 PH Random Variables

Since we only wish to rely on the connection to a Markov process, there is no need to restrict the transitions as in the examples of the previous section, and we may as well consider the general case. Thus, we define[3] a Markov process on the states $\{0, 1, \ldots, n\}$ with initial probability vector $(\tau_0, \boldsymbol{\tau})$ and infinitesimal generator

$$Q = \begin{bmatrix} 0 & \mathbf{0} \\ \boldsymbol{t} & T \end{bmatrix}, \tag{2.3}$$

where $\boldsymbol{\tau}$ is a row vector of size n, T is an $n \times n$ matrix, and $\boldsymbol{t}$ is a column vector of size n. Since Q is the generator of a Markov process, we have that

$$T_{ii} < 0, \qquad t_i \geq 0, \qquad T_{ij} \geq 0 \qquad \text{for } 1 \leq i \neq j \leq n \tag{2.4}$$

and

$$T\mathbf{1} + \boldsymbol{t} = \mathbf{0}, \tag{2.5}$$

where $\mathbf{1}$ is a column vector of 1's; we also have that

$$\tau_0 + \boldsymbol{\tau}\mathbf{1} = 1. \tag{2.6}$$

Definition 2.3.1 *The distribution of the time X till absorption into the absorbing state 0 is called the PH distribution with representation $(\boldsymbol{\tau}, T)$. We denote the distribution as PH$(\boldsymbol{\tau}, T)$ and use the shorthand X is PH$(\boldsymbol{\tau}, T)$.*

[3] Our notation is slightly different from that used in other work on phase-type random variables. The major difference is that we use the symbol $\boldsymbol{\tau}$ for the initial probability distribution, because we shall extensively use the symbol $\boldsymbol{\alpha}$ in another context later.

In view of (2.5), (2.6), $\boldsymbol{t}$ and τ_0 are implicitly given by T and $\boldsymbol{\tau}$, which is why we do not include them in the list of parameters providing a representation.

Example 2.3.2 Consider an equipment subject to three possible types of failures. When the equipment is new, the time to failure is exponentially distributed with mean 100 hours and the failure can be one of three types with probability 0.5, 0.4, and 0.1; of these the first two are nonfatal and the last is fatal.

A failure of Type 1 when repaired leaves the machine in its original new condition. A failure of type 2 when repaired decreases the mean time to next failure to 50 hours and also changes the respective probabilities of the next failure to be of types 1, 2, or 3 with probability 0.75, 0.2, 0.05, respectively. The time to repair either type of failure is exponentially distributed with mean one hour.

The lifetime of the equipment is a PH random variable with representation

$$\begin{aligned}\boldsymbol{\tau} &= \begin{bmatrix} 1 & 0 & 0 & 0 \end{bmatrix},\\ T &= \begin{bmatrix} -.01 & .005 & .004 & 0 \\ 1 & -1 & 0 & 0 \\ 0 & 0 & -1 & 1 \\ 0 & .015 & .004 & -.02 \end{bmatrix}.\end{aligned}$$

In state 1, the equipment functions like new. In state 4, it is functioning after the repair of a type 2 failure. In the states 2 and 3, it is undergoing repair, respectively, of a type 1 and a type 2 failure.

Example 2.3.3 The interoverflow time distribution from a c server queue with Poisson arrivals of rate λ, exponential servers of rate μ, and finite waiting room of size K (the $\mathrm{M/M}/c/c+K$ queue) is of phase-type. We show this by displaying the representation in the special case where $c = 3$ and $K = 2$:

$$\boldsymbol{\tau} = \begin{bmatrix} 0 & 0 & 0 & 0 & 0 & 1 \end{bmatrix},$$

$$T = \begin{bmatrix} -\lambda & \lambda & & & & \\ \mu & -\lambda-\mu & \lambda & & & \\ & 2\mu & -\lambda-2\mu & \lambda & & \\ & & 3\mu & -\lambda-3\mu & \lambda & \\ & & & 3\mu & -\lambda-3\mu & \lambda \\ & & & & 3\mu & -\lambda-3\mu \end{bmatrix}.$$

The transient states are numbered from 0 to $c + K$ and represent the number of customers in the system. The absorbing state represents the arrival of a customer who overflows because the system is full. Since we measure the time between two such overflows, the system is full at time 0.

2.4 Distribution and Moments

We shall now give a few basic properties of PH distributions.

Theorem 2.4.1 *Assume that X is PH($\boldsymbol{\tau}, T$). Its distribution function is given by*

$$F(x) = 1 - \boldsymbol{\tau} \exp\,(Tx)\mathbf{1} \qquad \textit{for } x \geq 0 \tag{2.7}$$

and its density function is given by

$$f(x) = \boldsymbol{\tau} \exp\,(Tx)\boldsymbol{t} \qquad \textit{for } x > 0, \tag{2.8}$$

where the matrix exponential is defined by

$$\exp(A) = \sum_{n \geq 0} \frac{1}{n!} A^n.$$

Proof. We associate with X a Markov process $\{J(x) : x \geq 0\}$ with the infinitesimal generator Q given in (2.3). The transition function $P(x)$ with elements $P_{ij}(x) = \mathrm{P}[J(x) = j | J(0) = i]$ is given by $P(x) = \exp(Qx)$ (see, e.g., Çinlar [12]), and it is easy to verify that

$$\exp(Qx) = \begin{bmatrix} 1 & \mathbf{0} \\ \mathbf{1} - \exp(Tx)\mathbf{1} & \exp(Tx) \end{bmatrix}.$$

We then immediately obtain that

$$\begin{aligned} F(x) &= \mathrm{P}[J(x)=0] = \sum_{0\le i\le n} \mathrm{P}[J(0)=i]\,\mathrm{P}[J(x)=0|J(0)=i] \\ &= \tau_0 + \sum_{1\le i\le n} \tau_i P_{i0}(x) = \tau_0 + \boldsymbol{\tau}\mathbf{1} - \boldsymbol{\tau}\exp(Tx)\mathbf{1}, \end{aligned}$$

which proves (2.7). Equation (2.8) follows from a direct computation of $F'(x)$ using $T\mathbf{1} = -\boldsymbol{t}$. □

The transition matrix $P(x) = \exp(Qx)$ is such that for every i and j, $P_{ij}(x)$ is either identically zero or strictly positive for all $x > 0$ (see Chung [14, Page 126]). Thus, since $\boldsymbol{\tau}$ and $\boldsymbol{t}$ are nonnegative vectors, the density $\boldsymbol{\tau}\exp(Tx)\boldsymbol{t}$ of a PH distribution is either identically 0 or strictly positive for all $x > 0$.

It must be observed that the representation is not unique; instead, there exist, for any PH distribution, infinitely many different representations. For instance, we have seen that the exponential distribution has the representation on page 39. Other representations of arbitrary order are given by pairs $(\boldsymbol{\tau}, T)$, where $\boldsymbol{\tau}$ is a probability vector and $T\mathbf{1} = -\lambda\mathbf{1}$ or $\boldsymbol{\tau}T = -\lambda\boldsymbol{\tau}$. To see this, expand the right-hand side of (2.7) and replace $\boldsymbol{\tau}T^n\mathbf{1}$ by $(-\lambda)^n$.

The distribution $\mathrm{PH}(\boldsymbol{\tau}, T)$ is nondefective if and only if $F(\infty) = 1$; that is, if and only if absorption occurs in a finite time almost surely (a.s.; i.e., with probability 1). Concerning nondefective PH distributions, we have the following theorem.

Theorem 2.4.2 *For any nondefective PH distribution, there exists a representation $(\boldsymbol{\tau}, T)$ satisfying the additional condition that starting from any phase i in $\{1, \ldots, n\}$, absorption into state 0 occurs with probability 1.*

Proof. Partition the set of phases $\{1, \ldots, n\}$ into two subsets A and B such that A consists of states from which absorption into 0 occurs almost surely (a.s.; i.e., with probability 1), while B consists of those states from which with positive probability absorption might not occur. We assert that $T_{ij} = 0$ if $i \in A$ and $j \in B$; otherwise, there is a positive probability that the Markov process moves from i to j and then fails to get absorbed, which contradicts the definition of A.

Furthermore, we have that $\tau_j = 0$ for all $j \in B$, otherwise the distribution would be defective.

It is now a trivial matter to verify that the distribution $\mathrm{PH}(\boldsymbol{\tau}_A, T_A)$ is the same as the distribution $\mathrm{PH}(\boldsymbol{\tau}, T)$, where $\boldsymbol{\tau}_A$ is the subvector of $\boldsymbol{\tau}$ corresponding to indices in A and T_A is the submatrix of T with row and column indices in A (verify, for example, that the distribution functions are equal). □

One may directly verify on the transition diagram whether a PH distribution is nondefective: it is necessary and sufficient that for every phase i such that $\tau_i > 0$, all the possible paths lead to the absorbing phase 0. There also exists a simple algebraic condition under which absorption into 0 in finite time occurs a.s.; it plays an important role as we shall see later, and we state it below as a theorem.

Theorem 2.4.3 *Consider a PH$\{\boldsymbol{\tau}, T\}$ distribution. Absorption into state 0 occurs with probability 1 from any phase i in $\{1, \ldots, n\}$ if and only if the matrix T is nonsingular.*

Moreover, $(-T^{-1})_{ij}$ is the expected total time spent in phase j during the time until absorption, given that the initial phase is i.

Proof. Let us denote by E_{ij} the expected sojourn time in phase j, starting from phase i, and let us assume that absorption occurs a.s. This implies that E_{ij} is finite for all i and j. By conditioning on the first phase visited after i, we readily obtain that

$$\begin{aligned} E_{ij} &= \sum_{k \neq i} \frac{T_{ik}}{-T_{ii}} E_{kj} \qquad \text{for } i \neq j, \\ E_{ii} &= \frac{1}{-T_{ii}} + \sum_{k \neq i} \frac{T_{ik}}{-T_{ii}} E_{ki} \end{aligned}$$

for $1 \leq i, j \leq n$, which is equivalent to $TE = -I$. Thus, T is nonsingular and $E = -T^{-1}$.

Assume now that T^{-1} exists, and define $\boldsymbol{\phi}(x) = \exp(Tx)\mathbf{1}$; note that $0 \leq \phi_i(x) \leq 1$ for all i and x since it is the probability, starting from the phase i, of being in any one of the phases 1 to n at time x.

Now, one has that

$$\exp(Tt) = I + \int_0^t T \exp(Tx) dx; \tag{2.9}$$

the proof of this is by differentiation of both sides and verification that both sides are equal at $t = 0$.

If we pre- and postmultiply both sides of (2.9), respectively, by T^{-1} and $\mathbf{1}$, we obtain that

$$T^{-1}\boldsymbol{\phi}(t) = T^{-1}\mathbf{1} + \left\{\int_0^t \exp(Tx)dx\right\}\mathbf{1}. \tag{2.10}$$

Since $\mathbf{0} \leq \boldsymbol{\phi}(t) \leq \mathbf{1}$, the left-hand side of (2.10) is bounded. Therefore, the increasing function $\int_0^t \exp(Tx)dx$ on the right must have a finite limit as t goes to infinity.

Since $\exp(Tx)_{ij}$ is the probability of being in phase j at time x, given that the initial phase is i, we have that E is equal to $\int_0^\infty \exp(Tx)dx$. Thus, E is finite, and therefore all states are transient. □

Remark 2.4.4 A direct consequence of the theorem above is that $-T^{-1} \geq 0$.

Remark 2.4.5 An algebraic proof of Theorem 2.4.3 goes as follows. A matrix T which satisfies (2.4), (2.5) is called a semistable matrix, and it can be shown that the real part of each eigenvalue is negative. Furthermore, T is invertible if and only if the real part of each eigenvalue is strictly negative, in which case T is called a stable matrix.

If T is stable, then $\lim_{t\to\infty} \exp(Tt) = 0$ (to see this, one expands $\exp(Tt)$ and replaces T by its Jordan form). Thus, taking the limit on both sides of (2.9), we find that

$$\int_0^\infty \exp(Tx)dx = -T^{-1}. \tag{2.11}$$

We shall henceforth only consider nondefective PH distributions and assume without loss of generality that the rate matrix T is nonsingular. With this assumption, using (2.8), one shows that the Laplace–Stieltjes transform of PH$(\boldsymbol{\tau}, T)$ is given by

$$\phi(s) = \tau_0 + \boldsymbol{\tau}(sI - T)^{-1}\boldsymbol{t} \qquad \text{for } \mathrm{Re}(s) \geq 0. \tag{2.12}$$

Differentiating k times with respect to s and setting $s = 0$, we obtain the moments

$$\mathrm{E}[X^k] = k!\,\boldsymbol{\tau}(-T^{-1})^k\mathbf{1} \qquad \text{for } k \geq 1. \tag{2.13}$$

These formulas lend themselves well to numerical computations, as we shall see in section 2.8. One may also recover known analytic expressions in simple cases. For instance, if X has the Erlang $E(n,\lambda)$ distribution, with representation

$$\begin{aligned}
\boldsymbol{\tau} &= \begin{bmatrix} 1 & 0 & \dots & 0 \end{bmatrix}, \\
T &= \begin{bmatrix} -\lambda & \lambda & & & \\ & -\lambda & \ddots & & \\ & & \ddots & \lambda & \\ & & & -\lambda & \lambda \\ & & & & -\lambda \end{bmatrix},
\end{aligned}$$

then $T^k_{i,i+j} = (-1)^{k+j}\binom{k}{j}\lambda^k$ for all $k \geq 0$, $1 \leq i \leq i+j \leq n$, and the other elements are equal to 0. From (2.8), we obtain that

$$\begin{aligned}
f(x) &= \sum_{k\geq 0} \frac{1}{k!} x^k \boldsymbol{\tau} T^k \boldsymbol{t} \\
&= \sum_{k\geq n-1} \frac{1}{k!}(-1)^{k+n+1} x^k \lambda^{k+1} \binom{k}{n-1} \\
&= \frac{\lambda^n}{(n-1)!} x^{n-1} e^{-\lambda x},
\end{aligned}$$

which agrees with (2.2). Similarly, since $T^{-1}_{i,i+j} = -\lambda^{-1}$ for $1 \leq i \leq i+j \leq n$ and 0 otherwise, one readily finds the well-known expressions for the moments of the Erlang distribution.

Example 2.4.6 This is a generalization of Example 2.1.1. Our purpose is to show how the fact that PH distributions are associated to Markov processes simplifies conditioning arguments. We consider the distribution $\{a_k : k \geq 0\}$ of the number of Poisson events (of rate λ) during a random interval X with PH$(\boldsymbol{\tau}, T)$ distribution. For simplicity, we assume that $\boldsymbol{\tau}\mathbf{1} = 1$, so that X is strictly positive a.s.

We shall consider X as the time till absorption in phase 0 for the Markov process $\{J(t) : t \geq 0\}$ with the generator Q in (2.3).

We define the vectors $\boldsymbol{\xi}(k)$ such that $\xi_i(k)$ $(k \geq 0, 1 \leq i \leq n)$ is the probability that k events occur during X given that $J(0) = i$. Clearly, $a_k = \boldsymbol{\tau}\boldsymbol{\xi}(k)$. We have that

$$\boldsymbol{\xi}(0) = (\lambda I - T)^{-1}\boldsymbol{t},$$

because the conditional probability of having no event given X is $e^{-\lambda X}$ and the density function of X given $J(0) = i$ is the ith component of $\exp(Tx)\boldsymbol{t}$, so that

$$\begin{aligned} \boldsymbol{\xi}(0) &= \int_0^\infty e^{-\lambda x} \exp(Tx)\boldsymbol{t}dx \\ &= \int_0^\infty \exp\{(-\lambda I + T)x\}\boldsymbol{t}dx \\ &= (\lambda I - T)^{-1}\boldsymbol{t} \end{aligned}$$

by (2.11).

For $k \geq 1$, we condition on the epoch U of the first event *and* the phase $J(u)$ of the Markov process at that time. We obtain that

$$\xi_i(k) = \int_0^\infty \lambda e^{-\lambda u} du \sum_{1 \leq j \leq n} (\exp(Tu))_{ij}\xi_j(k-1),$$

because we have k events in the interval X if and only if

> at the epoch u of the first event, the Markov process is still in a transient phase j, the probability of which is $(\exp Tu)_{ij}$,

and in addition,

> $k-1$ events occur in the remaining part of X. The probability of this, given that $J(U) = j$, is $\xi_j(k-1)$ by the strong Markov property.

In vector notations, we have that

$$\begin{aligned} \boldsymbol{\xi}(k) &= \int_0^\infty \lambda e^{-\lambda u} du \exp(Tu)\boldsymbol{\xi}(k-1) \\ &= \lambda(\lambda I - T)^{-1}\boldsymbol{\xi}(k-1). \end{aligned}$$

Observe that we used here the Markov property just as we used earlier the memoryless property.

2.5 Discrete PH Distributions

Just as we defined continuous PH distributions in the preceding section, we define a discrete PH distribution $\mathrm{PH}_d(\boldsymbol{\tau}, T)$ by considering the absorption time into state 0 in the discrete time Markov chain with initial probability vector $(\tau_0, \boldsymbol{\tau})$ and transition probability matrix

$$P = \begin{bmatrix} 1 & \mathbf{0} \\ \boldsymbol{t} & T \end{bmatrix}.$$

Here, we have that $T_{ij} \geq 0$ and $t_i \geq 0$ for $1 \leq i, j \leq n$ and that $\boldsymbol{t} + T\mathbf{1} = \mathbf{1}$. We omit the subscript d and write $\mathrm{PH}(\boldsymbol{\tau}, T)$ when the context clearly indicates that we are dealing with a discrete PH distribution.

The discrete analogues of the exponential, hyperexponential and Erlang distributions are, respectively, the geometric, mixture of geometric, and negative binomial distributions; these are clearly seen to be discrete PH distributions. To illustrate the modeling appeal of PH distributions, we present two simple examples below.

Example 2.5.1 In a poll with a large number of voters (who therefore can be considered to be an infinite population), two candidates A and B, respectively, obtain the fractions p and $1-p$ of the votes, where the value of p will not be known until all votes have been counted. A pollster sequentially polls voters coming out of the poll booth, and once one candidate has attained a lead of at least N votes more than the other, the pollster decides to declare her the winner. The number of voters that are polled before a declaration can be made is a discrete PH random variable with representation

$$\begin{aligned}
\boldsymbol{\tau} &= \begin{bmatrix} 0 & \ldots & 0 & 1 & 0 & \ldots & 0 \end{bmatrix}, \\
T &= \begin{bmatrix}
0 & p & & & & & \\
1-p & 0 & p & & & & \\
& 1-p & 0 & \ddots & & & \\
& & 1-p & \ddots & p & & \\
& & & \ddots & 0 & p & \\
& & & & 1-p & 0 & p \\
& & & & & 1-p & 0
\end{bmatrix};
\end{aligned}$$

there are $2N-1$ states, numbered from $-(N-1)$ to $N-1$, representing the difference between the numbers of polled voters who preferred candidate A over candidate B.[4]

Example 2.5.2 Communication channels are monitored continuously for transmission errors which occur due to weather conditions, electrical disturbances, and a variety of other causes which cannot be controlled. When the sustained error rate is high, use of the channel is abandoned and traffic is rerouted on alternate channels. A tool for error monitoring is the *leaky bucket algorithm* described below:

- Initialize a counter at value 0.

- If the unit of information (called a packet) is received in error, then increment the counter by an integer U.

- If it is received without error and the counter value is positive, then decrement the counter by 1.

- Stop using the channel when the counter reaches a threshold K.

We assume that each packet has a constant probability p of being in error independently of all other packets. Often, p is calculated by assuming that each packet contains a fixed number of bits, which may be in error independently of others, with a constant probability, called the bit error rate.

The total number of packets transmitted before taking the channel out of commission is a discrete PH random variable, of which we give below a representation in the case where $K = 7$ and $U = 2$. In the general case, the phases are numbered from 0 to $K-1$:

$$\boldsymbol{\tau} = \begin{bmatrix} 1 & 0 & \dots & 0 \end{bmatrix},$$

[4]The definition given earlier requires that the transient phases be numbered from 1 to $2N-1$, but the actual labeling of the phases is not important, and it is clearer, in the present case, to use the integers from $-(N-1)$ to $N-1$.

$$T = \begin{bmatrix} 1-p & \cdot & p & \cdot & \cdot & \cdot & \cdot \\ 1-p & \cdot & \cdot & p & \cdot & \cdot & \cdot \\ \cdot & 1-p & \cdot & \cdot & p & \cdot & \cdot \\ \cdot & \cdot & 1-p & \cdot & \cdot & p & \cdot \\ \cdot & \cdot & \cdot & 1-p & \cdot & \cdot & p \\ \cdot & \cdot & \cdot & \cdot & 1-p & \cdot & \cdot \\ \cdot & \cdot & \cdot & \cdot & \cdot & 1-p & \cdot \end{bmatrix}$$

(for further detail about this problem see Ramaswami and Wang [104]).

Let us now turn to some important properties of discrete PH distributions. From the fact that

$$P^k = \begin{bmatrix} 1 & \mathbf{0} \\ \mathbf{1} - T^k\mathbf{1} & T^k \end{bmatrix} \qquad \text{for } k \geq 0,$$

we immediately obtain the density and distribution function, as stated below.

Theorem 2.5.3 *Assume that X is $PH_d(\boldsymbol{\tau}, T)$. We have that*

$$\begin{aligned} \mathrm{P}[X=0] &= \tau_0, \\ \mathrm{P}[X=k] &= \boldsymbol{\tau} T^{k-1}\boldsymbol{t} \qquad \text{for } k \geq 1, \end{aligned}$$

and

$$\mathrm{P}[X \leq k] = 1 - \boldsymbol{\tau} T^k \mathbf{1} \qquad \text{for } k \geq 0.$$

□

Using arguments identical to the continuous case, we can show that for a nondefective discrete PH distribution, there exists a representation such that $I - T$ is nonsingular, which is equivalent to stating that absorption occurs a.s. from any phase. We then have that $(I-T)^{-1}_{ij}$ is the expected number of steps in phase j before absorption, given that the initial phase is i. The probability generating function of PH$(\boldsymbol{\tau}, T)$ is also easy to obtain and is given by

$$\begin{aligned} G(z) &= \sum_{k\geq 0} z^k \mathrm{P}[X=k] \\ &= \tau_0 + z\boldsymbol{\tau}(I - zT)^{-1}\boldsymbol{t} \qquad \text{for } |z| \leq 1. \end{aligned} \tag{2.14}$$

By differentiating this successively with respect to z and setting $z = 1$, we obtain the factorial moments

$$\mathrm{E}[X(X-1)\cdots(X-k+1)] = k!\boldsymbol{\tau}(I-T)^{-k}T^{k-1}\mathbf{1} \tag{2.15}$$

for $k \geq 1$.

Example 2.5.4 This is a continuation of Example 2.4.6, where we determined the distribution $\{a_k : k \geq 0\}$ of the number N of Poisson events (of rate λ) during a random interval X with continuous PH$(\boldsymbol{\tau}, T)$ distribution. This is in fact a discrete PH distribution with representation $(\boldsymbol{\sigma} S)$, where

$$\begin{aligned} \boldsymbol{\sigma} &= \boldsymbol{\tau} S, \\ S &= \lambda(\lambda I - T)^{-1}. \end{aligned}$$

To show this, we define the Markov process $\{J(t) : t \geq 0\}$ with the generator (2.3) and we consider its state at the epochs $\{t_1, t_2, \ldots\}$ of transition of the Poisson process.

The process $\{J(t_1), J(t_2), \ldots\}$ is a discrete time Markov chain with one absorbing state, and N is the number of transitions before absorption. Thus, N has a discrete PH distribution. To determine the matrix S of transition probabilities between transient states, we condition on the length U of the interval between two Poisson epochs and find that

$$\begin{aligned} S &= \int_0^\infty \lambda e^{-\lambda x} \exp(Tx)dx \\ &= \lambda \int_0^\infty \exp\{(-\lambda I + T)x\}dx \\ &= \lambda(\lambda I - T)^{-1}. \end{aligned}$$

Since $J(0)$ is chosen with the distribution $\boldsymbol{\tau}$, the distribution at the first Poisson epoch is $(1 - \boldsymbol{\tau} S\mathbf{1}, \boldsymbol{\tau} S)$.

2.6 Closure Properties

We have stated earlier that it is advantageous to use PH distributions because there is a connection with Markov processes. We use it here to show that the family of PH distributions is closed under certain operations such as convolution and convex mixture. We shall prove these properties in the discrete case, leaving the continuous case as an exercise.

Theorem 2.6.1 *Let X and Y be two independent random variables; assume that X is $PH_d(\boldsymbol{\tau}, T)$ with n phases and that Y is $PH_d(\boldsymbol{\beta}, S)$ with m phases. Then their sum $X+Y$ is $PH_d(\boldsymbol{\gamma}, C)$ with $n+m$ phases, where*

$$\boldsymbol{\gamma} = [\boldsymbol{\tau}, \tau_0 \boldsymbol{\beta}]$$

and

$$C = \begin{bmatrix} T & \boldsymbol{t} \cdot \boldsymbol{\beta} \\ 0 & S \end{bmatrix},$$

where $(\boldsymbol{t} \cdot \boldsymbol{\beta})_{ij} = t_i \beta_j$.

Comment. Since the distribution of $X + Y$ is the convolution of the distributions of X and of Y, this shows that the family of PH distributions is closed under the convolution operation.

Proof. A simple proof goes as follows. Assume that the transient phases of the Markov chain $\{X_k : k \geq 0\}$ associated with X are numbered from 1 to n and that those of the Markov chain $\{Y_k : k \geq 0\}$ associated with Y are numbered from $n+1$ to $n+m$. The Markov chain $\{Z_k : k \geq 0\}$ associated with $\text{PH}_d(\boldsymbol{\gamma}, C)$ is constructed by "catenating" the states of $\{X_k\}$ and of $\{Y_k\}$.

The chain $\{Z_k\}$ moves first along the paths of $\{X_k\}$ and then along the paths of $\{Y_k\}$. The total time spent by $\{Z_k\}$ in $\{1, \ldots, n+m\}$ is the sum of the time T_1 spent in the phases $\{1, \ldots, n\}$, which has the same distribution as X, and the time T_2 spent in $\{n+1, \ldots, n+m\}$. Since the transition probabilities from the first set of indices to the second are given by $\boldsymbol{t} \cdot \boldsymbol{\beta}$, this means that the first state in the path of T_2 is chosen with the distribution $\boldsymbol{\beta}$ always, so that T_2 has the same distribution as Y and T_1 and T_2 are independent. Thus the result is proved. □

A formal presentation of the argument above is given in Latouche [46, Theorem 3.1]. Another formal proof is given in Neuts [79, Theorem 2.2.2], where it is shown by direct computation using (2.14) that the probability generating function of the constructed PH distribution is indeed the product of the probability generating functions of X and Y.

Theorem 2.6.2 *Let X and Y be two PH random variables; assume that X is $PH_d(\boldsymbol{\tau}, T)$ with n phases and that Y is $PH_d(\boldsymbol{\beta}, S)$ with m phases; denote their distribution functions, respectively, by $F(\cdot)$ and $G(\cdot)$. The convex mixture $\theta F(\cdot) + (1-\theta)G(\cdot)$, $0 \le \theta \le 1$, is a PH distribution with representation $PH_d(\boldsymbol{\gamma}, C)$ with $n+m$ phases, where*

$$\boldsymbol{\gamma} = [\theta\boldsymbol{\tau}, (1-\theta)\boldsymbol{\beta}]$$

and

$$C = \begin{bmatrix} T & 0 \\ 0 & S \end{bmatrix}.$$

Proof. The proof is straightforward. Indeed, consider the Markov chain $\{Z_k\}$ associated with $\mathrm{PH}_d(\boldsymbol{\gamma}, C)$. The subsets $\{1, \ldots, n\}$ and $\{n+1, \ldots, n+m\}$ are disjoint. The total sojourn time in the transient phases is either

a) the sojourn time in the first subset with the probability θ, in which case it is distributed like X,

or

b) the sojourn time in $\{n+1, \ldots, n+m\}$ with probability $1-\theta$, in which case it is distributed like Y.

Its distribution is the mixture $\theta F(\cdot) + (1-\theta)G(\cdot)$ of the distributions of X and Y. □

The two theorems above extend in a natural way to convolutions and mixtures of finitely many PH distributions. They do not, in general, extend to infinitely many distributions. We shall give a counterexample later, after we present a set of cases where it is known that an infinite mixture yields a PH distribution.

Theorem 2.6.3 *Consider the mixture $G(\cdot) = \sum_{k \ge 0} p_k F^{*k}(\cdot)$, where $F(\cdot)$ is a continuous $PH(\boldsymbol{\tau}, T)$ distribution, $F^{*k}(\cdot)$ denotes the k-fold* *convolution of $F(\cdot)$ with itself, and $\{p_\cdot\}$ is a discrete $PH_d(\boldsymbol{\beta}, S)$ density. The distribution $G(\cdot)$ is of phase-type with representation given by $PH(\boldsymbol{\gamma}, C)$, where*

$$\boldsymbol{\gamma} = \boldsymbol{\tau} \otimes \boldsymbol{\beta}$$

and

$$C = T \otimes I + \boldsymbol{t} \cdot \boldsymbol{\tau} \otimes S, \tag{2.16}$$

where we denote by $\otimes$ the Kronecker product.

Proof. First of all, note that $G(\cdot)$ is the distribution of the random sum $Y = \sum_{0 \le k \le N} X_k$, where the X_k's are independent with distribution $F(\cdot)$ and N has the density $\{p_\cdot\}$. We shall use the Markov processes associated with the X_k's and with N by the phase-type representations and construct a Markov process associated with Y.

Let the orders of S and T be m and n, respectively. Also, for simplicity, assume that $\beta_0 = \tau_0 = 0$; the modifications needed in the general case are left to the reader. Define a two-dimensional, continuous time Markov process with state space $\{0\} \cup \{(i,j) : 1 \le i \le n, 1 \le j \le m\}$. The first and second components of the nonzero states will, respectively, correspond to a phase of the PH$(\boldsymbol{\tau}, T)$ and of the PH$_d(\boldsymbol{\beta}, S)$ distributions.

The dynamics are described as follows: the continuous time Markov chain PH$(\boldsymbol{\tau}, T)$ alone makes transitions (and thus the first component alone changes) until it gets absorbed, at which time one of the X_n's is determined. At this time, the discrete time Markov chain PH$_d(\boldsymbol{\beta}, S)$ makes a transition. If it gets absorbed, this means that N is determined and the process terminates; otherwise, a new component i is chosen with the initial distribution $\boldsymbol{\tau}$, and the continuous time Markov chain starts over again. At time 0, both the continuous and discrete Markov chains are initialized, respectively, with the distributions $\boldsymbol{\tau}$ and $\boldsymbol{\beta}$.

Thus, the probability of starting from (i,j) at time 0 is $\tau_i \beta_j$; the transition rates are given in the table below. As usual, we have that $\boldsymbol{t} = -T\mathbf{1}$ and $\boldsymbol{s} = \mathbf{1} - S\mathbf{1}$.

From	To	With rate
(i,j)	0	$t_i s_j$
(i,j)	$(r,j),\ r \neq i$	$T_{ir} + t_i \tau_r S_{jj}$
(i,j)	$(r,k),\ k \neq j$	$t_i \tau_r S_{jk}$

Clearly, the time to absorption of this process is equal to Y, the number of changes of the second component is equal to N, the interval of time between two such changes has distribution $F(\cdot)$, and the intervals are independent.

The representation given in the statement of the theorem is simply in terms of the vector of initial probabilities and the matrix of transition rates given in the table above. □

An algebraic proof of the above is through the computation of the Laplace–Stieltjes transform of the infinite mixture; see [79, Page 53]. A similar property holds if $F(\cdot)$ is a discrete PH distribution; we omit the formal statement and the proof thereof.

Theorem 2.6.4 *Let X and Y be two independent random variables; assume that X is $PH_d(\boldsymbol{\tau}, T)$ with n phases and that Y is $PH_d(\boldsymbol{\beta}, S)$ with m phases. The random variable $Z = \min(X, Y)$ has a PH distribution with representation $PH_d(\boldsymbol{\gamma}, C)$ with nm phases, where*

$$\boldsymbol{\gamma} = \boldsymbol{\tau} \otimes \boldsymbol{\beta}$$

and

$$C = T \otimes S.$$

Proof. As in the proof of Theorem 2.6.3, we define a two-dimensional Markov process of which the first and second components, respectively, are associated with X and Y. The possible transitions among transient states are from (i, j) to (r, k) with rate $T_{ir}S_{jk}$. □

More generally, all order statistics of a set of independent PH random variables are themselves PH. These results hold for continuous PH random variables as well.

An elementary but striking property is the following. It shows that the class of discrete PH distributions is dense[5] in the class of distributions on $\mathbb{N}$.

Theorem 2.6.5 *Any probability distribution with finite support on $\mathbb{N}$ is a discrete PH distribution.*

[5] The precise definition of "denseness" is beyond the scope of our discussion. Weak convergence of distribution functions is enough for our purpose.

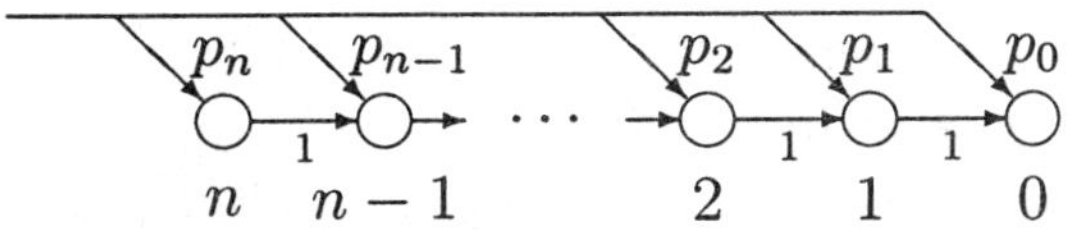

Figure 2.3: Diagrammatic representation of a discrete PH distribution with finite support.

Proof. Let the given distribution have the density $\{p_0, p_1, \ldots, p_n\}$. Take $\boldsymbol{\tau} = (p_1, p_2, \ldots, p_n)$ and T such that $T_{i,i-1} = 1$ for $2 \le i \le n$, $T_{ij} = 0$ otherwise (see Figure 2.3). □

2.7 Further Properties

We shall not discuss in greater detail further properties of PH distributions. They are still the subject of active research, and for completeness, we only cite here a few results without proof.

The Erlang distribution $E(n, n\lambda)$ approaches a point mass at $1/\lambda$ as $n \to \infty$. Given that discrete PH distributions are dense in the class of distributions on N, it is then clear that continuous PH distributions are dense in the class of distributions on $\mathbb{R}_+$. This property makes them a very versatile modeling tool.

Maier and O'Cinneide [64] have proved the following complete characterization of the family of PH distributions.

Theorem 2.7.1 *The continuous PH distributions form the smallest family of distributions on* $\mathbb{R}_+$ *which*

a) *contains the point mass at zero and all exponential distributions,*

b) *is closed under finite mixtures and convolutions, and*

c) *is closed under the infinite mixture* $G(\cdot) = \sum_{n\ge 0}(1-p)p^n F^{*n}(\cdot)$ *for* $0 \le p \le 1$. □

For the discrete PH distributions, the exponential distributions are replaced by the point mass at 1 in the characterization above.

Finally, one shows that the Laplace–Stieltjes transform $\psi(s)$ of a continuous PH distribution $F(\cdot)$ has a unique pole of maximal real part which is real (O'Cinneide [85]) and which we denote by $-\eta$. The following theorem then follows, where $k+1$ is the degree of the factor $(s+\eta)$ in the denominator of $\psi(s)$.

Theorem 2.7.2 *Let $F(\cdot)$ be a PH$(\boldsymbol{\tau}, T)$ distribution with density $f(\cdot)$. We have that*

$$f(x) = \gamma x^k e^{-\eta x} + o(x^k e^{-\eta x}) \qquad \text{as } x \to \infty$$

for some $\gamma > 0, k \geq 0$, where $-\eta$ is a real eigenvalue of T. □

This enables us to exhibit an example of an infinite mixture of PH distributions which is not itself a PH distribution (refer to Theorem 2.6.3). Consider the mixture of infinitely many Erlang distributions:

$$G(\cdot) = \sum_{n\geq 1} p_n E(\cdot; n, \lambda)$$

with $p_n = 1/n - 1/(n+1)$ for $n \geq 1$. Its density function is given by

$$\begin{aligned} g(x) &= \sum_{n\geq 1} \left(\frac{1}{n} - \frac{1}{n+1}\right) \frac{\lambda^n x^{n-1}}{(n-1)!} e^{-\lambda x} \\ &= e^{-\lambda x} \sum_{n\geq 1} \frac{\lambda^n x^{n-1}}{(n+1)!} \\ &= \frac{1}{\lambda x^2}(1 - e^{-\lambda x} - \lambda x e^{-\lambda x}), \end{aligned}$$

which is not of the form given by the theorem above.

2.8 Algorithms

In order to actually calculate the density or the cumulative distribution function of a continuous PH random variable, one may use differential equation techniques from numerical analysis. Indeed, the density function of a PH($\boldsymbol{\tau}, T$) distribution is given by $f(x) = \boldsymbol{\tau}\boldsymbol{\phi}(x)$ for $x \geq 0$, where $\boldsymbol{\phi}(x) = \exp(Tx)\boldsymbol{t}$; see (2.8). That function satisfies the differential equations

$$\boldsymbol{\phi}'(x) = T\boldsymbol{\phi}(x), \qquad \boldsymbol{\phi}(0) = \boldsymbol{t}.$$

Similarly, the distribution may be obtained as the solution of a linear differential equation: define $\boldsymbol{\xi}(x) = \exp(Tx)\mathbf{1}$ ($x \geq 0$); it satisfies the equation

$$\boldsymbol{\xi}'(x) = T\boldsymbol{\xi}(x), \qquad \boldsymbol{\xi}(0) = \mathbf{1},$$

and we have that $F(\cdot) = 1 - \boldsymbol{\tau}\boldsymbol{\xi}(\cdot)$. Linear differential equations have been well studied, and various techniques are available, in textbooks as well as program libraries.

In the following, we present an alternate method based on an interesting probabilistic construction. We have preferred that method in our own work and have found it to be extremely satisfactory. It is based on the *uniformization* technique, also known as the method of *randomization*, first described by Jensen [34].

Consider a Markov process $\{X(t) : t \geq 0\}$ with generator Q such that $|q_{ii}| \leq c < \infty$ (all i) for some constant c (this automatically holds when there are only finitely many states, as here). Then, the matrix $K = \frac{1}{c}Q + I$ is stochastic, as can easily be verified.

Now, let us define the stochastic process $\{Y(t) : t \geq 0\}$ as follows. Take a Poisson process with rate c and denote by $0 = t_0, t_1, t_2, \ldots$ the epochs of events in that process. Take a discrete time Markov chain $\{Z_n : n \geq 0\}$ with transition matrix K independent of the Poisson process. Define the process $\{Y(t) : t \geq 0\}$ such that $Y(t) = Z_n$ for $t_n \leq t < t_{n+1}$ for $n \geq 0$. Not surprisingly, $\{Y(t) : t \geq 0\}$ happens to be a Markov process, and furthermore, its generator is equal to Q; we refer to Çinlar [12, Chapter 8], for details.

Algebraically, if we define the transition matrix $P(t)$ where $P_{ij}(t) = \mathrm{P}[Y(t) = j | Y(0) = i]$, we obtain by a simple conditioning argument on the number of Poisson events in $(0, t]$ that

$$P(t) = \sum_{n \geq 0} e^{-ct} \frac{(ct)^n}{n!} K^n$$

$$
\begin{aligned}
&= \sum_{n\geq 0} e^{-ct}\frac{(ct)^n}{n!}\sum_{0\leq k\leq n}\binom{n}{k}\left(\frac{1}{c}Q\right)^k \\
&= \exp(Qt),
\end{aligned}
$$

which is the transition matrix of the process $\{X(t) : t \geq 0\}$.

This is a powerful method. On the one hand, it allows one to interpret a continuous time Markov process as a discrete time Markov chain, for which one merely replaces the constant unit of time between any two transitions by independent exponential random variables with the same parameter (hence the term *uniformization*). On the other hand, it allows us to evaluate the transition matrix $P(t)$ without recourse to any differential equations. A particularly appealing feature of the methods based on this idea is that it allows us to obtain global error control; see below.

Let us now turn to the computation of the density and distribution functions of a PH$(\boldsymbol{\tau}, T)$ random variable. We may uniformize the associated Markov process with generator

$$
Q = \begin{bmatrix} 0 & \mathbf{0} \\ \boldsymbol{t} & T \end{bmatrix}
$$

by choosing $c = \max(-T_{ii} : 1 \leq i \leq n)$ and

$$
K = (1/c)Q + I = \begin{bmatrix} 1 & \mathbf{0} \\ \boldsymbol{p} & P \end{bmatrix},
$$

where $P = (1/c)T + I$ and $\boldsymbol{p} = (1/c)\boldsymbol{t}$. Now, we readily obtain that

$$
\exp(Tx) = \sum_{k\geq 0} e^{-cx}\frac{(cx)^k}{k!}P^k,
$$

so that (2.7, 2.8) become

$$
F(x) = 1 - \sum_{k\geq 0} e^{-cx}\frac{(cx)^k}{k!}\boldsymbol{\tau}P^k\mathbf{1} \tag{2.17}
$$

and

$$
f(x) = c\sum_{k\geq 0} e^{-cx}\frac{(cx)^k}{k!}\boldsymbol{\tau}P^k\boldsymbol{p}. \tag{2.18}
$$

Equation (2.18) is directly obtained by noting that absorption occurs in $(x, x+dx)$ if and only if some $(k+1)$st Poisson event occurs in that interval *and* the discrete time Markov chain enters the absorbing state precisely at the $(k+1)$st transition. A similar argument leads to (2.17).

The two equations above lend themselves to easy computation. We shall discuss the evaluation of $F(\cdot)$, leaving it to the reader to adapt the procedure to $f(\cdot)$.

The series $\sum_{k\geq 0} \boldsymbol{\tau} P^k \mathbf{1}$ is finite since it is the expected number of steps before absorption for the discrete time Markov chain; it is equal to $\boldsymbol{\tau}(I-P)^{-1}\mathbf{1}$ by (2.15). Therefore, we may choose for any $\epsilon > 0$ a bound K_1 such that

$$\sum_{k\geq K_1+1} \boldsymbol{\tau} P^k \mathbf{1} < \epsilon,$$

so that

$$-\epsilon < F(x) - \left(1 - \sum_{0\leq k\leq K_1} e^{-cx}\frac{(cx)^k}{k!}a_k\right) \leq 0,$$

uniformly in x, where $a_k = \boldsymbol{\tau} P^k \mathbf{1}$ $(k \geq 0)$. Once the coefficients $(a_k : 0 \leq k \leq K_1)$ are determined, it is easy to compute the distribution function $F(x)$ for any given x. The procedure is described in Figure 2.4. The variable p in the inner loop successively takes the values of the Poisson probabilities $p_k = \exp(-cx)(cx)^k/k!$ given by the recursion

$$p_0 = e^{-cx}, \qquad p_{k+1} = p_k cx/(k+1)$$

for $k \geq 0$.

If cx is quite large, there is a risk of floating-point underflow in the computation of $\exp(-cx)$, in which case p_0 (and therefore all the p_{ks}'s) would be computed as being equal to 0. This may be handled in either of two ways. One may initialize the recursion with $p_0 = 1$ and, in order to limit the risk of floating-point overflow, occasionally renormalize the computed sequence so that the probabilities sum to 1. Alternatively, a second way around the problem is to compute the mode p_m of the Poisson distribution, which occurs at $m = \lfloor cx \rfloor$, as accurately as possible, using natural logarithms and Stirling's approximation

$$\ln\ n! \;=\; \frac{1}{2}\ln(2\pi) + \left(n+\frac{1}{2}\right)\ln\ n - n$$

Continuous PH Distribution Function
$M := \boldsymbol{\tau}(I-P)^{-1}\mathbf{1}$;
$a_0 := \boldsymbol{\tau}\mathbf{1}$;
$k := 0$;
$\boldsymbol{v} := P\mathbf{1}$;
repeat
  $k := k+1$;
  $a_k := \boldsymbol{\tau}\boldsymbol{v}$;
  $\boldsymbol{v} := P\boldsymbol{v}$
until $|\sum_{0\le i\le k} a_i - M| < \epsilon$;
$K_1 := k$;
for any x of interest **do**
  $p := \exp(-cx)$;
  $F := pa_0$;
  for $k := 1$ **to** K_1 **do**
    $p := cxp/k$;
    $F := F + pa_k$
  od
od

Figure 2.4: A simple algorithm to compute the distribution function of a continuous PH(τ, T) random variable. One uses $c = \max\{-T_{ii} : 1 \le i \le n\}$.

$$+\frac{1}{12n} - \frac{1}{360n^3} + \frac{1}{1260n^5} - \frac{1}{1680n^7} \cdots,$$

and then to use the recursion to compute terms successively on both sides of the mode.

The uniformization method is very useful. It often helps to avoid arduous numerical integrations and to obtain globally accurate algorithms. See Lucantoni and Ramaswami [62] and Ramaswami and Lucantoni [100] for details.

Chapter 3

Markovian Point Processes

Renewal processes provide simple models of point processes which may describe an ordered set of points (arrival instants, service completion epochs, equipment failures, etc.) on $[0, \infty)$. Their main simplifying feature is the independence and equidistribution of successive interrenewal intervals. We study in some detail renewal processes for which the intervals have a PH distribution.

In the second part of the chapter, we describe a Markovian point process which generalizes and unifies the Markovian processes commonly used in applied probability.

The results presented here illustrate anew the simplicity and the algorithmic tractability offered by the underlying Markovian structure in the method of phases.

3.1 The PH Renewal Process

We consider a renewal process with a PH distribution for the interrenewal intervals. We call such a process a PH renewal process. We shall primarily deal with the continuous case, stating the parallel results for discrete time only in summary form.

As with PH distributions, we may associate a Markov process with a PH renewal process. Assume as given a Markov process on $\{0, 1, \ldots, n\}$

with initial distribution $(0, \boldsymbol{\tau})$ and infinitesimal generator

$$Q = \begin{bmatrix} 0 & \mathbf{0} \\ \boldsymbol{t} & T \end{bmatrix},$$

which is associated with the PH distribution $F(\cdot)$; we assume that there is no atom at 0. Start the process at time 0 and let it evolve to absorption. Instantaneously upon absorption, restart the Markov process by choosing a new state with the distribution $\boldsymbol{\tau}$ and let it proceed to absorption again, etc.

Consider the set $\{0 = t_0 < t_1 < t_2 < \cdots\}$ of time points at which the Markov process is reinitialized with the distribution $\boldsymbol{\tau}$. This clearly forms a renewal process with interrenewal distribution $\text{PH}(\boldsymbol{\tau}, T)$.

If the state at time 0 is chosen with a different distribution $\boldsymbol{\beta}$, then we have a *delayed* renewal process where the first interval has the distribution $\text{PH}(\boldsymbol{\beta}, T)$. Unless otherwise stated, when we refer to a renewal process in what follows, we consider the nondelayed case, i.e., the one where there is a renewal at time 0.

With respect to renewal processes, some key quantities of interest are the renewal function $R(x)$, which is the expected number of renewals in the interval $[0, x]$, and the renewal density $r(x) = R'(x)$, which has the physical interpretation that $r(x)dx$ is the elementary probability of having a renewal in the interval $(x, x + dx)$.

Let us begin by observing that our construction above defines a Markov process $\{J(t) : t \geq 0\}$ on $\{1, \ldots, n\}$; we ignore the instantaneous visits to the state 0 and require that the trajectories be right continuous. This is the "phase process" associated with the PH renewal process. Its infinitesimal generator is given by the matrix

$$D = T + \boldsymbol{t} \cdot \boldsymbol{\tau}.$$

We see this by noting that in our construction of the PH renewal process, there may be a transition from i to j with $i \neq j$ in two ways: either directly with the rate T_{ij} or after absorption in the state 0, which occurs at the rate t_i instantaneously followed by a restart in the state j, which has probability τ_j. The diagonal elements of D are such that $D\mathbf{1} = \mathbf{0}$. This Markov process plays a key role in analyzing the PH renewal process.

Recall that a phase j is transient if there is a path from j to the absorbing phase 0. Symmetrically, the phase j is said to be useful if there exists a path to j from some phase i with $\tau_i > 0$; if there were no such i, then j would never be visited and it would be useless to include it in the model. It will be much simpler, from now on, to restrict ourselves to *irreducible* PH representations, i.e., to pairs $(\boldsymbol{\tau}, T)$ such that all phases are transient and all are useful. Clearly, there is no loss of generality if we only consider irreducible representations (see Theorem 2.4.2). The following property is immediate.

Lemma 3.1.1 *If the representation $(\boldsymbol{\tau}, T)$ is irreducible, then the matrix $D = T + \boldsymbol{t} \cdot \boldsymbol{\tau}$ is irreducible.* □

The renewal density of a PH renewal process is expressed as follows.

Theorem 3.1.2 *Consider the (possibly delayed) PH renewal process with representation $(\boldsymbol{\tau}, T)$ and initial phase distribution $\boldsymbol{\beta}$. Its renewal density is given by $r(x) = \boldsymbol{\beta} \exp(Dx)\boldsymbol{t}$, $x > 0$.*

Proof. A renewal occurs in $(x, x+dx)$ if and only if the phase process $\{J(t) : t \geq 0\}$ is in some state j at time x, and an absorption into 0 occurs in $(x, x+dx)$. Therefore,

$$\begin{aligned} r(x)dx &= \sum_{1\leq i\leq n} \sum_{1\leq j\leq n} \beta_i \mathrm{P}[J(x) = j | J(0) = i] t_j dx \\ &= \sum_{1\leq i\leq n} \sum_{1\leq j\leq n} \beta_i [\exp(Dx)]_{ij} t_j dx. \end{aligned}$$

□

Setting $\boldsymbol{\beta} = \boldsymbol{\tau}$ gives the renewal density of the (nondelayed) renewal process which has a renewal at time 0. With this, we may determine the renewal function (of the nondelayed renewal process) as

$$R(x) = 1 + \int_0^x r(u)du,$$

where $r(u) = \boldsymbol{\tau} \exp(Du)\boldsymbol{t}$. In order to evaluate this, we shall need the following algebraic property.

Lemma 3.1.3 *The matrix $D - \Pi$ is nonsingular, where*

$$\Pi = \mathbf{1} \cdot \boldsymbol{\pi}$$

and

$$\boldsymbol{\pi} = (\boldsymbol{\tau} T^{-1} \mathbf{1})^{-1} \boldsymbol{\tau} T^{-1}. \tag{3.1}$$

The vector $\boldsymbol{\pi}$ is the stationary probability vector of the phase process, i.e., the solution of the system $\boldsymbol{\pi} D = \mathbf{0}$, $\boldsymbol{\pi}\mathbf{1} = 1$. Furthermore, we have that $\boldsymbol{\pi}(D - \Pi)^{-1} = -\boldsymbol{\pi}$.

Proof. The proof that the vector $\boldsymbol{\pi}$ given in (3.1) is the solution of $\boldsymbol{\pi} D = \mathbf{0}$, $\boldsymbol{\pi}\mathbf{1} = 1$ is by direct verification.

Assume that there exists a nonzero vector $\boldsymbol{x}$ such that $\boldsymbol{x}(D - \mathbf{1} \cdot \boldsymbol{\pi}) = \mathbf{0}$. After postmultiplication by $\mathbf{1}$, we find that $\boldsymbol{x}\mathbf{1} = 0$ and therefore that $\boldsymbol{x} D = \mathbf{0}$. Since D is irreducible by Lemma 3.1.1, this implies that $\boldsymbol{x}$ is proportional to $\boldsymbol{\pi}$, which contradicts the earlier conclusion that $\boldsymbol{x}\mathbf{1} = 0$; thus the first statement is proved.

Finally, $\boldsymbol{\pi}$ may be written as $\boldsymbol{\pi} = (\boldsymbol{\pi}\mathbf{1})\boldsymbol{\pi} - \boldsymbol{\pi} D$, or $\boldsymbol{\pi} = \boldsymbol{\pi}(\Pi - D)$, from which we have the last statement. □

With this, we easily determine an expression for the renewal function.

Theorem 3.1.4 *The renewal function of the PH renewal process (starting with a renewal at time 0) with representation $(\boldsymbol{\tau}, T)$ is given by*

$$R(x) = 1 + xM^{-1} + \boldsymbol{\tau}(\exp(Dx) - I)(D - \Pi)^{-1}\boldsymbol{t},$$

where $M = \boldsymbol{\tau}(-T)^{-1}\mathbf{1}$ is the mean interrenewal time and $\Pi = \mathbf{1} \cdot \boldsymbol{\pi}$.

Proof. Since $R(x)$ is the expected number of renewals in $[0, x]$ under the assumption that the first renewal occurs at time 0, we have that

$$\begin{aligned} R(x) &= 1 + \int_0^x r(u)du \\ &= 1 + \int_0^x \boldsymbol{\tau} \exp(Du)\boldsymbol{t} du \\ &= 1 + \boldsymbol{\tau} \sum_{n \geq 0} \frac{x^{n+1}}{(n+1)!} D^n \boldsymbol{t}. \end{aligned}$$

If we write $\boldsymbol{t}$ as $(D-\Pi)(D-\Pi)^{-1}\boldsymbol{t}$, using the facts that $D^n\Pi = D^n\mathbf{1}\cdot\boldsymbol{\pi} = \mathbf{0}$ for $n \geq 1$ and that $\boldsymbol{\pi}(D-\Pi)^{-1} = -\boldsymbol{\pi}$, we obtain that

$$\begin{aligned} R(x) &= 1 - \boldsymbol{\tau}\Pi(D-\Pi)^{-1}\boldsymbol{t}x + \boldsymbol{\tau}\sum_{n\geq 1}\frac{x^n}{n!}D^n(D-\Pi)^{-1}\boldsymbol{t} \\ &= 1 + \boldsymbol{\pi t}x + \boldsymbol{\tau}(\exp(Dx) - I)(D-\Pi)^{-1}\boldsymbol{t}, \end{aligned}$$

which concludes the proof since

$$\boldsymbol{\pi t} = (\boldsymbol{\tau}T^{-1}\mathbf{1})^{-1}\boldsymbol{\tau}T^{-1}(-T)\mathbf{1} = M^{-1}$$

by (2.13). □

3.2 The Number of Renewals

Consider the PH renewal process defined by $(\boldsymbol{\tau}, T)$, and let $N(x)$ denote the number of renewals in $(0, x]$. Our construction clearly shows that $(N(x), J(x))$ is Markovian. Thus, to determine the distribution of $N(x)$, it is convenient to consider the joint distributions

$$P_{ij}(k,x) = \mathrm{P}[N(x) = k, J(x) = j | N(0) = 0, J(0) = i]$$

for $k \geq 0$, $1 \leq i, j \leq n$. We shall denote by $P(k,x)$ the $n \times n$ matrix $\{P_{ij}(k,x)\}$, and we also define the matrix generating function

$$P^*(z,x) = \sum_{k\geq 0} z^k P(k,x).$$

The marginal distribution of $N(x)$ is given by the scalar sequence $\{\boldsymbol{\beta}P(k,x)\mathbf{1} : k \geq 0\}$ if $\boldsymbol{\beta}$ is the initial phase distribution.

Theorem 3.2.1 *The matrices $P(\cdot,\cdot)$ satisfy the following system of differential equations:*

$$P'(0,x) = TP(0,x), \tag{3.2}$$

$$P'(k,x) = TP(k,x) + \boldsymbol{t}\cdot\boldsymbol{\tau}P(k-1,x) \qquad \textit{for } k \geq 1 \tag{3.3}$$

with

$$P(0,0) = I, \qquad P(k,0) = 0 \qquad \textit{for } k \geq 1.$$

Their generating function is given by

$$P^*(z,x) = \exp\{(T + z\boldsymbol{t}\cdot\boldsymbol{\tau})x\}. \tag{3.4}$$

Proof. The generator of the process $\{(N(x), J(x)) : x \geq 0\}$ is given in block form by the matrix

$$Q = \begin{bmatrix} T & \boldsymbol{t}\cdot\boldsymbol{\tau} & 0 & \cdots \\ 0 & T & \boldsymbol{t}\cdot\boldsymbol{\tau} & \cdots \\ 0 & 0 & T & \cdots \\ \vdots & \vdots & \vdots & \ddots \end{bmatrix}.$$

The matrix of transition functions is given by

$$P(x) = \begin{bmatrix} P(0,x) & P(1,x) & P(2,x) & \cdots \\ 0 & P(0,x) & P(1,x) & \cdots \\ 0 & 0 & P(0,x) & \cdots \\ \vdots & \vdots & \vdots & \ddots \end{bmatrix},$$

since we have that

$$P_{ij}(k,x) = \mathrm{P}[N(x) = k + k', J(x) = j | N(0) = k', J(0) = i]$$

for $k, k' \geq 0$, $1 \leq i, j \leq n$.

The proof of (3.2, 3.3) is immediate once we write in block form the Kolmogorov backward differential equation $P'(x) = QP(x)$. From there, we find that

$$\begin{aligned} \frac{\partial}{\partial x} P^*(z,x) &= \sum_{k\geq 0} z^k P'(k,x) \\ &= (T + z\boldsymbol{t}\cdot\boldsymbol{\tau})P^*(z,x), \end{aligned} \tag{3.5}$$

so that (3.4) follows. □

These equations provide matrix analogues of the corresponding results for the Poisson process (take $n = 1$, $T = -\lambda$, and $\boldsymbol{\tau} = 1$).

Remark 3.2.2 We now give an alternate, more elaborate derivation using a renewal argument.

Clearly, we have that $P(0,x) = \exp(Tx)$ since $N(x) = 0$ if and only if all the phase transitions up to time x are among the transient states. This immediately gives the differential equation for $P(0,x)$.

For $k \geq 1$, we condition on the first renewal epoch $x - u$ and write that

$$P(k,x) = \int_0^x \exp(T(x-u))\boldsymbol{t} \cdot \boldsymbol{\tau} P(k-1,u)du;$$

premultiplying both sides by $\exp(-Tx)$ and differentiating with respect to x, we obtain the differential equation for $P(k,x)$.

Multiplying by z^k and adding, we get

$$P^*(z,x) = \exp(Tx) + z\int_0^x \exp(T(x-u))\boldsymbol{t} \cdot \boldsymbol{\tau} P^*(z,u)du.$$

Now, premultiply by $\exp(-Tx)$ and differentiate; this yields (3.5).

With some experience, one can write the integral equation above without having to go through the elaborate steps of setting up the equations for $P(k,x)$; one uses a conditioning argument on the first arrival epoch $x - u$, at which one "throws in" a factor z to record the fact that there is a renewal at that epoch. Later, when we have to write equations for matrix generating functions, we shall often employ this device without explaining the details.

As in the preceding chapter, uniformization offers an alternative method to compute the probabilities $P(\cdot,\cdot)$. Its advantages relative to solving the system of differential equations by numerical integration are similar to those found earlier when we discussed the computation of the PH density and distribution function. With both methods, the actual computation of the probabilities is a nontrivial task, expensive both in memory and time requirements. Fortunately, in the models that will be discussed, many quantities of interest may be obtained without having to compute these quantities explicitly.

Lemma 3.2.3 *Define* $c = \max\{|T_{ii}| : 1 \leq i \leq n\}, P = c^{-1}T + I,$ *and* $\boldsymbol{p} = c^{-1}\boldsymbol{t} = \mathbf{1} - P\mathbf{1}$.

The matrices $P(\nu, x)$ $(\nu \geq 0,\ x \geq 0)$ *are given by*

$$P(\nu,x) = \sum_{m\geq\nu} e^{-cx}\frac{(cx)^m}{m!}K_{m\nu}, \tag{3.6}$$

where the matrices $K_{m,\nu}$ *of order* n *are defined as follows:*

$$\begin{aligned} K_{m,\nu} &= 0 \qquad \textit{for } m < \nu, &(3.7)\\ K_{m,0} &= P^m, &(3.8)\\ K_{m+1,\nu+1} &= PK_{m,\nu+1} + \boldsymbol{p\tau}K_{m,\nu} &(3.9)\end{aligned}$$

for $m, \nu \geq 0$.

Proof. Consider the uniformized version of the phase process $\{J(t) : t \geq 0\}$ associated with the PH renewal process. As noted in section 2.8, this allows us to consider the Markov process as made up of a Poisson process and a discrete time Markov chain. The Poisson process with rate c determines when an absorbing discrete time Markov chain makes its transitions. The transition matrix among transient states is P, and the absorption probabilities are given by the vector $\boldsymbol{p}$. When the Markov chain gets absorbed, a new transient state is instantaneously chosen with the distribution $\boldsymbol{\tau}$.

Now, given that the initial phase is i, let $(K_{m,\nu})_{ij}$ denote the conditional probability that the Markov chain is in the transient phase j immediately after the mth transition and that there have been ν instantaneous visits to the absorbing state during the first m transitions. The equations (3.7), (3.8) are then immediate, and (3.9) is obtained by a simple conditioning argument on the first transition.

The equation (3.6) follows immediately if we condition on the number of Poisson events in $(0, x]$. □

Example 3.2.4 This is a continuation of Example 2.1.2; it gives one more illustration of our general theme, that the Markovian structure in phase processes allows for a strong similarity to the Poisson process.

Consider the PH/G/∞ queue which is the infinite server queue with PH$(\boldsymbol{\tau}, T)$ renewal arrivals and general service time distribution $H(\cdot)$. Let $P_{ij}(n, x)$ denote the probability, given that at time 0 all servers are idle and that the arrival process is in phase i, that at time x, exactly n servers are busy and the arrival process is in phase j. Define the matrix generating function

$$P^*(z, x) = \sum_{n=0}^{\infty} z^n P(n, x).$$

As in Example 2.1.2, we condition on the first arrival; as in the proof of Theorem 3.2.1, we keep track of the phase at epochs of arrival and at time x. This gives

$$P^*(z, x) \quad = \quad \exp(Tx) + \int_0^x \exp(T(x-u))\boldsymbol{t} \cdot \boldsymbol{\tau} P^*(z, u) H(u) du$$

$$+z \int_0^x \exp(T(x-u))\boldsymbol{t} \cdot \boldsymbol{\tau} P^*(z,u)(1-H(u))du.$$

In the right side of the above equation, the first term corresponds to the case where there is no arrival in $(0, x]$. The second term corresponds to the case where an arrival occurs at time $x - u$ and leaves before time x; at time $x - u$ a new phase is chosen with the distribution $\boldsymbol{\tau}$. In the third case, the customer who arrives at time $x - u$ is still present at time x and contributes a factor z.

We proceed as in Remark 3.2.2, premultiply by $\exp(-Tx)$, differentiate, and obtain

$$\frac{\partial}{\partial x} P^*(z,x) = \{D + (z-1)(1-H(x))\boldsymbol{t} \cdot \boldsymbol{\tau}\} P^*(z,x) \tag{3.10}$$

with the initial condition $P^*(z, 0) = 1$.

From (3.10), one may numerically obtain the moments of the queue length at time x. For further details, see Ramaswami and Neuts [101].

Example 3.2.5 This is the continuation of Examples 2.4.6 and 2.5.4. We consider here a PH renewal process with representation $(\boldsymbol{\tau}, T)$ and define N as the number of renewals in the interval $(0, X)$, where X has an exponential distribution with rate λ, independent of the renewal process. The number of renewals has a geometric distribution, with parameter $p = \boldsymbol{\tau} S \mathbf{1}$, where $S = \lambda(\lambda I - T)^{-1}$.

To see this, we compare X and the first renewal epoch, say, Y, and obtain that $\mathrm{P}[N = 0] = \mathrm{P}[X < Y]$. If $X > Y$, then $N = 1 + N_1$, where N_1 is the number of renewals in the remainder of the interval (Y, X). Given the memoryless property of the exponential distribution, N_1 has the same distribution as N and we find that N is geometrically distributed with parameter $p = \mathrm{P}[N = 0]$. Now,

$$\begin{aligned} \mathrm{P}[N=0] &= \int_0^\infty \boldsymbol{\tau} \exp(Tx)\mathbf{1}\, \lambda e^{-\lambda x} dx \\ &= \lambda \boldsymbol{\tau} (\lambda I - T)^{-1} \mathbf{1} \\ &= \boldsymbol{\tau} S \mathbf{1}, \end{aligned}$$

which is easily computed.

3.3 The Stationary Process

Given a renewal process with interrenewal time distribution $F(.)$, its stationary version is obtained by imposing an initial delay with distribution

$$G(x) = \frac{1}{M}\int_0^x (1 - F(u))du \qquad \text{for } x \geq 0,$$

where M is the expected interval between two renewal epochs. The distribution $G(\cdot)$ is the asymptotic survival distribution. That is, if for fixed t we define the survival random variable V_t as the length of the interval from t to the next renewal, then one has that $G(x) = \lim_{t\to\infty} \mathrm{P}[V_t \leq x]$.

If $F(\cdot)$ is PH$(\boldsymbol{\tau}, T)$, then the distribution of the survival interval V_t is PH$(\boldsymbol{\beta}(t), T)$, where $\boldsymbol{\beta}(t) = \boldsymbol{\beta}\exp(Dt)$ is the distribution of the phase at time t for the process $\{J(x) : x \geq 0\}$. Knowing (by Lemma 3.1.3) that $\lim_{t\to\infty} \boldsymbol{\beta}(t) = \boldsymbol{\pi}$, and that $\boldsymbol{\pi}$ is given by (3.1), we have proved the following theorem.

Theorem 3.3.1 *If $F(\cdot)$ is PH$(\boldsymbol{\tau}, T)$, then the distribution*

$$G(x) = \frac{1}{M}\int_0^x (1 - F(u))du \qquad \textit{for } x \geq 0, \tag{3.11}$$

where M is the expected interval between two renewal epochs, is PH$(\boldsymbol{\pi}, T)$, where $\boldsymbol{\pi} = (\boldsymbol{\tau}T^{-1}\mathbf{1})^{-1}\boldsymbol{\tau}T^{-1}$. □

This means that the stationary version of the PH renewal process is the delayed renewal process obtained by choosing the initial phase according to the stationary distribution $\boldsymbol{\pi}$.

Example 3.3.2 Consider an M/PH/1 queue, i.e., an M/G/1 queue with PH service times. A consequence of the theorem above is that the stationary waiting time distribution $W(\cdot)$ is PH. If the service time distribution is PH$(\boldsymbol{\tau}, T)$, then $W(\cdot)$ is PH$((1-\rho)\boldsymbol{\pi}, T)$.

To prove this, we recall the Pollachek–Khinchine formula (Cohen [16, Page 255]) which states that

$$\tilde{w}(s) = \frac{s(1-\rho)}{s - \lambda + \lambda\phi(s)},$$

where $\widetilde{w}(s)$ and $\phi(s)$ are, respectively, the Laplace–Stieltjes transforms of the stationary waiting time and of the service time distributions and $\rho = \lambda M$ is the traffic coefficient; the expression for $\phi(s)$ is given in (2.12).

This may also be written as

$$\widetilde{w}(s) = \frac{1-\rho}{1-\rho\gamma(s)} = \sum_{i\geq 0}(1-\rho)\rho^i[\gamma(s)]^i,$$

where

$$\gamma(s) = \frac{1}{sM}(1-\phi(s))$$

is the Laplace–Stieltjes transform of the asymptotic survival distribution (3.11).

Therefore, the distribution function $W(\cdot)$ itself is such that

$$W(x) = \sum_{i\geq 0}(1-\rho)\rho^i G^{*i}(x),$$

i.e., it is a geometric mixture of the convolution powers of $G(\cdot)$. Since $G(\cdot)$ is PH by Theorem 3.3.1, and since the geometric distribution is discrete PH, $W(\cdot)$ is PH by Theorem 2.6.3.

3.4 Discrete Time PH Renewal Processes

We can use discrete time Markov chains very much like we used continuous time Markov processes in section 3.1. This allows us to define discrete time PH renewal processes with interrenewal intervals distributed as $\mathrm{PH}_d(\boldsymbol{\tau}, T)$.

We assume that $\boldsymbol{\tau}\mathbf{1} = 1$, so that renewals occur singly. If $\boldsymbol{\tau}\mathbf{1} < 1$, then renewals occur in batches of random size with a geometric distribution. This is a special case of the process which we define in the next section.

The generator of the phase process is $T^* = T + \boldsymbol{t} \cdot \boldsymbol{\tau}$ with $\boldsymbol{t} = \mathbf{1} - T\mathbf{1}$. We may, as before, assume without loss of generality that it is irreducible. We shall also assume that it is aperiodic. Under these conditions, the following results are easy to prove; we omit the details.

The stationary probability vector of T^* is $\boldsymbol{\pi} = (1/\mathrm{E}[X])\boldsymbol{\tau}(I-T)^{-1}$, where $\mathrm{E}[X] = \boldsymbol{\tau}(I-T)^{-1}\mathbf{1}$ is the expected interrenewal time. The stationary version of the renewal process is obtained by choosing the initial phase with the vector $\boldsymbol{\pi}$.

The renewal density $\{r(k) : k \geq 0\}$, where $r(k)$ is the probability of a renewal at epoch k, is given by $r(k) = \boldsymbol{\tau}(T + \boldsymbol{t}\cdot\boldsymbol{\tau})^{k-1}\boldsymbol{t}$ for $k \geq 1$.

The distribution of the remaining lifetime at k is $\mathrm{PH}(\boldsymbol{\beta}_k, T)$, where $\boldsymbol{\beta}_0$ is the initial phase distribution and $\boldsymbol{\beta}_k = \boldsymbol{\beta}_0(T + \boldsymbol{t}\cdot\boldsymbol{\tau})^k$. As k tends to infinity, this distribution converges to $\mathrm{PH}_d(\boldsymbol{\pi}, T)$.

The density of the elapsed lifetime Y_k at k is given by

$$\mathrm{P}[Y_k = j] = \boldsymbol{\beta}_0 T^k \mathbf{1}$$

for $j = k$ and

$$\mathrm{P}[Y_k = j] = \boldsymbol{\beta}_0(T + \boldsymbol{t}\cdot\boldsymbol{\tau})^{k-j-1}\boldsymbol{t}\cdot\boldsymbol{\tau}T^j\mathbf{1}$$

for $0 \leq j \leq k-1$; it converges to $\mathrm{PH}_d(\boldsymbol{\pi}, T)$ as k tends to infinity.

3.5 A General Markovian Point Process

Let us consider anew the process $\{(N(x), J(x)) : x \geq 0\}$ of section 3.2, where $N(x)$ is the number of renewals in $(0, x]$ and $J(x)$ is the phase at time x. Its infinitesimal generator is

$$\begin{bmatrix} T & \boldsymbol{t}\cdot\boldsymbol{\tau} & 0 & \cdots \\ 0 & T & \boldsymbol{t}\cdot\boldsymbol{\tau} & \cdots \\ 0 & 0 & T & \cdots \\ \vdots & \vdots & \vdots & \ddots \end{bmatrix} \tag{3.12}$$

and is very similar to the generator of the familiar Poisson process:

$$\begin{bmatrix} -\lambda & \lambda & 0 & \cdots \\ 0 & -\lambda & \lambda & \cdots \\ 0 & 0 & -\lambda & \cdots \\ \vdots & \vdots & \vdots & \ddots \end{bmatrix}.$$

Thus it appears that the PH renewal process may be seen as a generalization of the pure birth process. There, the birth process $N(x)$ is modulated by the phase process $J(x)$ whose state dictates the instantaneous birth rates.

If we partition the state space $\{(n, j) : n \geq 0, 1 \leq j \leq m\}$ into subsets

$$\ell(k) = \{(k, 1), \ldots, (k, m)\} \qquad \text{for } k \geq 0,$$

called *levels* , we see that the diagonal blocks in (3.12) are formed by the rates at which phase changes occur while the level remains the same. The blocks above the diagonal are formed by the rates at which a birth occurs: the process moves from one level to the next level above, and a change of phase may also occur. In a PH renewal process, these phase changes occur in a very restricted way since the new phase is chosen at each birth with the same distribution $\boldsymbol{\tau}$.

We may generalize this birth process in two ways: we may allow for multiple births so that a transition is possible in one step from each level $\ell(k)$ to some level $\ell(k + \nu)$ with $\nu \geq 1$; second, we may allow phase changes at birth epochs to occur without any restriction. We define in this way a general Markovian counting process with infinitesimal generator Q given below:

$$Q = \begin{bmatrix} D_0 & D_1 & D_2 & D_3 & \cdots \\ 0 & D_0 & D_1 & D_2 & \cdots \\ 0 & 0 & D_0 & D_1 & \cdots \\ 0 & 0 & 0 & D_0 & \cdots \\ \vdots & \vdots & \vdots & \vdots & \ddots \end{bmatrix}, \tag{3.13}$$

where $D_0, D_1, D_2, \ldots$ are matrices of order m with $D_k \geq 0$ for all $k \geq 1$, $D_0(i, j) \geq 0$ for $1 \leq i \neq j \leq m$, $D_0(i, i) < 0$ for $1 \leq i \leq m$, and $\sum_{k \geq 0} D_k \mathbf{1} = \mathbf{0}$.

The only restriction which we impose is that the transition rate from (k, i) to $(k + \nu, j)$ must be independent of k: the birth rate and the size of the increase may depend on the phase, but not on the current level.

This family of counting processes has received several names in the literature: *versatile* Markovian processes in Neuts [78] and *Neuts* processes in Ramaswami [92]; at present, the name MAP (Markovian arrival process) introduced in Lucantoni, Meier-Hellstern, and Neuts [61] is frequently used, as is the name BMAP (batch Markovian arrival process). We now give a few examples to illustrate the variety of models subsumed under it as special cases.

Example 3.5.1 Batch PH Renewal Processes. Consider a PH renewal process with representation $(\boldsymbol{\tau}, T)$. Assume that at the nth renewal epoch, a batch of size X_n arrives, where the X_n's are i.i.d. random variables with density $\{p_k : k \geq 1\}$. This process is a Markovian point process with generator of the form (3.13), where $D_0 = T$ and $D_k = p_k \boldsymbol{t} \cdot \boldsymbol{\tau}$ $(k \geq 1)$.

Alternatively, we may assume that the batch size depends both on the phase i from which absorption occurs and the phase j in which the process restarts with the distribution $\{p_k(i,j) : k \geq 1\}$. Then $D_k = P_k \circ (\boldsymbol{t} \cdot \boldsymbol{\tau})$, where we denote by $A \circ B$ the matrix C such that $C_{ij} = A_{ij}B_{ij}$.

Example 3.5.2 Markov Modulated Poisson Process. This is a Poisson process in which the instantaneous arrival rate depends on an auxiliary Markov process which serves as a random environment. Suppose that the auxiliary process has m states and generator G and that when the auxiliary state is i, arrivals occur at rate λ_i. The resulting process is called a Markov modulated Poisson process. It belongs to the family defined here with $D_0 = G - \Delta(\boldsymbol{\lambda})$, $D_1 = \Delta(\boldsymbol{\lambda})$, and $D_k = 0$ for $k \geq 2$, where $\Delta(\boldsymbol{\lambda})$ is the diagonal matrix with the λ_i values in the diagonal.

Markov modulated Poisson processes have been used extensively, for example, in modeling arrivals of packet streams to communication systems in which the rate of arrivals depends on the number of sources (terminals, video codecs, etc.) that are active. The *interrupted Poisson process* is a particular case with two phases and $\lambda_2 = 0$; in fact, it is a PH renewal process with hyperexponential interrenewal intervals.[6]

We may add the assumption that arrivals occur in batches with environment-dependent sizes: if the auxiliary variable is i, then the batch is of size k with probability $p_i(k)$ $(k \geq 1)$. Then $D_k = \Delta(\boldsymbol{\lambda})\Delta(\boldsymbol{p}(k))$ for $k \geq 1$.

[6]This is one more indication of the fact that PH processes do not have a unique representation.

Example 3.5.3 PH Semi-Markov Processes. Consider a semi-Markov process $\{(X_n, J_n) : n \geq 0\}$ with PH interval distributions: the conditional distribution of X_n given that $J_n = i$ is $\mathrm{PH}(\boldsymbol{\tau}_i, T_i)$ and $\mathrm{P}[J_{n+1} = j | J_n = i] = p_{ij}$. This is also a Markovian point process; we leave it to the reader to write the matrices D_k, $k \geq 0$.

Example 3.5.4 Thinning and Superpositions. It is elementary to verify that the superposition of two Markovian point processes is again a Markovian point process (use the same construction as in Theorem 2.6.4 and consider a two-dimensional phase process composed of the individual phases). Similarly, if the points of increase are discarded with probability p independently of each other, then the resulting thinned process remains a Markovian point process.

Example 3.5.5 Departure Processes. This is a continuation of Example 2.3.3. Consider an M/M/c/c+K queue (Poisson arrivals, exponential services, c servers, and a buffer of size K). The overflow process is a Markovian point process, of which the matrix D_0 is given below in the special case where $c = 3$ and $K = 2$:

$$D_0 = \begin{bmatrix} -\lambda & \lambda & & & & \\ \mu & -\lambda-\mu & \lambda & & & \\ & 2\mu & -\lambda-2\mu & \lambda & & \\ & & 3\mu & -\lambda-3\mu & \lambda & \\ & & & 3\mu & -\lambda-3\mu & \lambda \\ & & & & 3\mu & -\lambda-3\mu \end{bmatrix};$$

the only nonzero element of D_1 is λ in the lower right corner.

The departure process is also a Markovian point process, as one may easily verify. If we assume that waiting customers become impatient and renege independently of each other and if their waiting time exceeds an exponentially distributed interval, then the process of reneging customers is also a Markovian point process.

3.6 Analysis of the General Process

We show in this section that the analysis of the general Markovian point process is very similar to that of the PH renewal process; naturally, multiple births add an element of complexity. The basic random variables of interest are the phase $J(x)$ at time x and $N(x)$, the number of steps up (or arrivals, or events, etc.) during the interval $(0, x]$. Equivalently, $N(x)$ is the level reached at time x, assuming that the process starts in the level 0 at time 0.

We assumed earlier that the intervals between renewals had a nondefective distribution. As we saw with the Theorems 2.4.2 and 2.4.3, this is equivalent to the assumption that the matrix T on the diagonal of (3.12) is nonsingular. Here, we shall require that $N(x)$ increases without bounds as x tends to infinity. In other words, the process never gets stuck in a level; for every pair $n' > n$, starting from any state (n, i), it will move to some state (n', j) in a finite time a.s. This holds if and only if the matrix D_0 on the diagonal of (3.13) is nonsingular.

Second, we need to assume that the expected size of the jumps to higher levels is finite so that the expected value of $N(x)$ remains finite over finite intervals. This holds if and only if the vector $\boldsymbol{d} = \sum_{k\geq 1} kD_k\mathbf{1}$ is finite. This automatically holds for the PH renewal process, naturally.

The phase process $\{J(x) : x \geq 0\}$ is a Markov process with infinitesimal generator $D = \sum_{k\geq 0} D_k$. Our third assumption is that the matrix D is irreducible. We saw in section 3.1 that, for a PH renewal process, this is equivalent to requiring that the phases are all useful. Without this assumption, the discussion of the limiting behavior of our process becomes more involved.

Let us denote by $\Lambda(x)$ $(x \geq 0)$ the expected value of $N(x)$ and by $\lambda(x)$ its derivative. The instantaneous rate $\lambda(x)$ has the physical interpretation that $\lambda(x)dx$ is the expected increase in the level during $(x, x + dx)$. The functions $\Lambda(\cdot)$ and $\lambda(\cdot)$ are a generalization of the renewal function $R(\cdot)$ and the renewal density $r(\cdot)$ studied in section 3.2.

Throughout this section, we shall assume that the initial state is $(N(0) = 0,\ J(0) = j)$ with probability β_j with $\boldsymbol{\beta}\mathbf{1} = 1$.

Theorem 3.6.1 *The instantaneous rate of increase is given by*

$$\lambda(x) = \boldsymbol{\beta}\exp(Dx)\boldsymbol{d}; \tag{3.14}$$

its asymptotic value is

$$\lambda^* = \boldsymbol{\pi}\boldsymbol{d}, \tag{3.15}$$

where $\boldsymbol{\pi}$ is the stationary probability vector of D.

The expected increase $\Lambda(x) = \mathrm{E}[N(x)]$ over the interval $(0, x]$ is given by

$$\Lambda(x) = x\lambda^* + \boldsymbol{\beta}(\exp(Dx) - I)(D - \Pi)^{-1}\boldsymbol{d}.$$

Proof. The proof mostly repeats the arguments of Theorems 3.1.2 and 3.1.4. To prove (3.15), we recall that (under our general assumptions) the limit of $\mathrm{P}[J(x) = j | J(0) = i]$ as $x \to \infty$ is π_j independently of i so that $\boldsymbol{\beta}\exp(Dx)$ converges to $\boldsymbol{\pi}$. □

We leave it to the reader to develop computational procedures for evaluating the intensity function. This may be done by solving a suitable system of linear differential equations. Indeed, we see from (3.14) that $\lambda(x) = \boldsymbol{\beta y}(x)$, where the vector $\boldsymbol{y}(x)$ is a solution of the system $\boldsymbol{y}'(x) = D\boldsymbol{y}(x)$, $\boldsymbol{y}(0) = \boldsymbol{d}$.

Alternately, one may proceed by uniformization: (3.14) may also be written as

$$\lambda(x) = \sum_{n \geq 0} e^{-cx} \frac{(cx)^n}{n!} \boldsymbol{\beta} D^{*n} \boldsymbol{d},$$

where $c \geq \max |D_{ii}|$ and $D^* = I + 1/cD$.

The stationary version of the Markovian point process is obtained by choosing $\boldsymbol{\beta} = \boldsymbol{\pi}$. We then have that the marginal distribution of the phase is $\boldsymbol{\pi}\exp(Dx) = \boldsymbol{\pi}$ at any time x and that the instantaneous increase rate is constant: $\lambda(x) \equiv \lambda^*$; furthermore, the expected increase over the interval $(0, x]$ is linear: $\Lambda(x) = \lambda^* x$. Also, the increase probabilities $\mathrm{P}[N(t + x) - N(x) = k]$ are indeed independent of x since we have that

$$\begin{aligned} &\mathrm{P}[N(t + x) - N(x) = k] \\ &\quad = \sum_{1 \leq j \leq m} \mathrm{P}[J(x) = j]\mathrm{P}[N(t + x) - N(x) = k | J(x) = j] \\ &\quad = \sum_{1 \leq j \leq m} \pi_j \mathrm{P}[N(t) = k | J(0) = j]. \end{aligned}$$

To conclude this chapter, we shall consider the time-dependent conditional distribution of the Markovian point process. Define

$$P_{ij}(k, x) = \mathrm{P}[N(x) = k, J(x) = j | N(0) = 0, J(0) = i]$$

for $k \geq 0$, $1 \leq i, j \leq m$; denote by $P(k,x)$ the matrix $\{P_{ij}(k,x)\}$ and by $P^*(z,x)$ the generating function

$$P^*(z,x) = \sum_{k\geq 0} z^k P(k,x).$$

The following theorem is a direct generalization of Theorem 3.2.1.

Theorem 3.6.2 *The matrices $P(\cdot,\cdot)$ satisfy the following system of differential equations:*

$$P'(k,x) = \sum_{0\leq j\leq k} D_j P(k-j,x) \tag{3.16}$$

for $k \geq 0$ with $P(0,0) = I$ and $P(k,0) = 0$ for $k \geq 1$. The generating function $P^*(\cdot,\cdot)$ satisfies the integral equation

$$\begin{aligned} P^*(z,x) &= \exp(D_0 x) \\ &+ \int_0^x \exp(D_0(x-u)) \sum_{k\geq 1} z^k D_k P^*(z,u)du \end{aligned} \tag{3.17}$$

and is given by

$$P^*(z,x) = \exp[D(z)x], \tag{3.18}$$

where $D(z) = \sum_{k\geq 0} z^k D_k$.

Proof. The equation (3.16) is just a reformulation of Kolmogorov's backward equations.

The equation (3.17) is obtained by conditioning on the first epoch when there is a jump in level. The first term corresponds to the case where the process remains in the level 0 during the whole interval $[0,x]$. The summand in the second term corresponds to the case where the process jumps from level 0 to level k at time $x-u$, the epoch of the first jump.

Premultiplying both sides of (3.17) by $\exp(-D_0 x)$ and differentiating with respect to x gives

$$\frac{\partial}{\partial x} P^*(z,x) = D(z)P^*(z,x),$$

which, together with $D(z,0) = I$, gives (3.18). □

Considering the stationary version of the Markovian point process, the same approach may be used to determine the *joint* generating function of $N(t)$ and $N(x+t)$ (along with the necessary phases), from which one obtains the auto-covariance function of the process. We refer the reader to Neuts [82, Chapter X].

Needless to say, the material in this section may be adapted to the discrete time case. The resulting processes have been called DMAPs (discrete time Markovian arrival processes) by some authors.

Part III

The Matrix-Geometric Distribution

Chapter 4

Birth-and-Death Processes

QBDs are matrix generalizations of simple birth-and-death processes on the nonnegative integers in the same way as PH distributions are matrix generalization of the exponential distribution. For this reason, a review of birth-and-death processes is useful before we consider the more general QBDs. Indeed, our presentation will be along lines that set the stage for the chapters on QBDs proper. It will be seen later that the arguments (based on renewal theory) for birth-and-death processes carry over to more general processes in an entirely parallel manner and are based on Markov renewal theory.

In our study of the simpler birth-and-death process, we shall recall a few well-known properties, sometimes from a slightly unusual perspective, and we may appear in places to belabor the obvious. Our intention is to facilitate the understanding of the material in later chapters, which will then be seen to constitute a direct generalization, where scalar quantities are replaced by matrices. If the reader should feel impatient with the pace we impose, he might skim through the material in the first sections, returning to it as the need arises at a later stage.

In applications, Markov processes come in two flavors: discrete time Markov chains and regular continuous time Markov processes.[7] The analysis of either kind proceeds along the same lines as the other, but the final equations are slightly different, as we have demonstrated in Chapters 2 and 3 on phase processes. In this chapter we give the

[7]Loosely stated, a continuous time Markov process is regular if it undergoes only a finite number of transitions in any interval of finite length.

detailed proofs in both cases in order to illustrate once more their similarity as well as their differences. In later chapters, we shall give the proofs in the discrete time case only, despite the fact that most applications (and indeed all the examples in Chapter 1) are continuous time processes, because the arguments for discrete time processes are slightly simpler.

4.1 Terminating Renewal Processes

We shall frequently invoke limit theorems for terminating renewal processes, and these are briefly reviewed in this section; the reader is referred to Çinlar [12] or Feller [21] for further details.

A renewal process is said to be *terminating* if $F(\infty) < 1$, where $F(\infty)$ is the limit $\lim_{x\to\infty} F(x)$ of the associated distribution function $F(\cdot)$ of the interrenewal times. For a terminating process, the renewal interval is infinite with positive probability $1 - F(\infty)$. In that case, the total number of renewals in $[0, \infty)$ is finite a.s. Indeed, denoting by N the total number of renewals, it is easily seen that

$$\mathrm{P}[N = n] = F(\infty)^{n-1}(1 - F(\infty)) \qquad \text{for } n \geq 1. \tag{4.1}$$

The *renewal function* $R(\cdot)$ is defined as follows:

$$R(x) = \sum_{i \geq 0} F^{\star i}(x) \qquad \text{for } x \geq 0,$$

where $F^{\star i}(\cdot)$ is the i-fold convolution of $F(\cdot)$ with itself and $F^{\star 0}(.)$ is the one point distribution at 0. If the first renewal occurs at time 0, then $R(x)$ is also the expected number of renewals in $[0, x]$.

From the point of view of our applications, there are basically two cases of interest:

(i) F is a discrete distribution on the positive integers, or a subset thereof, with density $\{f_n\}$;

(ii) F is an absolutely continuous distribution on the positive reals with density f.

In the discrete case, the renewal function has a renewal density $\{r(n)\}$ such that $R(n) = \sum_{0 \le k \le n} r(k)$ with $r(k)$ being the probability that a renewal occurs at time k. In the continuous case, the renewal function has a density $r(x)$ such that $R(x) = 1 + \int_0^x r(u)du$. The quantity $r(x)dx$ may be interpreted as the elementary probability that a renewal occurs in the interval $(x, x + dx)$.

A *renewal equation* is an equation of the form $g = h + F \star g$, where h is known and g is unknown. Its unique solution is given by $g = R \star h$.

In the discrete case, the renewal equation and its solution are, respectively,

$$g(x) = h(x) + \sum_{0 \le \nu \le x} f(\nu) g(x - \nu)$$

and

$$g(x) = \sum_{0 \le \nu \le x} r(\nu) h(x - \nu).$$

Observe that while the renewal equation is typically obtained by conditioning on the *first* epoch of renewal, its solution in terms of the renewal function is obtained by considering all the renewal epochs up to time x. Indeed, if one thinks of $g(x)$ as the total reward up to time x accruing in a renewal reward process in which each renewal results in a reward of $h(u)$ over an interval of length u beginning with the renewal, then the two equations above give two different methods of evaluating the total reward. The first one expresses the total reward as the sum of the reward from the initial event and the sum total of the remaining ones, if any. The second one expresses the total reward as the sum of the rewards from each individual event up to time x.

The same observation holds in the continuous case, where the renewal equation and its solution are as follows:

$$g(x) = h(x) + \int_{[0,x]} g(x - u) dF(u)$$

and

$$\begin{aligned} g(x) &= \int_{[0,x]} h(x - u) dR(u) \\ &= h(x) + \int_0^x r(u) h(x - u) du. \end{aligned}$$

In many applications, the limit as x tends to infinity of the solution of a renewal equation is of interest. In the case of a terminating renewal process, that limit depends on the expected total number of renewals, which from (4.1) is easily seen to be given by

$$R(\infty) = 1/(1 - F(\infty)) < \infty. \tag{4.2}$$

Indeed, the following result holds.

Theorem 4.1.1 *Suppose that the distribution function $F(\cdot)$ is such that $F(\infty) < 1$, that $h(\cdot)$ is bounded over any compact set, and that $h(\infty) = \lim_{x\to\infty} h(x)$ exists. Then,*

$$\lim_{x\to\infty} h \star R(x) = h(\infty)R(\infty) = h(\infty)/(1 - F(\infty)). \tag{4.3}$$

Proof. In the continuous case, we extend the definition of $h(x)$ so that it is equal to 0 for negative values of x. Then we have that

$$\begin{aligned} h \star R(x) &= \int_{[0,x]} h(x-u)dR(u) \\ &= \int_{[0,\infty)} h(x-u)dR(u). \end{aligned}$$

Since $h(\infty)$ is finite, there exists a constant M such that $|h(x)| \leq M$ for all x. Thus, we may apply the dominated convergence theorem and write that

$$\begin{aligned} \lim_{x\to\infty} h \star R(x) &= \lim_{x\to\infty} \int_{[0,\infty)} h(x-u)dR(u) \\ &= \int_{[0,\infty)} \lim_{x\to\infty} h(x-u)dR(u) \\ &= \int_{[0,\infty)} h(\infty)dR(u) \\ &= h(\infty)R(\infty). \end{aligned}$$

Now, $R(\infty)$ being the expected total number of renewals in $[0,\infty)$, it is equal to $(1 - F(\infty))^{-1}$ by (4.1).

The proof in the discrete case is exactly analogous and is left to the reader. □

If, instead of being set equal to zero, the first renewal epoch is a random variable, then we have a *delayed* renewal process and the renewal function is defined as $R(\cdot) = \sum_{i\geq 0} G \star F^{\star i}(\cdot)$, where $G(\cdot)$ is the distribution function of the first renewal epoch. Theorem 4.1.1 is generalized as follows; we omit the details.

Theorem 4.1.2 *Consider a delayed, terminating renewal process with interrenewal distribution $F(\cdot)$ and initial delay distribution $G(\cdot)$. Define the delayed renewal function $R(\cdot) = \sum_{i\geq 0} G \star F^{\star i}(\cdot)$. Suppose that $h(\cdot)$ is bounded over any compact set and that $h(\infty) = \lim_{x\to\infty} h(x)$ exists. Then,*

$$\lim_{x\to\infty} h \star R(x) = h(\infty)R(\infty) = h(\infty)G(\infty)/(1 - F(\infty)). \tag{4.4}$$

□

4.2 Discrete Time Birth-and-Death Processes

A discrete time birth-and-death process is a Markov chain $\{X_k : k \geq 0\}$ on the nonnegative integers; it represents a random walk with a boundary at 0 and is skip-free: from any state, the chain can move up or down to an adjacent state only. Its transition matrix is of the form

$$P = \begin{bmatrix} a_{00} & a_{01} & 0 & 0 & \cdots \\ a_{10} & a_{11} & a_{12} & 0 & \cdots \\ 0 & a_{21} & a_{22} & a_{23} & \cdots \\ 0 & 0 & a_{32} & a_{33} & \cdots \\ \vdots & \vdots & \vdots & \vdots & \ddots \end{bmatrix}. \tag{4.5}$$

Clearly, it is irreducible if and only if $a_{i,i+1} > 0$ and $a_{i,i-1} > 0$ for all i, which we shall assume throughout. Then we know that the limit

$$\pi_n = \lim_{k\to\infty} \mathrm{P}[X_k = n] \qquad \text{for } n \geq 0$$

exists and is independent of the initial state. It is equal to zero if the Markov chain is transient or null recurrent, and it is strictly positive for all n if the Markov chain is positive recurrent.[8] In the positive recurrent case, the vector $\boldsymbol{\pi}$ is a true probability distribution, called the *stationary* or *steady state* distribution; it is the unique solution of the linear system $\boldsymbol{\pi} P = \boldsymbol{\pi}$, $\boldsymbol{\pi}\mathbf{1} = 1$.

We shall determine the stationary distribution of the Markov chain (4.5) through an argument which may appear more involved than those usually given in the literature; our reason is that we wish to use and illustrate in simple circumstances the method which we shall use later in more general settings.

Consider the Markov chain $\{X_k : k \geq 0\}$ with transition matrix (4.5), and denote by $P_{ij}(k) = \mathrm{P}[X_k = j | X_0 = i]$ its k-step transition probability. Further, define

$$ {}_nP_{ij}(k) = \mathrm{P}[X_k = j, X_1 > n, \ldots, X_{k-1} > n | X_0 = i]; $$

this is the taboo probability that the Markov chain is in state j at time k, starting from the state i, and that it has not visited any of the states in $\{0, \ldots, n\}$ during the interval $[1, k-1]$.

Since the stationary distribution is independent of the initial state, we assume without loss of generality that $X_0 = 0$. Thus, the Markov chain is forced to pass through n before it can reach the state $n+1$, and, conditioning on the last epoch of a visit to n, we can write

$$ P_{0,n+1}(k) = \sum_{0 \leq \nu \leq k} P_{0n}(k - \nu)\, {}_nP_{n,n+1}(\nu). \qquad (4.6) $$

If the chain starts in $n+1$, then the successive epochs of visit to $n+1$ avoiding $\{0, \ldots, n\}$ form a renewal process; that process is terminating since, at each visit, the probability of a return is at most equal to $1 - a_{n+1,n} < 1$. If the chain starts in n, then the successive epochs of visits to $n+1$, avoiding $\{0, \ldots, n\}$, form a delayed, terminating renewal process and ${}_nP_{n,n+1}(\cdot)$ is the renewal density for that process.

Taking the limit in (4.6) as $k \to \infty$, using Theorem 4.1.2, we finally obtain that $\pi_{n+1} = \pi_n R_{n+1}$, where R_{n+1} is the expected total number of renewals in the delayed renewal process. We restate this as a theorem, leaving it to the reader to verify that the definition of R_n given below is equivalent to the one given here.

---

[8] If the Markov chain is periodic, then one must consider the Cesaro limit.

Theorem 4.2.1 *Assume that the Markov chain with transition matrix (4.5) is irreducible and aperiodic. Its limit probabilities satisfy the relation*

$$\pi_n = \pi_{n-1} R_n \qquad \text{for all } n \geq 1, \tag{4.7}$$

where R_n is the expected total number of visits to the state n between two successive visits to $n-1$.

The Markov chain is positive recurrent if and only if the series $\sum_{k\geq 1} R_1 R_2 \cdots R_k$ converges, in which case

$$\pi_0 = \left(1 + \sum_{k\geq 1} R_1 R_2 \cdots R_k\right)^{-1}.$$

□

We have so far established the form of the stationary distribution. Its determination requires a more precise characterization of the quantities R_n, and that issue is taken up next. It would be easy to use (4.7) and algebraically solve the system $\boldsymbol{\pi} P = \boldsymbol{\pi}$; we shall, however, follow a more intricate path in order to illustrate in the scalar case an argument which we shall use later for matrix equations.

Let us consider an n, arbitrary but fixed. By Theorem 4.1.2, we have that $R_n = G(\infty)/(1 - F(\infty))$, where $G(\infty)$ is the probability, starting from $n-1$, of visiting n before returning to $n-1$; this is clearly equal to $a_{n-1,n}$ since any other transition will eventually lead to a return to $n-1$ before a visit to n. As for the probability $F(\infty)$ of returning to n before visiting $n-1$, starting from n, we shall henceforth denote it by U_n; it is given by

$$U_n = \sum_{k\geq 1} {}_nP_{nn}(k),$$

and we have that

$$R_n = a_{n-1,n}(1 - U_n)^{-1}. \tag{4.8}$$

Consider now the circumstances where the chain returns to n, starting from n, avoiding all lower states. Either the chain enters n immediately at its first step, or it enters $n+1$ at the first step, and then makes a first passage from $n+1$ to n. We therefore may write that

$$U_n = a_{n,n} + a_{n,n+1} G_{n+1},$$

where G_{n+1} is the probability that, starting from $n+1$, the Markov chain moves down to n in a finite time.

Thus, in order to characterize the quantities R_n, we are led to the problem of analyzing downward passage probabilities G_n. For now, the following result is adequate for our purpose.

Lemma 4.2.2 *The irreducible Markov chain with transition matrix* (4.5) *is recurrent if and only if* $G_1 = 1$.

Proof. Since the Markov chain is irreducible, it is recurrent if and only if one of its states is recurrent. The probability $U_0 = a_{00} + a_{01}G_1$ of returning to 0, starting from 0, in a finite amount of time is equal to 1—and the chain is recurrent—if and only if $G_1 = 1$. □

Equivalently, we may state that the Markov chain is recurrent if and only if $G_n = 1$ for all $n \geq 1$, since in a recurrent Markov chain, the first passage probability in a finite time is equal to 1 for all pairs of states. Conversely, $G_n < 1$ for all n if the states are transient. This yields the following result.

Theorem 4.2.3 *The Markov chain with transition matrix* (4.5) *is recurrent if and only if* $R_n = a_{n-1,n}/a_{n,n-1}$ *for all* $n \geq 1$. *It is positive recurrent if and only if*

$$\sum_{k\geq 1} \frac{a_{01}\cdots a_{k-1,k}}{a_{10}\cdots a_{k,k-1}} < \infty.$$

In that case, one has that

$$\begin{aligned}
\pi_0 &= \left(1 + \sum_{k\geq 1} \frac{a_{01}\cdots a_{k-1,k}}{a_{10}\cdots a_{k,k-1}}\right)^{-1}, \\
\pi_n &= \pi_0 \frac{a_{01}\cdots a_{n-1,n}}{a_{10}\cdots a_{n,n-1}} \qquad \textit{for all } n \geq 1.
\end{aligned}$$

Proof. The Markov chain is recurrent if and only if $G_n = 1$ for all n, i.e., if and only if $U_n = a_{n,n} + a_{n,n+1}$ for all n. Since $a_{n,n-1} + a_{n,n} + a_{n,n+1} = 1$, we conclude from (4.8) that this holds if and only if $R_n = a_{n-1,n}/a_{n,n-1}$. The remainder of the theorem is easy to prove. □

If we substitute the value for R_n in $\pi_n = \pi_{n-1}R_n$, we find that $\pi_n a_{n,n-1} = \pi_{n-1}a_{n-1,n}$: this asserts the equivalence of the stationary flows across the cut which separates the states in $\{0, \dots, n-1\}$ from those in $\{n, n+1, \dots\}$. It is a highly intuitive property.

In the special case where the transition probabilities $a_{i,i+1}$ and $a_{i+1,i}$ do not depend on i, we call the birth-and-death process *spatially homogeneous* and we obtain the following simpler result.

Theorem 4.2.4 *Consider an irreducible, aperiodic Markov chain with transition matrix*

$$P = \begin{bmatrix} b_0 & a_0 & 0 & 0 & \cdots \\ a_2 & a_1 & a_0 & 0 & \cdots \\ 0 & a_2 & a_1 & a_0 & \cdots \\ 0 & 0 & a_2 & a_1 & \cdots \\ \vdots & \vdots & \vdots & \vdots & \ddots \end{bmatrix}, \qquad (4.9)$$

and let $\boldsymbol{\pi} = (\pi_0, \pi_1, \dots)$ *denote the vector of its limit probabilities. Then,* $\pi_n = \pi_{n-1}R$ *for all* $n \geq 1$*, where* R *is equal to the expected number of visits to* n *between two successive visits to* $n-1$ *and is independent of* n*. In particular, if the chain is positive recurrent, the stationary distribution is given by the geometric density* $\pi_n = (1-R)R^n$ *with* $R = a_0/a_2$*.*

Proof. Traditionally, in order to prove the theorem, one writes the system

$$\begin{aligned} -\pi_0 b_0 + \pi_1 a_2 &= 0, && (4.10) \\ \pi_{n-1}a_0 - \pi_n a_1 + \pi_{n+1}a_2 &= 0 \quad \text{for } n \geq 2, && (4.11) \\ \sum_{i \geq 0} \pi_i &= 1 && (4.12) \end{aligned}$$

and performs a few simple algebraic manipulations. The approach which we follow in this proof is more involved, but it brings out the structure of the steady state distribution and it will be found to be more fruitful when we deal with the more general models of interest to us.

Take an arbitrary but fixed value of n. Replace in the Markov chain (4.9) the states $\{0, 1, \ldots, n\}$ by an absorbing state A, replace the transition from $n+1$ to n by a transition from $n+1$ to A, and keep everything else unchanged. The transition matrix for this new Markov chain is

$$P = \left[\begin{array}{c|c} 1 & 0 \quad 0 \quad 0 \quad \ldots \\ \hline a_2 & \\ 0 & \\ 0 & \widetilde{P}_n \\ \vdots & \end{array}\right],$$

with

$$\widetilde{P}_n = \begin{bmatrix} a_1 & a_0 & 0 & \cdots \\ a_2 & a_1 & a_0 & \cdots \\ 0 & a_2 & a_1 & \cdots \\ \vdots & \vdots & \vdots & \ddots \end{bmatrix}. \tag{4.13}$$

The probability G_{n+1} is equal to the probability of eventually being absorbed in this new Markov chain, starting from the first nonabsorbing state. This is clearly independent of n since $\widetilde{P}_n$ does not depend on n. Thus, we may write that $G_n \equiv G$. Since $R_n = a_0(1 - U_n)^{-1}$ with $U_n = a_1 + a_0 G_{n+1}$, this proves that $R_n = R$ for all n.

For a recurrent chain, $G_n \equiv 1$, implying that $R_n \equiv a_0/a_2$. We use Theorem 4.2.1 to prove the remaining statements. □

In the spatially homogeneous case, the classification of the Markov chain becomes easy, as shown in the theorem below.

Theorem 4.2.5 *The Markov chain with transition matrix* (4.9) *is recurrent if and only if* $a_0 \leq a_2$*; it is positive recurrent if and only if* $a_0 < a_2$.

The probability $G \equiv G_n$ *of a first passage from state* n *to state* $n-1$ *for* $n \geq 1$ *is the minimal nonnegative solution of the equation*

$$G = a_2 + a_1 G + a_0 G^2. \tag{4.14}$$

Proof. Let γ_n denote the probability that, starting from n, the Markov chain eventually visits 0, and let $\boldsymbol{\gamma} = (\gamma_1, \gamma_2, \gamma_3, \ldots)^t$; we set $\gamma_0 = 1$. We note that $\boldsymbol{\gamma} = \sum_{i \geq 0} \widetilde{P}_1^i \boldsymbol{a}$, where $\boldsymbol{a} = (a_2, 0, 0, \ldots)^t$, for the ith term in the series is the probability that the first visit to 0 occurs at time $i+1$.

The vector $\boldsymbol{\gamma}$ is the minimal nonnegative solution of the equation

$$\boldsymbol{\gamma} = \widetilde{P}_1\boldsymbol{\gamma} + \boldsymbol{a}. \tag{4.15}$$

To show this, we assume that there is another nonnegative vector $\boldsymbol{z}$ solution of (4.15). If we iteratively replace $\boldsymbol{z}$ on the right-hand side by $\widetilde{P}_1\boldsymbol{z} + \boldsymbol{a}$, we find that

$$\begin{aligned}\boldsymbol{z} &= \sum_{0\le i\le k} \widetilde{P}^i\boldsymbol{a} + \widetilde{P}^{k+1}\boldsymbol{z} \\ &\ge \sum_{0\le i\le k} \widetilde{P}^i\boldsymbol{a}\end{aligned}$$

for all $k \ge 0$, so that $\boldsymbol{z} \ge \lim_{k\to\infty} \sum_{0\le i\le k} \widetilde{P}^i\boldsymbol{a} = \boldsymbol{\gamma}$.

Conditioning on the first passage from n to $n-1$, we find that $\gamma_n = G_n\gamma_{n-1} = G\gamma_{n-1}$, so that $\gamma_n = G^n$ for $n \ge 1$. Replacing γ_n by G^n in (4.15) yields that G is the minimal solution of (4.14), so that the second statement is proved.

The two roots of (4.14) are 1 and a_2/a_0. If $a_0 \le a_2$, then $G = 1$ and the chain is recurrent. We know by Theorem 4.2.3 that the homogeneous birth-and-death process is positive recurrent if and only if $a_0 < a_2$. This completes the proof. □

More generally, if we assume that $a_{n,n-1} = a_2$, $a_{n,n} = a_1$, and $a_{n,n+1} = a_0$ for n greater than some finite value c, then we can show that the Markov chain is positive recurrent if and only if $a_0 < a_2$ and that it has a modified geometric distribution with $\pi_{c+k} = \pi_c R^k$ for $k \ge 0$. We omit the details of the arguments which are similar to those used above.

Remark 4.2.6 Another proof of the first statement in Theorem 4.2.5 is based on the theory of random walks. We briefly give its outline because that is the approach we will follow in Chapter 7; for details, see Chung [15, Chapter 8].

Consider the random walk defined by $S_n = S_{n-1} + Y_n$, where $\{Y_n : n \ge 1\}$ is a sequence of i.i.d. random variables such that Y_n is equal to 1 with probability a_0, 0 with probability a_1, and -1 with probability a_2.

It is obvious that the first passage time from the state 1 to the state 0 in the Markov chain has the same distribution as the hitting time of the negative half-line for the random walk, starting from 0.

This hitting time is finite a.s. if and only if $E[Y_n] \leq 0$, so that $g = 1$ if and only if $a_0 \leq a_2$. Furthermore, the expected hitting time is finite if and only if the inequality is strict (Chung [15, Theorems 8.3.4 and 8.4.4]).

Remark 4.2.7 Note that finite birth-and-death processes on the interval $\{0, 1, \ldots, K\}$ may be handled as special cases of the infinite processes: it suffices to replace $p_{K,K+1}$ by zero in our equations. Finite processes are inherently nonhomogeneous, however, and we therefore miss simplifications like those in Theorem 4.2.4.

4.3 The Basic Nonlinear Equations

In analyzing birth-and-death processes, we used the downward passage probability G_n from n to $n-1$. More generally, we define

$$g_n(k) = {}_{n-1}P_{n,n-1}(k) \qquad \text{for } k \geq 1;$$

i.e., $g_n(k)$ is the probability that, starting in n at time 0, the chain makes its first visit to $n-1$ at time k.

We also used the renewal density of visits to n, starting from $n-1$, before the first return to $n-1$. In order to simplify notations, we shall define

$$r_n(k) = {}_{n-1}P_{n-1,n}(k) \qquad \text{for } k \geq 1; \tag{4.16}$$

i.e., $r_n(k)$ is the probability that, starting in $n-1$, the chain is in n at time k and has not returned to $n-1$ yet.

Finally, we define

$$u_n(k) = {}_nP_{n,n}(k) \qquad \text{for } k \geq 1;$$

i.e., $u_n(k)$ is the probability of returning to n at time k for the first time, starting from n, avoiding $n-1$.

We shall denote, respectively, by $G_n(\cdot)$, $R_n(\cdot)$, and $U_n(\cdot)$ the generating functions of these probability sequences:

$$\begin{aligned} G_n(z) &= \sum_{k\geq 1} z^k g_n(k), \\ R_n(z) &= \sum_{k\geq 1} z^k r_n(k), \\ U_n(z) &= \sum_{k\geq 1} z^k u_n(k). \end{aligned}$$

We observe that $G_n = G_n(1)$, $R_n = R_n(1)$, and $U_n = U_n(1)$.

The densities defined above are closely interrelated, as we show next.

Theorem 4.3.1 *The following relations hold between $G_n(\cdot)$, $U_n(\cdot)$, and $R_n(\cdot)$ for $n \geq 1$; any one of these sequences determines the other two:*

$$\begin{aligned} U_n(z) &= za_{n,n} + za_{n,n+1}G_{n+1}(z) && (4.17) \\ &= za_{n,n} + R_{n+1}(z)za_{n+1,n}, && (4.18) \\ G_n(z) &= (1 - U_n(z))^{-1} za_{n,n-1}, && (4.19) \end{aligned}$$

and

$$R_n(z) = za_{n-1,n}(1 - U_n(z))^{-1} \tag{4.20}$$

for $|z| < 1$.

Proof. Consider the density $\{u_n(k) : k \geq 1\}$ and its generating function $U_n(z)$. Starting in n, the chain may return to n in two ways, while avoiding $n-1$: it may remain in n at the very next transition (the probability of which event is $a_{n,n}$) or it may move up to $n+1$ at the first transition (with probability $a_{n,n+1}$) and then eventually move down from $n+1$ to n, after a random interval of time, the generating function of which is $G_{n+1}(z)$. We thus obtain the equation (4.17).

Equation (4.18) is similarly obtained by conditioning on the *last* step before the return to n instead of the first step after the departure from n.

With respect to the density $\{g_n(k) : k \geq 1\}$, we decompose the event according to the number of times the process returns to n (with the density $\{U_n(\cdot)\}$) before it eventually moves down from n to $n-1$ (with probability $a_{n,n-1}$). We thus find that

$$G_n(z) = za_{n,n-1} + U_n(z)za_{n,n-1} + U_n^2(z)za_{n,n-1} + \cdots,$$

which is identical to (4.19) since $|U_n(z)| \le U_n(1) < 1$ (here and in what follows, we write $U_n^k(z)$ for $(U_n(z))^k$).

Finally, we prove (4.20) as follows. In order for the chain to be in n at some time, having started in $n-1$ and having avoided $n-1$ in intermediate steps, it is necessary that the first transition should be from $n-1$ to n (contributing the factor $za_{n-1,n}$) and that the chain subsequently visits n, possibly several times, avoiding $n-1$; each such visit contributes a factor $U_n(z)$.

We leave it to the reader to give a formal proof. □

Corollary 4.3.2 *For the spatially homogeneous process with transition matrix* (4.9), *the functions $U_n(\cdot)$, $G_n(\cdot)$, and $R_n(\cdot)$ are all independent of n and their values $U(\cdot)$, $G(\cdot)$, and $R(\cdot)$ are such that*

$$\begin{aligned} U(z) &= za_1 + za_0 G(z) \\ &= za_1 + R(z)za_2, \\ G(z) &= (1-U(z))^{-1}za_2, \end{aligned}$$

and

$$R(z) = za_0(1-U(z))^{-1}.$$

Any one of these functions determines the other two.

Proof. We have noted earlier that the distribution of the first passage time from n to $n-1$ is governed by the matrix $\widetilde{P}$ in (4.13), obtained by deleting the rows and columns corresponding to the states 0 to $n-1$. Since this matrix does not depend on n, the distribution $\{g_n(k) : k \ge 1\}$ and the generating function $G_n(\cdot)$ do not depend on n. This implies by Theorem 4.3.1 that $U_n(\cdot)$ and $R_n(\cdot)$ are also independent of n. □

We have shown that the sequences of functions $\{U_n(\cdot)\}$, $\{G_n(\cdot)\}$, and $\{R_n(\cdot)\}$ are interrelated. They may also be characterized separately, and we have that

$$U_n(z) = za_{nn} + za_{n,n+1}(1-U_{n+1}(z))^{-1}za_{n+1,n}, \tag{4.21}$$
$$G_n(z) = za_{n,n-1} + za_{n,n}G_n(z) + za_{n,n+1}G_{n+1}(z)G_n(z), \tag{4.22}$$

and

$$R_n(z) = za_{n-1,n} + R_n(z)za_{n,n} + R_n(z)R_{n+1}(z)za_{n+1,n}. \tag{4.23}$$

This may be proved by simple algebraic manipulations; for instance, in order to prove (4.21), we replace $G_{n+1}(z)$ in (4.17) by the expression provided by (4.19). It is, however, more fruitful to consider the probabilistic justification, which is based on the following lemma.

Lemma 4.3.3 *The distribution of the first passage time from $n+r$ to n has the generating function $G_{n+r}(\cdot)\cdots G_{n+1}(\cdot)$. In particular, in the spatially homogeneous case of* (4.9), *this generating function is equal to $(G(\cdot))^r$.*

For upward movements, consider the probability that the Markov chain is in $n+r$ at time k, starting from n at time 0 and avoiding n in $[1\dots k-1]$; the generating function with respect to k of these probabilities is $R_{n+1}(\cdot)R_{n+2}(\cdot)\dots R_{n+r}(\cdot)$. For the spatially homogeneous chain, this is $(R(\cdot))^r$.

Proof. A first passage time from $n+r$ to n is really the sum of a first passage time from $n+r$ to $n+r-1$, one from $n+r-1$ to $n+r-2$, and so on, and finally one from $n+1$ to n. By the strong Markov property, these random times are conditionally independent, given that they are finite, so that the first statement follows.

The second statement is proved by induction after noting that, starting from n, the chain can get to $n+r$ only after visiting $n+1$, $n+2$, $\dots$, $n+r-1$; the formulas are obtained by conditioning on the last epoch of such visits prior to visiting $n+r$. □

Equation (4.22) may now be obtained by conditioning on the first step of the Markov chain. If that first step happens to be into $n+1$, then it is followed by a first passage from $n+1$ to $n-1$ with generating function $G_{n+1}(\cdot)G_n(\cdot)$. The other two terms correspond to the cases where the first step is to $n-1$ or to n.

To prove (4.23), we condition on the state from which n is entered at some fixed time. The Markov chain may enter n from $n-1$, n, or $n+1$; these cases, respectively, give the three terms. Observe that the factor $R_n(z)R_{n+1}(z)$ is the generating function for the visits to $n+1$.

The proof of (4.21) is immediate when we write it like this:

$$U_n(z) = za_{n,n} + za_{n,n+1} \sum_{\nu \geq 0} U_{n+1}^{\nu}(z) za_{n+1,n};$$

indeed, the first return to n, avoiding lower states, may either occur immediately, or it may happen after a jump to $n+1$, followed by some number ν of returns to $n+1$ and an eventual transition from $n+1$ to n.

In the spatially homogeneous case, the equations (4.21)–(4.23) become

$$U(z) = za_1 + za_0(1 - U(z))^{-1} za_2, \tag{4.24}$$
$$G(z) = za_2 + za_1 G(z) + za_0 G^2(z), \tag{4.25}$$

and

$$R(z) = za_0 + R(z) za_1 + R^2(z) za_2. \tag{4.26}$$

These equations are important, but they do not fully characterize the generating functions of interest. As an elementary verification, we consider the equation (4.25) for $z = 1$. It is clear that 1 is always a solution, yet we have seen that the first passage probability $G = G(1)$ is strictly less than 1 if the Markov chain is transient. Thus, we need a further criterion to determine G, or $G(z)$, etc. We shall see in Chapter 8 that the quantities we are interested in are such that for $0 \leq z \leq 1$, their values are *minimal nonnegative solutions* of their respective equations. Here, we shall merely show it in the special case where $z = 1$.

Theorem 4.3.4 *For the spatially homogeneous chain with matrix* (4.9), *the quantities* $G(1)$ *and* $R(1)$ *are given by*

$$G(1) = \min\{1, a_2/a_0\}$$

and

$$R(1) = \min\{1, a_0/a_2\}.$$

Proof. The equation $G(1) = a_2 + a_1 G(1) + a_0 G^2(1)$ has two solutions: 1 and a_2/a_0. By Theorem 4.2.5, $G = 1$ when $a_0 \leq a_2$ and $G < 1$ when $a_0 > a_2$, so that the first statement is proved.

By (4.17) and (4.20), we have that $R(1) = a_0(1 - a_1 - a_0 G(1))^{-1}$, from which the second statement readily follows. □

4.4 Transient Distributions

The functions discussed in the preceding section are also useful to characterize the transient behavior of the Markov chain. Let us define the probabilities

$$x_n(k) = \mathrm{P}[X_k = n | X_0 = 0],$$

and their generating functions $X_n(z) = \sum_{i \geq 0} z^i x_n(i)$, for $n \geq 0$. It immediately results from (4.6) and (4.16) that

$$\begin{aligned} X_{n+1}(z) &= X_n(z) R_{n+1}(z) \\ &= X_0(z) \prod_{1 \leq j \leq n+1} R_j(z). \end{aligned}$$

The generating function $U_0(z)$ of the first return time to 0 is equal to $z(a_{00} + R_1(z)a_{10})$ by (4.18), so that

$$\begin{aligned} X_0(z) &= \sum_{\nu \geq 0} U_0^\nu(z) \\ &= (1 - U_0(z))^{-1} \\ &= [1 - z(a_{00} + R_1(z)a_{10})]^{-1}. \end{aligned}$$

When the birth-and-death process is spatially homogeneous, these equations become

$$\begin{aligned} X_{n+1}(z) &= X_0(z) R^{n+1}(z), \\ X_0(z) &= [1 - z(b_0 + R(z)a_2)]^{-1}. \end{aligned} \qquad (4.27)$$

4.5 Continuous Time Birth-and-Death Processes

In continuous time, a birth-and-death process $\{X(t) : t \geq 0\}$ has a tridiagonal generator Q:

$$Q = \begin{bmatrix} q_{00} & q_{01} & 0 & 0 & \cdots \\ q_{10} & q_{11} & q_{12} & 0 & \cdots \\ 0 & q_{21} & q_{22} & q_{23} & \cdots \\ 0 & 0 & q_{32} & q_{33} & \cdots \\ \vdots & \vdots & \vdots & \vdots & \ddots \end{bmatrix}. \qquad (4.28)$$

In the irreducible case, we have that $q_{i,j} > 0$ for $i, j \geq 0$, $j = i - 1$ or $j = i + 1$, $q_{i,i} < 0$, and the row sums are equal to zero.

We need to make the assumption that the process is not explosive, i.e., that it cannot make infinitely many transitions in a finite amount of time. For instance, this assumption will be needed when we shall condition on the last visit to some state before a given time. Simple sufficient conditions for this are that $\sup_i(-q_{i,i}) \leq \lambda$ for some finite λ or that $\sum_{n\geq 0}(-1/q_{n,n})$ diverges. For further detail, we refer to Çinlar [12]. In most applications, it is easy to verify that one of these conditions holds.

For nonexplosive birth-and-death processes, the limiting probabilities $\pi_n = \lim_{t\to\infty} \mathrm{P}[X(t) = n]$ always exist, independently of the initial state; they are equal to zero in the transient and null recurrent cases.

The spatially homogeneous birth-and-death process has the following generator:

$$Q = \begin{bmatrix} b_0 & q_0 & 0 & 0 & \cdots \\ q_2 & q_1 & q_0 & 0 & \cdots \\ 0 & q_2 & q_1 & q_0 & \cdots \\ 0 & 0 & q_2 & q_1 & \cdots \\ \vdots & \vdots & \vdots & \vdots & \ddots \end{bmatrix} \tag{4.29}$$

with $q_0, q_2 > 0$, $b_0 = -q_0 < 0$, $q_1 = -q_0 - q_2 < 0$. The M/M/1 queue is of this type, while the M/M/∞ queue is a simple example of a level-dependent birth-and-death process like (4.28).

Theorem 4.5.1 *In the recurrent case, the stationary distribution of the Markov process given by* (4.28) *satisfies*

$$\pi_n = \pi_{n-1}\frac{q_{n-1,n}}{q_{n,n-1}} \qquad \text{for } n \geq 1.$$

The process is ergodic if and only if

$$\sum_{k\geq 1} \frac{q_{01}\cdots q_{k-1,k}}{q_{10}\cdots q_{k,k-1}} < \infty.$$

In the homogeneous case of (4.29)*, the recurrence becomes*

$$\pi_n = \pi_{n-1}q_0/q_2 \qquad \text{for } n \geq 1$$

and the process is ergodic if and only if $q_0 < q_2$.

Proof. We basically repeat the argument which led to Theorem 4.2.1. Assume without loss of generality that $X(0) = 0$. The Markov chain is forced to pass through $n-1$ before it can reach the state n. Upon conditioning on the epoch $t-u$ of the last entrance into n from $n-1$ before time t, we obtain that

$$\begin{aligned} &\mathrm{P}[X(t) = n|X(0) = 0] \\ &= \int_0^t \mathrm{P}[X(t-u) = n-1|X(0) = 0] q_{n-1,n}\, {}_{n-1}P_{n,n}(u)du \\ &= \int_0^t \mathrm{P}[X(t-u) = n-1|X(0) = 0] q_{n-1,n} d\tilde{V}_n(u), \end{aligned}$$

where, as before, ${}_nP_{i,j}(t)$ is the taboo probability that, starting from i at time 0, the Markov process is in j at time t and has not visited any of the states in $\{0, \dots, n\}$ during the interval $(0, t)$, and

$$\tilde{V}_n(t) = \int_0^t {}_{n-1}P_{n,n}(u)du$$

is the expected time spent in n during the interval $(0, t)$ before the first passage to $\{0, \dots, n-1\}$, given that $X(0) = n$.

Now, since the Markov process is irreducible, the process of successive visits to n under taboo of $\{0, \dots, n-1\}$ is terminating, and we have that

$$\begin{aligned} &\lim_{t\to\infty} \int_0^t \mathrm{P}[X(t-u) = n-1|X(0) = 0] d\tilde{V}_n(u) \\ &= \lim_{t\to\infty} \mathrm{P}[X(t) = n-1|X(0) = 0] M_n, \end{aligned}$$

where

$$M_n = \int_0^\infty {}_{n-1}P_{n,n}(u)du$$

is the expected total time spent in n before the first passage to $\{0, \dots, n-1\}$, given that the initial state is n, so that

$$\pi_n = \pi_{n-1} q_{n-1,n} M_n.$$

A simple probabilistic argument, conditioning on the first transition of the Markov process, gives that

$$M_n = \frac{1}{|q_{n,n}|} + \frac{q_{n,n+1}}{|q_{n,n}|} G_{n+1} M_n,$$

where G_{n+1} is the probability that, starting from $n+1$, the Markov chain reaches down to n in a finite time. Noting that $G_{n+1} = 1$ for a recurrent process, and using the fact that $q_{n,n} = -q_{n,n-1} - q_{n,n+1}$, we solve the above equation and obtain that $M_n = 1/q_{n,n-1}$. The remainder of the proof is elementary. □

Corollary 4.5.2 *The total time spent in state n starting from n before the first visit to $\{0, \dots, n-1\}$ is exponentially distributed with parameter U_n, where*

$$U_n = |q_{n,n}| - q_{n,n+1}G_{n+1} \tag{4.30}$$

and G_{n+1} is the probability that, starting from $n+1$, the Markov process reaches n in a finite time.

Proof. Let us denote by S the total sojourn time in n. It is equal to the sum $\sum_{1 \le i \le N} Y_i$, where N is the number of visits to n and Y_1, $\dots$, Y_N are the lengths of the intervals of sojourn. The Y_i's are i.i.d. random variables, exponentially distributed with parameter $|q_{n,n}|$, and N is geometrically distributed with parameter $(q_{n,n+1}/|q_{n,n}|)G_{n+1}$, so that S is exponentially distributed with parameter U_n. □

We shall henceforth use the notation

$$\begin{aligned} R_n &= q_{n-1,n} \int_0^\infty {}_{n-1}P_{n,n}(u)du \\ &= q_{n-1,n} M_n = q_{n-1,n} U_n^{-1}, \end{aligned} \tag{4.31}$$

so that the same relation $\pi_n = \pi_{n-1} R_n$ holds as in the discrete time case.

4.6 Passage and Sojourn Times

We shall examine here the continuous time analogue of the relations which we determined in section 4.3.

We first define the distribution functions $\tilde{G}_n(\cdot)$ of the first passage time from n to $n-1$. This is given by

$$\tilde{G}_n(t) = \int_0^t {}_{n-1}P_{n,n-1}(u)du \qquad \text{for } t \ge 0, \tag{4.32}$$

and we have that $G_n = \tilde{G}_n(\infty)$. We denote by $G_n(\cdot)$ the Laplace–Stieltjes transform of $\tilde{G}_n(\cdot)$,

$$G_n(s) = \int_0^\infty e^{-sy} d\tilde{G}_n(y),$$

and we also have that $G_n = G_n(0)$.

We next define the functions $\tilde{R}_n(t) = q_{n-1,n}\tilde{V}_n(t)$, which are such that $R_n = \lim_{t\to\infty} \tilde{R}_n(t)$. Observe that this is slightly different from the discrete time case, where we directly used the delayed renewal process of visits to n, *starting from the state* $n-1$. Finally, we denote by $V_n(\cdot)$ and $R_n(\cdot)$ the Laplace–Stieltjes transforms of $\tilde{V}_n(\cdot)$ and $\tilde{R}_n(\cdot)$, respectively.

Theorem 4.6.1 *The following relations hold:*

$$G_n(s) = V_n(s) q_{n,n-1} \tag{4.33}$$

and

$$V_n(s) = [s - q_{n,n} - q_{n,n+1} G_{n+1}(s)]^{-1}, \tag{4.34}$$

so that

$$R_n(s) = q_{n-1,n}[s - q_{n,n} - q_{n,n+1} G_{n+1}(s)]^{-1}. \tag{4.35}$$

Proof. Since ${}_{n-1}P_{n,n-1}(u) = {}_{n-1}P_{n,n}(u) q_{n,n-1}$, we find from (4.32) that $\tilde{G}(t) = \tilde{V}(t) q_{n,n-1}$, which proves the first statement.

To prove the second statement, we proceed as follows. Starting from n, there are two ways in which the process may be in n at time y without having visited $n-1$: either it does not leave n in $[0,y]$, or it moves to $n+1$ at some time $\tau < y$ and returns to n at some time $\tau + z < y$. This gives

$${}_{n-1}P_{n,n}(y) = e^{q_{n,n}y} + \int_o^y \int_0^{y-\tau} e^{q_{n,n}\tau} q_{n,n+1} d\tau \, d\tilde{G}_{n+1}(z) \, {}_{n-1}P_{n,n}(y-\tau-z),$$

so that, multiplying by e^{-sy} and integrating, we obtain that

$$V_n(s) = \frac{1}{s - q_{n,n}} + \frac{q_{n,n+1}}{s - q_{n,n}} G_{n+1}(s) V_n(s),$$

from which (4.34) is immediate. □

Taking $s = 0$ in (4.35), we find again that

$$R_n = q_{n-1,n}(-q_{n,n} - q_{n,n+1}G_{n+1})^{-1} = q_{n-1,n}U_n^{-1}.$$

The three sequences of transforms may also be separately characterized, and we have that

$$\begin{aligned} q_{n,n-1} + (q_{n,n} - s)G_n(s) + q_{n,n+1}G_{n+1}(s)G_n(s) &= 0, & (4.36)\\ q_{n-1,n} + R_n(s)(q_{n,n} - s) + R_n(s)R_{n+1}(s)q_{n+1,n} &= 0, & (4.37) \end{aligned}$$

and

$$1 + V_n(s)(q_{n,n} - s) + V_n(s)q_{n,n+1}V_{n+1}(s)q_{n+1,n} = 0. \qquad (4.38)$$

These equations may be either proved algebraically, using Theorem 4.6.1, or proved by a probabilistic argument like equations (4.21)–(4.23). The proof of (4.36) is by conditioning on the state visited when the Markov process leaves n for the first time. To prove (4.38), we define $v_n(t) = {}_{n-1}P_{n,n}(t)$, verify that

$$v_n(t) = e^{q_{n,n}t} + \int_0^t {}_{n-1}P_{n,n+1}(u)q_{n+1,n}e^{q_{n,n}(t-u)}du,$$

and

$${}_{n-1}P_{n,n+1}(u) = \int_0^t v_n(t-u)q_{n,n+1}v_{n+1}(u)du,$$

and then take transforms. We omit the details.

Just as in the discrete time case, and for precisely the same reason, the functions $G_n(s)$, $R_n(s)$ and $V_n(s)$ do not depend on n for the spatially homogeneous process (4.29), and equations (4.33)–(4.36) become

$$\begin{aligned} G(s) &= V(s)q_2,\\ V(s) &= (s - q_1 - q_0G(s))^{-1},\\ R(s) &= q_0(s - q_1 - q_0G(s))^{-1}, \end{aligned}$$

and

$$q_2 + q_1G(s) + q_0G(s)^2 = 0.$$

The solutions to these equations are not unique, but the solutions of interest to us have a minimality property. For instance, for $s = 0$, the probability $G = G(0)$ of a finite first passage time from 1 to 0 is $\min(1, q_2/q_0)$. Indeed, G is also the first passage probability for the Markov chain embedded at epochs of transitions. This is a spatially homogeneous, discrete time birth-and-death process with $a_0 = q_0/|q_1|$, $a_1 = 0$, and $a_2 = q_2/|q_1|$; it remains to apply Theorem 4.3.4.

We conclude this chapter with a discussion of the probabilistic significance of R_n. Different interpretations have been proposed in the literature, but we favor the following one, which holds in continuous as well as in discrete time.

Proposition 4.6.2 *For all $n \geq 1$, the factor R_n is equal to the expected sojourn time in n per unit of sojourn time in $n-1$.*

This is clearly the case in discrete time: we have seen that R_n is the expected total number of visits to n between two successive visits to $n-1$ (see Theorem 4.2.1), and the Markov chain spends exactly one unit of time between two transitions.

In continuous time, we have various expressions for R_n, such as $\pi_n = \pi_{n-1} R_n$. If we multiply both sides by some large T, since $\pi_{n-1}T$ and $\pi_n T$ are asymptotically the expected times spent in $n-1$ and n, respectively, we see that their ratio $R_n = \pi_n T / \pi_{n-1} T$ is asymptotically the expected time in n per unit of sojourn time in $n-1$ during $[0, T]$.

An alternative justification is based on the defining equation (4.31) which we recall:

$$R_n = q_{n-1,n} \int_0^\infty {}_{n-1}P_{n,n}(u)du = q_{n-1,n} M_n. \tag{4.39}$$

Assume that the Markov chain is in state $n-1$ at some arbitrary time, and consider the events which may occur during an interval of time of length h, where h is arbitrarily small. At the end of this interval, three cases may occur:

a. with probability $q_{n-1,n}h$, the Markov chain has moved to n, where it will spend M_n units of time before it is in $n-1$ again;

b. with probability $q_{n-1,n-2}h$, it has moved to $n-2$, from where it will not be able to visit n until it visits $n-1$ again;

c. with probability $1 + q_{n,n}h$, the Markov chain is still in $n - 1$.

Thus, the Markov chain spends on the average $q_{n-1,n}hM_n = R_n h$ units of time in n for each interval of length h spent in $n - 1$.

We can rewrite (4.39) as

$$R_n = \left(-\frac{q_{n-1,n}}{q_{n-1,n-1}} M_n\right) \Big/ \left(-\frac{1}{q_{n-1,n-1}}\right).$$

The denominator is the expected sojourn time in state $n - 1$ and $-q_{n-1,n}/q_{n-1,n-1}$ is the conditional probability of moving to state n upon leaving state $n - 1$. Thus, we may also interpret R_n in continuous time as the expected sojourn time spent in state n within a return time to $n - 1$, expressed in units of mean sojourn time in $n - 1$; this interpretation is the one given originally in Neuts [79].

Chapter 5

Processes Under a Taboo

The whole argument in Chapter 4 revolves around the behavior of the birth-and-death process when it is in one subset of states (n or higher), until it visits another subset of states (0 to $n-1$). We shall repeatedly use this type of argument in what follows. Instead of giving the details in each case, we present them here in a general setting, so that we shall be able to simply refer to them when needed. This material plays a very important role in the rest of the book; for that reason, we give detailed proofs even for simple and well-established results.

We consider an irreducible, aperiodic Markov chain $\{X_n : n \geq 0\}$ on a denumerable state space S with transition matrix P. We denote by $\boldsymbol{\pi}$ its limiting probability vector with $\boldsymbol{\pi} > \mathbf{0}$ when the Markov chain is positive recurrent. We initially assume that the process evolves in discrete time, the continuous case being examined later.

The key word here is *taboo*. We consider a taboo subset of states $T \subset S$ and examine the behavior of the system during an interval of time during which T is avoided. We define

$$\tau = \inf\{n \geq 1 : X_n \in T\}$$

as the first epoch after time 0, when the Markov chain is in one of the states in T. If $X_0 \notin T$, we shall call τ the first passage time to T, or the sojourn time in the complement $S \setminus T$; if $X_0 \in T$, τ will also be called the first return time to T.

One objective is to generalize the relation $\pi_n = \pi_{n-1} R_n$ of Theorem 4.2.1 and to establish the relation $\boldsymbol{\pi}_D = \boldsymbol{\pi}_T R^*$, where D is a subset of states disjoint from T, R^* is a matrix of expected values to be defined later, and $\boldsymbol{\pi}_T$ and $\boldsymbol{\pi}_D$ are the subvectors of limiting probabilities, respectively, for the states in T and in D.

In the last section, we establish the same property for continuous time Markov processes. We present this material in full detail in order to make it clear that the results for discrete and continuous time processes have the same probabilistic significance, despite some formal and superficial differences. This will allow us in later chapters to concentrate on proving the results for discrete time Markov chains and to present them only briefly for the continuous time case.

5.1 Expected Time to Exit

Let D be a proper subset of S, let $P_D = \{P_{ij} : i, j \in D\}$ be the submatrix of transition probabilities between states of D, and let us denote by N_D the matrix of expected sojourn times in the states of D before the first visit to any state outside of D. One has that

$$[N_D]_{ij} = \mathrm{E}\Big[\sum_{n\geq 0} \mathcal{I}\{X_n = j \;\&\; S_D \geq n | X_0 = i\}\Big],$$

where $\mathcal{I}\{\cdot\}$ is the indicator function and S_D is the total sojourn time in D. We use the symbol "&" to represent the operator "and," instead of the more customary $\cap$ or $\wedge$, in order to make the expressions easier to read.

The random variable S_D is greater than or equal to n if and only if $X_i \in D$ for all i in $\{0, \dots, n\}$. We immediately have that

$$\begin{aligned} [N_D]_{ij} &= \sum_{n\geq 0} \mathrm{P}[X_n = j \;\&\; S_D \geq n | X_0 = i] \\ &= \sum_{n\geq 0} [P_D^n]_{ij}. \end{aligned}$$

If D has finitely many states, then the following lemma applies (Ortega and Rheinboldt [89, Lemma 2.3.1]) and $N_D = (I - P_D)^{-1}$ as we already saw in (2.15).

Lemma 5.1.1 *Let M be a matrix of finite order. The series $\sum_{n\geq 0} M^n$ converges if and only if the spectral radius of M is strictly less than 1. In that case, $\sum_{n\geq 0} M^n = (I - M)^{-1}$.* □

In the infinite dimensional case, N_D is the minimal nonnegative solution of the system $(I - P_D)N = I$ (Çinlar [12] and Kemeny, Snell, and Knapp [40]). In this sense, the equation $(I - P_D)N = I$ characterizes the matrix N_D in the infinite dimensional case as well.

In order to have uniform notations, we shall adopt the following convention and write $\sum_{n\geq 0} M^n = (I - M)^{-1}$ for any substochastic matrix M such that the series converges, even when M is infinite dimensional.

With this convention, we summarize the above discussion as a lemma.

Lemma 5.1.2 *The matrix N_D of expected sojourn times in the subset D, before the first passage to the complementary subset $S \setminus D$, is given by*

$$N_D = (I - P_D)^{-1},$$

where

$$(I - P_D)^{-1} = \sum_{n\geq 0} P_D^n$$

is the minimal nonnegative solution of the systems $(I - P_D)N_D = I$ and $N_D(I - P_D) = I$. □

From now on, we shall freely interchange expressions like $(I - P_D)^{-1}$ and $\sum_{n\geq 0} P_D^n$ and interpret both as a matrix of expected sojourn times. Note that we have, with the lemma above, the first of several cases where a matrix is characterized as the minimal solution of a matrix equation.

5.2 Linking Subsets of States

We need to introduce several notations. We denote by $P_{ij}(k)$ the k-step transition probability from i to j,

$$P_{ij}(k) = \mathrm{P}[X_k = j | X_0 = i],$$

and by $P_{TD}(k)$, where T and D are two disjoint subsets of states, the matrix $\{P_{ij}(k) : i \in T, j \in D\}$; the rows of this matrix are indexed by the states in T and the columns by the states in D. This definition is extended in an obvious manner to several matrices which we shall define later.

We also define ${}_TP_{ij}(k)$ as the k-step *taboo* transition probability from i to j under taboo of the subset T,

$$ {}_TP_{ij}(k) = \mathrm{P}[X_1 \notin T, \ldots, X_{k-1} \notin T, X_k = j | X_0 = i], $$

or more concisely,

$$ {}_TP_{ij}(k) = \mathrm{P}[X_k = j \;\&\; \tau \geq k | X_0 = i], $$

where τ is the first passage time to T.

Conditioning on the last state visited in T before time k, we may write for all $i \in T$ and $j \in D$ that

$$ P_{ij}(k) = \sum_{0 \leq \nu \leq k} \sum_{i' \in T} P_{ii'}(k-\nu)\, {}_TP_{i'j}(\nu), $$

which we write in matrix notations as

$$ P_{TD}(k) = \sum_{0 \leq \nu \leq k} P_{TT}(k-\nu)\; {}_TP_{TD}(\nu) \tag{5.1} $$

for all $k \geq 0$.

For $j \in D$, ${}_TP_{jj}(\cdot)$ is the renewal density of the renewal process of successive returns to j until the first visit to the set T. For $i, j \in D$, $i \neq j$, we may think of ${}_TP_{ij}(\cdot)$ as the *delayed* renewal density of visits to j, starting from i. Instead of considering individual states, however, it will be more fruitful to consider globally ${}_TP_{DD}(\cdot)$ as the density of the *Markov renewal process* $\{(Y_n, S_n) : n \geq 0\}$ of successive visits to D: starting from $X_0 \in D$ at time 0, we set $Y_0 = X_0$ and $S_0 = 0$. Then, denoting by θ_1, θ_2, ... the successive epochs of visit to D, we define $Y_n = X_{\theta_n}$ and $S_n = \theta_n$, as long as $\theta_n < \tau$, the first passage time to T. We shall also make use of the delayed Markov renewal process where the initial state is in T; this has density ${}_TP_{TD}(\cdot)$.

We now introduce the convolution operator $\star$ for matrices. Let $A(\cdot)$ be a matrix of functions defined on $\mathbb{N}$, and let $B(\cdot)$ be a matrix of mass functions, i.e., $B(k) = \sum_{0 \le n \le k} F_B(n)$ for $k \ge 0$ with $F_B(n) \ge 0$ for all n; we define

$$A \star B(k) = \sum_{0 \le \nu \le k} A(k-\nu) F_B(\nu) \tag{5.2}$$

and

$$B \star A(k) = \sum_{0 \le \nu \le k} F_B(k-\nu) A(\nu) \tag{5.3}$$

for $k \ge 0$. If both $A(\cdot)$ and $B(\cdot)$ are matrices of mass functions, so that $A(k) = \sum_{0 \le n \le k} F_A(n)$ for $k \ge 0$ with $F_A(n) \ge 0$ for all n, we may indifferently define $A \star B(\cdot)$ by (5.2) or by

$$A \star B(k) = \sum_{0 \le \nu \le k} F_A(k-\nu) B(\nu),$$

since the right-hand side of the two equations are equal. Note that unlike for scalar functions, the matrix convolution operation $\star$ is not commutative and $A \star B \neq B \star A$.

Now, returning to (5.1), we may write it as follows,

$$P_{TD}(k) = P_{TT} \star {}_T R_{TD}(k), \tag{5.4}$$

if we define

$${}_T R_{ij}(k) = \sum_{0 \le n \le k} {}_T P_{ij}(n)$$

for $k \ge 0$.

Since the Markov chain is irreducible, it may not return indefinitely to any state j without visiting T: either the states are transient, in which case the total number of visits to j is finite a.s., or they are persistent and the Markov chain reaches T in a finite time a.s. Thus, these processes under taboo of T are terminating, and the series

$${}_T R_{ij} = \sum_{n \ge 0} {}_T P_{ij}(n), \tag{5.5}$$

which has the same interpretation as $R(\infty)$ in a terminating delayed renewal process, converges for all i and j; it is equal to the expected number of visits to j, starting from i, before the first visit to T.

We shall, for brevity, state that the matrix ${}_T R_{TD} = \{ {}_T R_{ij} : i \in T, j \in D\}$ *records* the expected number of visits to D between two visits to T.

Theorem 5.2.1 *Assume that the Markov chain is irreducible. Its limiting probability vector satisfies*

$$\boldsymbol{\pi}_D = \boldsymbol{\pi}_T \, {}_TR_{TD} \tag{5.6}$$

for all disjoint subsets of states T and D, where the matrix ${}_TR_{TD}$ records the expected numbers of visits to the states of D between two visits to T.

Remark 5.2.2 Before we prove this property, we observe that it constitutes a direct generalization of Theorem 4.2.1. Indeed, consider the birth-and-death chain with transition matrix (4.5), and choose $T = \{0, 1, \ldots, n-1\}$ and $D = \{n\}$. By (5.6), we have that

$$\pi_n = \begin{bmatrix} \pi_0 & \pi_1 & \cdots & \pi_{n-1} \end{bmatrix} \begin{bmatrix} R_{0n} \\ R_{1n} \\ \vdots \\ R_{n-1n} \end{bmatrix}, \tag{5.7}$$

where R_{in} is the expected number of visits to n, starting from i, before the first return to T. For $i \leq n-2$, it is not possible to visit n before visiting some state in T, so that $R_{in} = 0$ for $0 \leq i \leq n-2$ and (5.7) reduces to $\pi_n = \pi_{n-1} R_{n-1n}$, which is identical to (4.7) since R_n is defined as the expected number of visits to n between two visits to $n-1$.

Proof of Theorem 5.2.1 As we prove below, the fact that the Markov renewal process of visits to D is terminating implies that

$$\lim_{k\to\infty} P_{TD}(k) = \lim_{k\to\infty} P_{TT}(k) \lim_{k\to\infty} {}_TR_{TD}(k). \tag{5.8}$$

Since the Markov chain is irreducible, limiting probabilities are independent of the initial state, so that $\lim_{k\to\infty} P_{TD}(k) = \mathbf{1} \cdot \boldsymbol{\pi}_D$ and $\lim_{k\to\infty} P_{TT}(k) = \mathbf{1} \cdot \boldsymbol{\pi}_T$. As $\lim_{k\to\infty} {}_TR_{TD}(k) = {}_TR_{TD}$, we eventually obtain (5.6).

Thus, we need to prove that (5.4) implies (5.8). For i and j fixed, $i \in T$ and $j \in D$, we write (5.4) in detailed form as follows:

$$P_{ij}(k) = \sum_{s\in T} P_{is} \star {}_TR_{sj}(k). \tag{5.9}$$

For each s, the product $P_{is} \star {}_TR_{sj}(\cdot)$ is a scalar expression for which Theorem 4.1.2 applies, and

$$\begin{aligned}\lim_{k\to\infty} P_{is} \star {}_TR_{sj}(k) &= \lim_{k\to\infty} P_{is}(k) \lim_{k\to\infty} {}_TR_{sj}(k) \\ &= \pi_s \, {}_TR_{sj}.\end{aligned}$$

If T has finitely many states, the theorem is proved; otherwise, we need to proceed in several steps.

First, we apply Fatou's lemma (Feller [21, Page 108]) to (5.9) and find that

$$\pi_j \geq \sum_{s\in T} \pi_s \, {}_TR_{sj} \qquad \text{for all } j \notin T.$$

In order to prove equality, we proceed by contradiction. If there is one index for which strict inequality holds, then we have that

$$\sum_{j\notin T} \pi_j > \sum_{j\notin T} \sum_{s\in T} \pi_s \, {}_TR_{sj}$$

or, writing $T^c = S \setminus T$,

$$\boldsymbol{\pi}_{T^c}\mathbf{1} > \boldsymbol{\pi}_T \, {}_TR_{TT^c}\mathbf{1}. \tag{5.10}$$

Now, we have that

$${}_TR_{TT^c} = P_{TT^c}N_{T^c}, \tag{5.11}$$

which we show as follows: take $i \in T$, $j \notin T$, define Y_j as the total number of visits to j before the first return to T, and condition on the state visited at time 1. It is clear that

$$\begin{aligned}{}_TR_{ij} &= \mathrm{E}[Y_j|X_0 = i] \\ &= \sum_{k\notin T} \mathrm{P}[X_1 = k|X_0 = i]\, \mathrm{E}[Y_j|X_0 = k] \\ &= \sum_{k\notin T} P_{ik}\, (N_{T^c})_{kj}\end{aligned}$$

by Lemma 5.1.2. Combining (5.10) and (5.11), we find that

$$\boldsymbol{\pi}_{T^c}\mathbf{1} > \boldsymbol{\pi}_T P_{TT^c} N_{T^c}\mathbf{1}. \tag{5.12}$$

If we partition the transition matrix as

$$P = \begin{bmatrix} P_T & P_{TT^c} \\ P_{T^cT} & P_{T^c} \end{bmatrix}, \tag{5.13}$$

the equation $\boldsymbol{\pi} P = \boldsymbol{\pi}$ is decomposed as

$$\begin{aligned} \boldsymbol{\pi}_T P_T + \boldsymbol{\pi}_{T^c} P_{T^cT} &= \boldsymbol{\pi}_T, \\ \boldsymbol{\pi}_T P_{TT^c} + \boldsymbol{\pi}_{T^c} P_{T^c} &= \boldsymbol{\pi}_{T^c}, \end{aligned}$$

so that $\boldsymbol{\pi}_T P_{TT^c} = \boldsymbol{\pi}_{T^c}(I - P_{T^c})$. Replacing in (5.12), we obtain that

$$\begin{aligned} \boldsymbol{\pi}_{T^c}\mathbf{1} &> \boldsymbol{\pi}_{T^c}(I - P_{T^c})N_{T^c}\mathbf{1} \\ &= \boldsymbol{\pi}_{T^c}\mathbf{1} \end{aligned}$$

by Lemma 5.1.2, which is a contradiction. □

In order to make effective use of the fundamental equation (5.6), we need to be able to compute ${}_TR_{TD}$ for various sets T and D. For a fixed state j in D, let ${}_TH_{ij}$ denote the taboo probability of returning to j, starting from i, under taboo of T:

$${}_TH_{ij} = \mathrm{P}[\delta_j < \infty \;\&\; \delta_j < \tau | X_0 = i] \tag{5.14}$$

for all i and j, where

$$\delta_j = \inf\{n \geq 1 : X_n = j\} \tag{5.15}$$

is the first return time to j and τ is the first passage time to T.

A direct application of (4.2) yields the equation

$${}_TR_{jj} = \sum_{n\geq 0} ({}_TH_{jj})^n = (1 - {}_TH_{jj})^{-1}, \tag{5.16}$$

and we have by Theorem 4.1.2 that

$${}_TR_{ij} = {}_TH_{ij}(1 - {}_TH_{jj})^{-1} \tag{5.17}$$

for an initial state $i \neq j$.

To proceed along these lines would require that we consider individually each of the renewal processes which constitutes ${}_TR_{TD}$. Instead of doing so, we examine all of them together as a Markov renewal process. This is a characteristic feature of matrix analytic methods, and the results to follow are essential in illustrating our approach.

Let ${}_Tf_{ij}(n)$ be the probability that, starting from i, the Markov chain reaches D for the first time at the nth step, under taboo of T, and that the Markov chain is in j at time n:

$$ {}_Tf_{ij}(n) = \mathrm{P}[\theta = n \ \& \ \theta < \tau \ \& \ X_\theta = j | X_0 = i] $$

for $n \geq 1$, where

$$ \theta = \inf\{n \geq 1 : X_n \in D\} $$

is the first passage time to D after time 0 and τ is the first passage time to T. We set ${}_Tf_{ij}(0) = 0$ for all i and j.

Let us further define

$$ {}_TF_{ij}(k) = \sum_{0 \leq \nu \leq k} {}_Tf_{ij}(\nu) $$

and ${}_TF_{ij} = {}_TF_{ij}(\infty)$. By (5.15), we may write that $\theta = \inf\{\delta_j : j \in D\}$ and that

$$ {}_TF_{ij} = \mathrm{P}[\theta = \delta_j \ \& \ \theta < \infty \ \& \ \theta < \tau | X_0 = i]; \tag{5.18} $$

notice the difference with (5.14).

For $j \in D$, we condition on the first visit in D and easily obtain that

$$ {}_TP_{ij}(n) = \sum_{0 \leq \nu \leq n} \sum_{s \in D} {}_Tf_{is}(n - \nu) \, {}_TP_{sj}(\nu) $$

for all i and for $n \geq 1$. Naturally, for $n = 0$, ${}_TP_{ij}(0)$ is equal to 0 for $i \neq j$ and equal to 1 for $i = j$. Summing over n, we get

$$ {}_TR_{ij}(k) = \sum_{0 \leq \nu \leq k} \sum_{s \in D} {}_Tf_{is}(k - \nu) \, {}_TR_{sj}(\nu) $$

for $i \in T$ or, more compactly,

$$ \begin{aligned} {}_TR_{TD}(k) &= \sum_{0 \leq \nu \leq k} {}_Tf_{TD}(k - \nu) \, {}_TR_{DD}(\nu) \\ &= {}_TF_{TD} \star {}_TR_{DD}(k). \end{aligned} \tag{5.19} $$

For $i \in D$, since ${}_T P_{DD}(0) = I$, we also obtain the formula

$$ {}_T R_{DD}(k) = I + {}_T F_{DD} \star {}_T R_{DD}(k). \tag{5.20} $$

These equations establish a relation between the first passage distributions and the Markov renewal kernels of the process of successive visits to D.

Consider the limits of these various matrices as k tends to infinity. The matrix ${}_T R_{TD} = {}_T R_{TD}(\infty)$ was introduced in (5.5) and records the expected number of visits to the states of D between two visits to T. The matrix ${}_T R_{DD} = {}_T R_{DD}(\infty)$ records the expected number of returns to D before the first passage to T. Since the Markov chain is irreducible, the two limits are finite.

The matrix ${}_T F_{DD} = \{ {}_T F_{ij} : i, j \in D\}$ records the return probabilities to D under taboo of T; similarly, the matrix ${}_T F_{TD}$ records the probabilities, starting from T, of hitting the set D in a finite time and before the first return to T.

In the theorem below, the equations (5.21) and (5.22) are generalizations of (4.4) and (4.2), respectively, to the Markov renewal processes of interest here.

Theorem 5.2.3 *The following relations hold for any disjoint subsets T and D:*

$$ {}_T R_{TD} = {}_T F_{TD} \, {}_T R_{DD} \tag{5.21} $$

and

$$ {}_T R_{DD} = \sum_{\nu \geq 0} ({}_T F_{DD})^{\nu} = (I - {}_T F_{DD})^{-1}. \tag{5.22} $$

Proof. To prove (5.21), we apply the same reasoning to (5.19) as in the proof of Theorem 5.2.1; the details are left to the reader.

By taking the limit in (5.20), we similarly obtain that

$$ {}_T R_{DD} = I + {}_T F_{DD} \, {}_T R_{DD}, $$

so that ${}_T R_{DD}$ is a solution of the linear system $(I - {}_T F_{DD})X = I$. By Çinlar [12, Proposition 10.2.20] , it is the minimal solution, so that (5.22) is proved as well. □

One special case of interest is when $D = T^c$. Then, if we partition the transition matrix as in (5.13), we see that ${}_TF_{TT^c} = P_{TT^c}$ and that ${}_TF_{T^cT^c} = P_{T^c}$. Therefore, we get (5.11) as a consequence of Lemma 5.1.2, of Theorem 5.2.3, and of the relation below:

$$ {}_TR_{T^cT^c} = N_{T^c} = \sum_{\nu \geq 0} P_{T^c}^{\nu}. \tag{5.23}$$

Furthermore, for every subset D disjoint of T, we have that

$$ {}_TR_{ij} = ({}_TR_{DD})_{ij} = ({}_TR_{T^cT^c})_{ij} \qquad \text{for } i, j \in D,$$

so that we may also characterize the matrix ${}_TR_{DD}$ as

$$ {}_TR_{DD} = \{(N_{T^c})_{ij} : i, j \in D\}, \tag{5.24}$$

i.e., it is the submatrix of N_{T^c} with rows and columns indexed by D.

Another result which will be useful is the following. Consider a subset T of states, and assume that the initial state is $i \notin T$. What is the probability distribution of the first state visited in T?

Defining as before τ as the first passage time to T after time 0, we write the first passage probabilities as

$$F_{ij} = \mathrm{P}[\tau < \infty \;\&\; X_\tau = j | X_0 = i],$$

and we are interested in determining these quantities for the various states i in a given set D.

Lemma 5.2.4 *For any disjoint subsets T and D, the matrix F_{DT} which records the first passage probabilities from the states of D to T is such that*

$$F_{DT} = (I - {}_TF_{DD})^{-1} \, {}_DF_{DT} = {}_TR_{DD} \, {}_DF_{DT}. \tag{5.25}$$

In particular, if $D = T^c$, then we have that

$$F_{T^cT} = N_{T^c} P_{T^cT}. \tag{5.26}$$

Proof. We only prove the first equation; the second is an immediate consequence.

Starting from a state in D, two cases may occur exclusively. Either the Markov chain reaches T before returning to D, or it returns to a state of D before it reaches T, in which case it still has to reach T from D. This shows that

$$F_{DT} = {}_DF_{DT} + {}_TF_{DD}F_{DT}$$

or

$$(I - {}_TF_{DD})F_{DT} = {}_DF_{DT}.$$

We premultiply by ${}_TR_{DD}$, apply Theorem 5.2.3, and are done. □

5.3 Local Time

Theorem 5.2.1 tells us how to compute the whole stationary distribution if we know the stationary distribution $\boldsymbol{\pi}_T$ over a subset T of states. Here, we show how one may compute $\boldsymbol{\pi}_T$ by itself.

We define the *epochs of visits* to the subset T as the sequence $\{\tau_0, \tau_1, \tau_2, \ldots\}$, where τ_n is the nth time that the Markov chain visits a state in T. More formally,

$$\begin{aligned} \tau_0 &= \inf\{i \geq 0 : X_i \in T\}, \\ \tau_{n+1} &= \inf\{i \geq \tau_n + 1 : X_i \in T\} \qquad \text{for } n \geq 0. \end{aligned}$$

We also define the *restricted* process $\{X_n^T, n \geq 0\}$ as follows:[9] $X_n^T = X_{\tau_n}$ for $n \geq 0$, so that X_n^T is the actual state visited when the Markov chain is for the nth time in the subset T.

Theorem 5.3.1 *Let $\{X_n, n \geq 0\}$ be an irreducible, homogeneous, positive recurrent Markov chain on a denumerable state space S with transition matrix P. Let S be partitioned into two arbitrary subsets T and T^c.*

The restricted process $\{X_n^T, n \geq 0\}$ is an irreducible, homogeneous, positive recurrent Markov chain on the state space T. Its transition matrix $P(T)$ is given by

$$P(T) = P_T + P_{TT^c}(I - P_{T^c})^{-1}P_{T^cT}.$$

Its stationary probability vector is proportional to $\boldsymbol{\pi}_T$; i.e., we have that $\boldsymbol{\pi}_T P(T) = \boldsymbol{\pi}_T$.

[9] We follow Freedman [22] in calling $\{X_n^T, n \geq 0\}$ the restricted process. Other authors call it the *censored* process or the *embedded* process.

Proof. The detailed proof may be found in Kemeny, Snell, and Knapp [40, Chapter 6, Proposition 6.4 and Lemmas 6.6 and 6.7]. We only briefly indicate here why this holds.

The instants $\{\tau_n, n \geq 0\}$ are stopping times, so that the strong Markov property implies that the restricted process is a Markov chain.

For $i, j \in T$, if $X_0 = i$, then $X_{\tau_1} = j$ either if $X_1 = j$ or if $X_1 = k$ for some k in T^c *and* if the first state visited upon return to T is j, so that

$$\begin{aligned} P(T) &= P_T + P_{TT^c}F_{T^cT} \\ &= P_T + P_{TT^c}N_{T^c}P_{T^cT} \qquad \text{by (5.26)} \\ &= P_T + P_{TT^c}(I - P_{T^c})^{-1}P_{T^cT} \end{aligned}$$

by (5.23). If we premultiply both sides by $\boldsymbol{\pi}_T$, we find that

$$\begin{aligned} \boldsymbol{\pi}_T P(T) &= \boldsymbol{\pi}_T P_T + \boldsymbol{\pi}_T P_{TT^c}(I - P_{T^c})^{-1}P_{T^cT} \\ &= \boldsymbol{\pi}_T P_T + \boldsymbol{\pi}_T\, {}_TR_{TT^c}P_{T^cT} \qquad \text{by (5.11)} \\ &= \boldsymbol{\pi}_T P_T + \boldsymbol{\pi}_{T^c}P_{T^cT} \end{aligned}$$

by Theorem 5.2.1. The last expression is clearly equal to $\boldsymbol{\pi}_T$ since $\boldsymbol{\pi}P = \boldsymbol{\pi}$ and P is partitioned as (5.13). □

This theorem may be generalized as follows. Take two disjoint subsets T and D. The transition matrix of the restricted process on T satisfies the relation

$$\begin{aligned} P(T) &= {}_DF_{TT} + {}_TF_{TD}\,{}_TR_{DD}\,{}_DF_{DT} \\ &= {}_DF_{TT} + {}_TF_{TD}(I - {}_TF_{DD})^{-1}\,{}_DF_{DT}. \end{aligned} \tag{5.27}$$

The proof proceeds along the same lines; we omit the details.

Remark 5.3.2 We conclude this section by indicating another probabilistic interpretation of the matrix ${}_TR_{TD} = {}_TF_{TD}\,{}_TR_{DD}$. It generalizes the one given in Proposition 4.6.2 for the scalar R_n. In expanded form, we have that

$${}_TR_{ij} = \sum_{k \in D} {}_TF_{ik}\,{}_TR_{kj} \qquad \text{for all } i \in T,\ j \in D. \tag{5.28}$$

Let us define two clocks. The first one measures the time of the Markov chain and is called the global clock. The second clock measures the time spent in the set T and increases by only one unit between τ_n and τ_{n+1}, starting from 0 at time τ_0; it is called the local clock. It is like the stopwatch at a basketball game: when it advances, it does so at the same speed as the global clock, but it stops whenever the ball is out of bounds.

Now, consider the restricted process $\{X_n^T : n \geq 0\}$, and assume that $X_n^T = i$ at some local instant n. At the next local instant, we have that $X_{n+1}^T = i'$ with probability $[P(T)]_{ii'}$. In the global time, the Markov chain may have avoided the set D altogether with the probability ${}_DF_{ii'}$ or it may have moved to some state k in D with the probability ${}_TF_{ik}$ and then spent in j an expected amount of global time equal to ${}_TR_{kj}$ before returning to T.

Summing over k, we get the right-hand side of (5.28) and we see that for all i in T and all j outside of T, ${}_TR_{ij}$ is the expected amount of global time spent in j per unit of the local time[10] of T (which happens to be) spent in i.

5.4 Gaussian Elimination

Theorems 5.2.1 and 5.3.1 may be used to design an algorithm for the calculation of the stationary probability vector $\boldsymbol{\pi}$ of a finite, irreducible Markov chain. We assume that the state space is $\{1, \dots, n\}$.

We first define the partition $D = \{1\}$, $T = \{2, \dots, n\}$, and decompose the matrix P into four blocks as follows:

$$P = \begin{bmatrix} p_{11} & \boldsymbol{p}_{1\bullet} \\ \boldsymbol{p}_{\bullet 1} & P_{\bullet\bullet} \end{bmatrix},$$

where $P_{\bullet\bullet}$ is the submatrix of transition probabilities between the states in $\{2, \dots, m\}$. We similarly write $\boldsymbol{\pi}$ as $(\pi_1, \boldsymbol{\pi}_\bullet)$. From (5.6), (5.11), we

[10]The 1 "local time" of a Markov chain is usually defined as a function of the global time: the local time $\gamma(t)$ is the time spent in T up to time t. Thus, technically, we have that $\gamma(t) = n$ for $n = \tau_n, \tau_n + 1, \dots, \tau_{n+1} - 1$. Here, we use the term in a looser sense, as the time marked on the local clock.

obtain that

$$\pi_1 = \boldsymbol{\pi}_\bullet \boldsymbol{p}_{\bullet 1}(1 - p_{11})^{-1}, \tag{5.29}$$

which expresses π_1 in function of $(\pi_2, \ldots, \pi_n)$. From Theorem 5.3.1, $\boldsymbol{\pi}_\bullet$ is determined by

$$\boldsymbol{\pi}_\bullet = \boldsymbol{\pi}_\bullet P_2^*,$$

where

$$P_2^* = P_{\bullet\bullet} + \boldsymbol{p}_{\bullet 1}(1 - p_{11})^{-1}\boldsymbol{p}_{1\bullet} \tag{5.30}$$

is the transition matrix of the censored Markov chain on the states $\{2, \ldots, n\}$ and is an irreducible stochastic matrix.

We may now partition $\{2, \ldots, n\}$ into the subsets $\{2\}$ and $\{3, \ldots, n\}$ and obtain an expression for π_2 in function of $(\pi_3, \ldots, \pi_n)$, etc.

This is exactly how one proceeds when solving the linear system $\boldsymbol{\pi}(P - I) = \mathbf{0}$, $\boldsymbol{\pi}\mathbf{1} = 1$, by Gaussian elimination: the matrix $P - I$ is decomposed into the product LU, where U is unit upper triangular and L is of the form

$$L = \begin{bmatrix} L_* & \mathbf{0} \\ \boldsymbol{\ell} & 0 \end{bmatrix}$$

with L_* lower triangular and nonsingular. Since U is nonsingular, the system above is equivalent to $\boldsymbol{\pi}L = \mathbf{0}$, $\boldsymbol{\pi}\mathbf{1} = 1$.

The decomposition is performed one row and one column at a time. Starting from

$$P - I = \begin{bmatrix} p_{11} - 1 & \boldsymbol{p}_{1\bullet} \\ \boldsymbol{p}_{\bullet 1} & P_{\bullet\bullet} - I \end{bmatrix},$$

where $p_{11} - 1 \neq 0$ since the Markov chain is irreducible, we use this diagonal element as a pivot and zero out the remainder of the first line, thereby obtaining the matrix

$$\begin{bmatrix} p_{11} - 1 & \mathbf{0} \\ \boldsymbol{p}_{\bullet 1} & P_2^* - I \end{bmatrix},$$

where the first column is equal to the first column of the matrix L and P_2^* is given by (5.30).

The first equation in $\boldsymbol{\pi}L = \mathbf{0}$ will therefore be $\pi_1(p_{11}-1)+\boldsymbol{\pi}_\bullet \boldsymbol{p}_{\bullet 1} = 0$ and is identical to (5.29).

As we mentioned earlier, the stochastic matrix P_2^* is irreducible. Therefore, its first diagonal element is different from 1 and it is an acceptable pivot to zero out the first line of $P_2^* - I$. The same argument may be repeated at each step of the decomposition: at the ith step, one has to decompose the submatrix $P_i^* - I$, where P_i^* is the irreducible transition matrix of the restricted Markov chain on the subset $\{i, \ldots, n\}$.

In order to solve the full system $\boldsymbol{\pi} L = \mathbf{0}$, $\boldsymbol{\pi}\mathbf{1} = 1$, we may set the last component of the solution vector equal to one, solve the resultant lower triangular system, and normalize, thereby obtaining that

$$\boldsymbol{\pi} = (-\boldsymbol{\ell} L_*^{-1}\mathbf{1} + 1)^{-1}[-\boldsymbol{\ell} L_*^{-1}, 1].$$

Grassmann, Taksar, and Heyman [30] suggest that one should never use the diagonal element itself as a pivot, but rather use the sum of the off-diagonal elements. For instance, at the first step, instead of computing P_2^* as $P_2^* = P_{\bullet\bullet} + \boldsymbol{p}_{\bullet 1}(1 - p_{11})^{-1}\boldsymbol{p}_{1\bullet}$, one should compute it as $P_2^* = P_{\bullet\bullet} + s^{-1}\boldsymbol{p}_{\bullet 1}\boldsymbol{p}_{1\bullet}$, where $s = \sum_{2 \le j \le n} p_{ij} = \boldsymbol{p}_{1\bullet}\mathbf{1}$. The two expressions are algebraically equivalent, but the second has the advantage of reducing the influence of round-off errors, as demonstrated in Stewart [111] and O'Cinneide [87].

The detailed algorithm is given in Figure 5.1. Since we never use the diagonal elements in the forward decomposition phase, we dispense with the identity matrix altogether. At the start, the array A is assigned a copy of the transition matrix P; at the end, its lower part has been assigned the matrix L in the LU decomposition.

5.5 Continuous Time Processes

Entirely analogous results hold for regular continuous time Markov processes; they may be obtained via various arguments. The easiest way is to use uniformization when that is possible, i.e., when $\sup\{|Q_{ii}| : i \in S\} < \infty$, where Q is the generator. If the diagonal elements of Q are unbounded, one may use the Markov chain embedded at transition epochs. Finally, one may always directly consider the process in continuous time. This is the approach we use below to prove the main properties which will be needed later.

We denote by $P_{ij}(t)$ the transition function

$$P_{ij}(t) = \mathrm{P}[X(t) = j | X(0) = i]$$

```
GTH Algorithm
A := P;
for i := 1 to n − 1 do
    s := 0;
    for j := i + 1 to n do s := s + A_ij od;
    A_ii := s;
    for k := i + 1 to n do
        t := A_ki/s;
        for j := i + 1 to n do A_kj := A_kj + tA_ij od
    od
od;
π_n := 1;
for j := n − 1 downto 1 do
    s := 0;
    for i := j + 1 to n do s := s + π_i A_ij od;
    π_j := s/A_jj
od;
s := 0;
for i := 1 to n do s := s + π_i od;
for i := 1 to n do π_i := π_i/s od
```

Figure 5.1: Algorithm to solve $\pi(P - I) = 0$, $\pi\mathbf{1} = 1$. The array A initially holds the matrix P; at the end, the lower part of A is equal to the matrix L of the decomposition $P - I = LU$. In the decomposition phase, the pivots are computed from the off-diagonal elements in order to reduce the effect of round-off errors. The vector π is computed in the back-substitution phase by setting π_n equal to one; it is then normalized.

---

and, for $i \notin T$, by ${}_TP_{ij}(t)$ the taboo transition function

$${}_TP_{ij}(t) = \mathrm{P}[X(t) = j \;\&\; \tau \geq t | X(0) = i],$$

where $\tau = \inf\{t > 0 : X(t) \in T\}$ is the first passage time to T.

We observe here the major difference with the discrete time case. If the initial state is in T, then the process will remain in T for an interval of time, so that $\tau = 0$ a.s. For that reason, we may not condition on the last time when the Markov chain is *in* T, as in (5.1), but we must condition on the last time when the Markov process *leaves* the subset T, and we find that

$$P_{ij}(t) = \int_0^t \sum_{i' \in T} \sum_{j' \notin T} P_{ii'}(t-u) Q_{i',j'} \, {}_T P_{j',j}(u) \, du \qquad \text{for } i \in T, \, j \notin T,$$

or in matrix notations,

$$\begin{aligned} P_{TT^c}(t) &= \int_0^t P_{TT}(t-u) Q_{TT^c} \, {}_T P_{T^cT^c}(u) \, du \\ &= P_{TT} \star {}_T R_{TT^c}(t), \end{aligned}$$

where

$${}_T R_{TT^c}(t) = \int_0^t Q_{TT^c} \, {}_T P_{T^cT^c}(u) \, du.$$

Taking the limit as $t \to \infty$, we obtain the following property.

Theorem 5.5.1 *Assume that the Markov process is irreducible. Its limiting probability vector satisfies the relation*

$$\boldsymbol{\pi}_{T^c} = \boldsymbol{\pi}_T \, {}_T R_{TT^c} \tag{5.31}$$

for all subsets T, *where*

$${}_T R_{TT^c} = Q_{TT^c} N_{T^c} \tag{5.32}$$

and the matrix

$$N_{T^c} = \int_0^\infty {}_T P_{T^cT^c}(u) \, du$$

records the expected total sojourn time in the states of T^c *given an initial state in* T^c, *before the first visit to* T. □

The proof is very much similar to that of Theorem 5.2.1 and is therefore left to the reader. Observe that the definition of ${}_T R_{TT^c}$ here is slightly different from that of ${}_T R_{TD}$ in (5.9); we shall return to this point later.

The taboo transition functions ${}_TP_{T^cT^c}(t)$ may also be written as ${}_TP_{T^cT^c} = \exp(Q_{T^c}t)$, where Q_{T^c} is the southeast corner of the partitioned generator

$$Q = \begin{bmatrix} Q_T & Q_{TT^c} \\ Q_{T^cT} & Q_{T^c} \end{bmatrix};$$

if T^c is finite, we have that $N_{T^c} = -Q_{T^c}^{-1}$, as we already mentioned in Theorem 2.4.3. In the infinite dimensional case, it is the minimal solution of the system $-Q_{T^c}N = I$, as it results from Çinlar [12, Equation (10.5.17)]. We restate this as a lemma similar to Lemma 5.1.2.

Lemma 5.5.2 *For any subset of states D, the matrix of expected sojourn times in D, before the first passage to the complementary subset $S \setminus D$, is given by*

$$N_D = \int_0^\infty \exp(Q_D t)dt = -Q_D^{-1}.$$

It is the minimal nonnegative solution of the systems $-Q_D N_D = I$ and $N_D(-Q_D) = I$. □

We want to use a restricted form of (5.31) and (5.32), similar to (5.6) and (5.21), where we only consider a subset $D \subset T^c$. We may immediately write

$$\boldsymbol{\pi}_D = \boldsymbol{\pi}_T \, {}_TR_{TD}, \tag{5.33}$$

where

$$ {}_TR_{TD} = \int_0^\infty Q_{TT^c} \, {}_TP_{T^cD}(u)\, du. \tag{5.34}$$

In order to distinguish states in D from among the states of T^c, we write that

$$ {}_TR_{TD} = \int_0^\infty Q_{TD} \, {}_TP_{DD}(u)\, du + \int_0^\infty Q_{T\bullet} \, {}_TP_{\bullet D}(u)\, du,$$

where we represent by $\bullet$ the set of states which are neither in D nor in T (this makes the equations somewhat simpler to read).

If we define

$$ {}_TF_{ij}(t) = \mathrm{P}[\theta \le t \;\&\; \theta < \tau \;\&\; X_\theta = j | X_0 = i],$$

where θ and τ are, respectively, the first passage times to D and T, we easily find that

$$_T P_{\bullet D}(t) = \int_0^t {}_T F_{\bullet D}(du)\ {}_T P_{DD}(t-u),$$

so that we eventually may replace (5.34) by

$$_T R_{TD} = (Q_{TD} + Q_{T\bullet}\ {}_T F_{\bullet D})\ {}_T N_{DD}, \tag{5.35}$$

where

$$_T N_{DD} = \int_0^\infty {}_T P_{DD}(u)\, du.$$

That matrix records the total expected time spent in the states of D, starting from a state in D, before the first passage to T, and $_T F_{\bullet D} = {}_T F_{\bullet D}(\infty)$ records the first passage probabilities from the states in $T^c \backslash D$ to D under taboo of T. A simple argument shows that

$$_T F_{\bullet D}(t) = \int_0^t \exp(Q_{\bullet} u) Q_{\bullet D}\, du,$$

so that

$$_T F_{\bullet D} = -Q_{\bullet}^{-1} Q_{\bullet D} \tag{5.36}$$

by Lemma 5.5.2.

To define the restricted process $\{X^T(t) : t \geq 0\}$ of the subset T, we proceed as in Freedman [22] and define the local clock which increases by one unit per unit of global time during those intervals when $X(t)$ is in T and remains constant when $X(t)$ is not in T. Then, we define $X^T(t) = X(t')$ if t' is the global time corresponding to the local time t. One shows the following continuous time version of Theorem 5.3.1.

Theorem 5.5.3 *Let $\{X(t) : t \geq 0\}$ be an irreducible, homogeneous, positive recurrent Markov process on a discrete state space S with generator Q. Let T be an arbitrary subset of S.*

The restricted process $\{X^T(t) : t \geq 0\}$ is an irreducible, homogeneous, positive recurrent Markov process on T. Its generator $Q(T)$ is given by

$$Q(T) = Q_T + Q_{TT^c}(-Q_{T^c})^{-1} Q_{T^cT}.$$

Its stationary probability vector is proportional to π_T; i.e., we have that $\pi_T Q(T) = 0$.

Proof. A short proof goes as follows. Consider two states i and j in T and a short interval $(t, t+h)$ in the local time of T. Assume that $X^T(t) = i$, which means that $X(t') = i$ at the corresponding global instant t'. The event $[X^T(t+h) = j]$ can occur in two ways. Either $X(t'+h)$ is equal to j in the global time with probability $Q_{ij}h + o(h)$, or it is equal to some $k \notin T$ with probability $Q_{ik}h + o(h)$, the first state visited upon return to T being j with probability ${}_TF_{kj}$. Thus,

$$[Q(T)]_{ij}h = Q_{ij}h + \sum_{k \in T^c} Q_{ik}h \, {}_TF_{kj}$$

and the remainder of the proof follows, since ${}_TF_{T^cT} = -Q_{T^c}^{-1}Q_{T^cT}$. □

We conclude this chapter by giving a probabilistic interpretation of the matrix ${}_TR_{TD}$ which holds in continuous as well as in discrete time.

Proposition 5.5.4 *For any subset of states T for all $i \in T$ and $j \notin T$, ${}_TR_{ij}$ is the expected sojourn time in j per unit of the local time of T spent in i.*

Proof. For a discrete time Markov chain, we already gave this interpretation at the end of section 5.3. In continuous time, we use the same argument as in Proposition 4.6.2. Assume that the restricted process is in $i \in T$ at some arbitrary time, and consider the events which may occur during an interval of local time of length h, arbitrarily small. During that interval, the process may have moved to some state $k \notin T$ with the probability $Q_{ik}h$. If that did happen, then the Markov process has spent in $j \in T^c$ an expected amount of global time equal to $(N_{T^c})_{kj}$. Summing over all k in T^c, we find that the expected global time spent in j during a local interval of length h is $\sum_{k \in T^c} Q_{ik}h(N_{T^c})_{kj}$, which is equal to ${}_TR_{ij}h$ by (5.32). □

Chapter 6

Homogeneous QBDs

We analyze the stationary distribution of an ergodic QBD and prove by a direct application of the material in Chapter 5 that the stationary distribution is matrix-geometric; i.e., we show that the stationary distribution, like that of a birth-and-death process, is of the form $\boldsymbol{\pi}_n = \boldsymbol{\pi}_{n-1} R$, where $\boldsymbol{\pi}_{n-1}$ and $\boldsymbol{\pi}_n$ are vectors instead of scalars and R is a matrix of taboo expected values.

We indicate in section 6.3 how the stationary distribution of the boundary states is determined once R is known. At this point, the distribution is characterized, but we do not know yet how to compute it. This is not quite as easy as for simple birth-and-death processes, and we shall devote the whole of Chapter 8 to that issue. We determine in Chapter 7 the conditions under which the process is ergodic.

Taken together, the matrix R and the boundary distribution constitute a compact characterization of the whole stationary distribution. We will show later, in section 9.1, that it also provides some information about the dynamic behavior of the process.

6.1 Definitions

We consider a discrete time Markov chain $\{X_t, t \in \mathbb{N}\}$ on the two-dimensional state space $\{(n,i) : n \geq 0, 1 \leq i \leq m\}$, which we partition as $\cup_{n \geq 0} \ell(n)$, where $\ell(n) = \{(n,1), (n,2), \ldots, (n,m)\}$ for $n \geq 0$. The first coordinate n is called the *level*, and the second coordinate j is called the *phase* of the state (n, j). We shall also use the word *level* to denote the whole subset $\ell(n)$. The number m of states in each level may be either finite or infinite.

The Markov chain is called a QBD if one-step transitions from a state are restricted to states in the same level or in the two adjacent levels: it is possible to move in one step from (n, j) to (n', j') only if $n' = n$, $n+1$, or $n-1$ (provided in the last case that $n \geq 1$, of course).

The transition probabilities are assumed to be level-independent. More precisely, for n or n' greater than or equal to 1, the probability $\mathrm{P}[X_1 = (n', j) | X_0 = (n, i)]$ may depend on i, j, and $n' - n$, but not on the specific values of n and n'. Thus, the transition matrix is block-tridiagonal and has the following form:

$$P = \begin{bmatrix} B & A_0 & 0 & 0 & \cdots \\ A_2 & A_1 & A_0 & 0 & \cdots \\ 0 & A_2 & A_1 & A_0 & \cdots \\ 0 & 0 & A_2 & A_1 & \cdots \\ \vdots & \vdots & \vdots & \vdots & \ddots \end{bmatrix}, \tag{6.1}$$

where A_0, A_1, A_2, and B are square matrices of order m.[11]

We shall consider in Chapter 12 the more general case where the transition probabilities depend on the level and where the levels may have different numbers of phases.

We assume throughout this monograph that the process is irreducible, so that there exists a path between any two states.

6.2 The Matrix-Geometric Property

We temporarily assume that the QBD is aperiodic and positive recurrent and denote by $\boldsymbol{\pi}$ its stationary probability vector. It is the unique solution of the system $\boldsymbol{\pi} = \boldsymbol{\pi} P$, $\boldsymbol{\pi}\mathbf{1} = 1$.

[11]The notation A_0, A_1, A_2 became standard very soon after the beginning of work on QBDs by Neuts and his coworkers. It is not felicitous, and we have considered alternate notations. We have not found one so good that it would compensate for the resulting disadvantage to the reader, who would need a lexicon when comparing the material here to the existing literature on QBDs. We recommend the mnemonic device that the matrices are numbered in their order of occurrence in each column.

We partition the vector $\boldsymbol{\pi}$ by levels into subvectors $\boldsymbol{\pi}_n, n \geq 0$, where $\boldsymbol{\pi}_n$ has m components. The defining system may be decomposed as

$$\begin{aligned} \boldsymbol{\pi}_0(B - I) + \boldsymbol{\pi}_1 A_2 &= \mathbf{0}, \\ \boldsymbol{\pi}_{n-1} A_0 + \boldsymbol{\pi}_n(A_1 - I) + \boldsymbol{\pi}_{n+1} A_2 &= \mathbf{0} \qquad \text{for } n \geq 1, \\ \sum_{n \geq 0} \boldsymbol{\pi}_n \mathbf{1} &= 1 \end{aligned}$$

and is so strikingly similar to the system (4.10)–(4.12) that one might guess the following theorem.

Theorem 6.2.1 *If the QBD is positive recurrent, then there exists a nonnegative matrix N of order m such that*

$$\boldsymbol{\pi}_{n+1} = \boldsymbol{\pi}_n A_0 N \qquad \textit{for } n \geq 0.$$

The matrix N is such that N_{ij} $(1 \leq i, j \leq m)$ is equal to the expected number of visits to the state (n, j), starting from the state (n, i), before the first visit to any of the states in $\ell(n-1)$ and is independent of $n \geq 1$.

We may also write

$$\boldsymbol{\pi}_n = \boldsymbol{\pi}_0 R^n \qquad \textit{for } n \geq 0, \tag{6.2}$$

where $R = A_0 N$ is such that, for any $n \geq 0$, R_{ij} $(1 \leq i, j \leq m)$ is the expected number of visits to $(n+1, j)$ before a return to $\ell(0) \cup \cdots \cup \ell(n)$, given that the process starts in (n, i).

Proof. Fix $n \geq 0$ and partition the state space as $T \cup T^c$, where $T = \ell(0) \cup \cdots \cup \ell(n)$, $T^c = \ell(n+1) \cup \ell(n+2) \cup \cdots$. Using the same notations as in section 5.1, we have that

$$P_{T^c} = \begin{bmatrix} A_1 & A_0 & 0 & 0 & \cdots \\ A_2 & A_1 & A_0 & 0 & \cdots \\ 0 & A_2 & A_1 & A_0 & \cdots \\ 0 & 0 & A_2 & A_1 & \cdots \\ \vdots & \vdots & \vdots & \vdots & \ddots \end{bmatrix}$$

and

$$P_{TT^c} = \begin{bmatrix} 0 & 0 & 0 & \cdots \\ \vdots & \vdots & \vdots & \\ 0 & 0 & 0 & \cdots \\ A_0 & 0 & 0 & \cdots \end{bmatrix}.$$

By Theorems 5.2.1 and 5.2.3, we have that $\boldsymbol{\pi}_{T^c} = \boldsymbol{\pi}_T P_{TT^c}(I - P_{T^c})^{-1}$, or in expanded form,

$$(\boldsymbol{\pi}_{n+1}, \boldsymbol{\pi}_{n+2}, \cdots) = (\boldsymbol{\pi}_0, \cdots, \boldsymbol{\pi}_n) P_{TT^c}(I - P_{T^c})^{-1},$$

where $(I - P_{T^c})^{-1} = \sum_{i \geq 0} P_{T^c}^i$ converges since we have assumed that the QBD is irreducible.

We decompose the matrix $N_{T^c} = (I - P_{T^c})^{-1}$ into blocks $\{N_{kk'}; k, k' \geq 1\}$, where $N_{kk'}$ is the matrix of expected number of visits to the states in $\ell(n + k')$, starting from a state in $\ell(n + k)$, before the first visit to any of the states in $T = \ell(0) \cup \cdots \cup \ell(n)$.

Because of the extremely sparse structure of the matrix P_{TT^c}, we have that

$$P_{TT^c}(I - P_{T^c})^{-1} = \begin{bmatrix} 0 & 0 & 0 & \cdots \\ \vdots & \vdots & \vdots & \\ 0 & 0 & 0 & \cdots \\ A_0 N_{11} & A_0 N_{12} & A_0 N_{13} & \cdots \end{bmatrix}$$

and that

$$(\boldsymbol{\pi}_{n+1}, \boldsymbol{\pi}_{n+2}, \cdots) = (\boldsymbol{\pi}_n A_0 N_{11}, \boldsymbol{\pi}_n A_0 N_{12}, \cdots). \tag{6.3}$$

In particular, we have that $\boldsymbol{\pi}_{n+1} = \boldsymbol{\pi}_n A_0 N_{11} = \boldsymbol{\pi}_n A_0 N$ if we define $N = N_{11}$.

We now make the first of many uses of the homogeneity of the matrix P. Irrespective of the value chosen for n in defining the partition $T \cup T^c$, the matrix P_{T^c} is the same. Therefore, the matrix $N_{11} = N$ is independent of n and it records the expected number of visits to $\ell(n+1)$, starting from $\ell(n+1)$, before the first visit to $\ell(0) \cup \cdots \cup \ell(n)$ for all values of $n \geq 0$, as we claim in the statement of the theorem.

Furthermore, the *structure* of P_{TT^c} is independent of n; thus the argument above applies to every level, and we have that $\boldsymbol{\pi}_{n+1} = \boldsymbol{\pi}_n A_0 N$

for all $n \geq 0$, which may obviously be written as $\boldsymbol{\pi}_n = \boldsymbol{\pi}_0 R^n$ for all $n \geq 0$ if we set $R = A_0 N$. Clearly, $A_0 N$ records the expected number of visits to $\ell(n+1)$ starting from $\ell(n)$, avoiding $\cup_{0 \leq k \leq n} \ell(k)$, which concludes the proof. □

Remark 6.2.2 The matrix-geometric distribution first appeared in Evans [17]. The author considers QBDs for which A_0 is a scalar matrix and presents some of the key ideas which have been fully explored later by Neuts and others.

Remark 6.2.3 The matrix $R = A_0 N_{11}$ is also interpreted as recording the expected rate of visit to the states of $\ell(n+1)$ per unit of time spent in the states of $\ell(n)$ measured in the local time of the whole subset $T = \ell(0) \cup \cdots \cup \ell(n)$ (see Remark 5.3.2). Because of the structure of the QBD, the Markov chain is forced to pass through the level n when it leaves the complementary set T^c. Therefore, we may restrict ourselves to the local time of the level n only and interpret R_{ij} as the expected rate of visit to $(n+1, j)$ per unit of the local time of $\ell(n)$ spent in (n, i).

Theorem 6.2.1 has the following immediate consequence.

Corollary 6.2.4 *If the QBD is positive recurrent, and if m is finite, then the spectral radius* $\mathrm{sp}(R)$ *of the matrix R is strictly less than* 1.

Proof. If the QBD is positive recurrent, then the series $\sum_{n \geq 0} \boldsymbol{\pi}_n$ must converge. Since $\boldsymbol{\pi}_n = \boldsymbol{\pi}_0 R^n$ for $n \geq 0$, the series $\sum_{n \geq 0} R^n$ must converge, and this implies that $\mathrm{sp}(R) < 1$ (see Remark 5.1.2). □

Remark 6.2.5 When m is infinite, it is proved in Tweedie [115] that there exists $\theta > 1$ such that $\sum_{n \geq 0} \theta^n (R^n)_{ij}$ diverges for some i and j and $\sum_{n \geq 0} r^n R^n$ converges whenever $r < \theta$. The quantity $\theta^{-1} < 1$ is the natural analogue of the spectral radius for finite matrices.

The *matrix-geometric* property (6.2) leads to very simple expressions for various marginal distributions and moments. For instance, the vector $\boldsymbol{\varphi} = \sum_{n \geq 0} \boldsymbol{\pi}_n = \boldsymbol{\pi}_0 (I - R)^{-1}$ is the marginal probability distribution of the phase. The marginal distribution $\{p_n, n \geq 0\}$ of the level is given by

$$p_n = \boldsymbol{\pi}_n \mathbf{1} = \boldsymbol{\pi}_0 R^n \mathbf{1} \qquad \text{for } n \geq 0 \tag{6.4}$$

and has a discrete phase-type representation as we show below (see also Sengupta [110]).

Theorem 6.2.6 *The marginal distribution $\{p_n, n \geq 0\}$ of the level has the PH representation $(\boldsymbol{\xi}, V)$ with m phases, $\boldsymbol{\xi} = \boldsymbol{\pi}_0 R(I-R)^{-1}$, and $V = \Delta^{-1} R^{\mathrm{t}} \Delta$, where $\Delta = \mathrm{diag}(\boldsymbol{\xi})$ and R^{t} is the transpose of R.*

Proof. The vector $\boldsymbol{\xi}$ is clearly strictly positive, and we have that $\boldsymbol{\xi}\mathbf{1} = 1 - \boldsymbol{\pi}_0\mathbf{1} < 1$. Thus, $\boldsymbol{\xi}$ satisfies the conditions for being the initial probability vector in a phase-type representation. Furthermore, V is nonnegative and

$$\begin{aligned} \mathbf{1} - V\mathbf{1} &= \Delta^{-1}(I - R^{\mathrm{t}})\Delta\mathbf{1} \\ &= \Delta^{-1}(I - R^{\mathrm{t}})\boldsymbol{\xi}^{\mathrm{t}} \\ &= \Delta^{-1}R^{\mathrm{t}}\boldsymbol{\pi}_0^{\mathrm{t}} \\ &> \mathbf{0}, \end{aligned}$$

so that the matrix V is substochastic. Thus, $(\boldsymbol{\xi}, V)$ is the representation of a phase-type distribution. Since

$$p_n = \boldsymbol{\pi}_0 R^n \mathbf{1} = \boldsymbol{e}(R^{\mathrm{t}})^n \boldsymbol{\pi}_0^{\mathrm{t}},$$

where $\boldsymbol{e} = (1, 1, \ldots, 1)$, we have that

$$\begin{aligned} p_n &= \boldsymbol{e}\Delta V^n \Delta^{-1}\boldsymbol{\pi}_0^{\mathrm{t}} \\ &= \boldsymbol{\xi} V^{n-1}(V\Delta^{-1}\boldsymbol{\pi}_0^{\mathrm{t}}) \\ &= \boldsymbol{\xi} V^{n-1}(\Delta^{-1}R^{\mathrm{t}}\boldsymbol{\pi}_0^{\mathrm{t}}) \\ &= \boldsymbol{\xi} V^{n-1}(\mathbf{1} - V\mathbf{1}), \end{aligned}$$

which proves that $\{p_n\}$ is the density of the PH$(\boldsymbol{\xi}, V)$ distribution. □

By (2.15), we may directly evaluate the factorial moments M_k of the distribution $\{p_n : n \geq 0\}$ as

$$M_k = k!\boldsymbol{\xi}(I - V)^{-k}V^{k-1}\mathbf{1} \qquad \text{for } k \geq 1.$$

The perceptive reader will have noticed that there is more information in (6.3) than we have used. We have, for instance, that $\boldsymbol{\pi}_{n+2} = \boldsymbol{\pi}_n A_0 N_{12}$ and, more generally, that $\boldsymbol{\pi}_{n+k} = \boldsymbol{\pi}_n A_0 N_{1k}$ for all $k \geq 1$. We also find from (6.2) that $\boldsymbol{\pi}_{n+k} = \boldsymbol{\pi}_n (A_0 N_{11})^k$. This suggests that a simple relation may exist among the matrices N_{1k}, $k \geq 1$. This is indeed the case, as we show in the next theorem.

Theorem 6.2.7 *The matrices* $\{N_{1k}, k \geq 1\}$ *satisfy the following recurrence equation:*

$$N_{1,k+1} = N_{11}A_0N_{1k} \qquad \textit{for } k \geq 1.$$

Hence,

$$N_{1,k} = N_{11}(A_0N_{11})^{k-1} = N_{11}R^{k-1} \qquad \textit{for } k \geq 1.$$

Proof. We assume that $n \geq 0$ is arbitrary but fixed and that the initial state is in $\ell(n+1)$. We define Z_j as the number of visits to the state $(n+k+1, j)$ before the first visit to any state in $\ell(n)$ (or, equivalently, in $\ell(0) \cup \cdots \cup \ell(n)$). As seen above,

$$(N_{1k+1})_{ij} = \mathrm{E}[Z_j|X_0 = (n+1, i)].$$

We denote by τ the epoch of the first visit to $\ell(n)$ and by $0 = \theta_0 < \theta_1 < \theta_2 < \cdots < \tau$ the successive epochs of visit to $\ell(n+1)$, before the first visit to $\ell(n)$. We shall simply compute $(N_{1k+1})_{ij}$ by counting the number of visits to $(n+k+1, j)$ during each interval $(\theta_\nu, \theta_{\nu+1})$ and the number of such intervals. In order to do this, we define the random variables K_s as the number of visits to $(n+1, s)$ before the first visit to $\ell(n)$; we have that $\mathrm{E}[K_s|X_0 = (n+1, i)] = (N_{11})_{is}$.

We finally define the random variables W_j as the number of visits to $(n+k+1, j)$ before the first visit to either $\ell(n+1)$ or $\ell(n)$. Since $\mathrm{E}[W_j|X_0 = (n+2, i)] = (N_{1k})_{ij}$, we obtain by a simple argument that $E[W_j|X_0 = (n+1, i)] = (A_0N_{1k})_{ij}$.

By the strong Markov property, we have that

$$\mathrm{E}[Z_j|X_0 = (n+1, i)] = \sum_{1 \leq s \leq m} \mathrm{E}[K_s|X_0 = (n+1, i)]\mathrm{E}[W_j|X_0 = (n+1, s)],$$

from which the theorem follows. □

Remark 6.2.8 We have given in Theorem 6.2.1 one interpretation of the matrix R. We may now give one to its successive powers: we repeat the argument at the end of the proof above and find that for any $n \geq 0$, $k \geq 1$, $1 \leq i, j \leq m$, $(R^k)_{ij}$ is the expected number of visits to $(n+k, j)$

before the first return to $\ell(0) \cup \cdots \cup \ell(n)$ given that the process starts in (n, i).

The quantity $(R^k)_{ij}$ may also be interpreted as the expected rate of visit to $(n + k, j)$ per unit of the local time of $\ell(n)$ spent in (n, i); in the next section, this interpretation will be seen, unlike the former, to hold in the continuous time case as well.

The characterization which we have obtained so far for the stationary distribution $\boldsymbol{\pi}$ is not very satisfactory. Indeed, it is expressed in terms of $N = N_{11}$, i.e., one block in the infinite matrix $N_{\mathcal{T}^c}$, which is the minimal nonnegative solution of the infinite linear system $N_{\mathcal{T}^c}(I - P_{\mathcal{T}^c}) = I$. We look therefore for other characterizations, better suited for the numerical evaluation of R to be developed in Chapter 8. This leads us to introduce two other matrices, which are named U and G.

In the discussion which follows, we emphasize the fact that the matrix R and the two new matrices U and G are not just mathematical artifacts, but are related in a fundamental way to the dynamics of QBDs.

Let us assume that X_0 is in $\ell(n)$. Define τ as the first epoch of visit to the level $\ell(n-1)$ and θ as the first epoch of return to the level $\ell(n)$. The matrices U and G are defined as follows,

$$U_{ij} = \mathrm{P}[\theta < \tau \text{ and } X_\theta = (n, j) | X_0 = (n, i)]; \tag{6.5}$$

the matrix U records the probability, starting from $\ell(n)$, of returning to $\ell(n)$ before visiting $\ell(n-1)$;

$$G_{ij} = \mathrm{P}[\tau < \infty \text{ and } X_\tau = (n-1, j) | X_0 = (n, i)]; \tag{6.6}$$

the matrix G records the probability, starting from $\ell(n)$, of visiting $\ell(n-1)$ in a finite time. In view of the homogeneity of the matrix P, the values of U and of G do not depend on $n \geq 1$.

The three matrices R, U, and G are related by a collection of equations. For instance, by conditioning on the number of returns to $\ell(n)$ before the first visit to $\ell(n-1)$, we find that

$$N = \sum_{i \geq 0} U^i = (I - U)^{-1} \tag{6.7}$$

and therefore that

$$R = A_0(I - U)^{-1};$$

similarly,

$$G = \sum_{i \geq 0} U^i A_2 = (I - U)^{-1} A_2.$$

If we condition on the state after the first transition, we find that

$$U = A_1 + A_0 G;$$

indeed, in order to return to $\ell(n)$, avoiding $\ell(n-1)$, the QBD may either remain in $\ell(n)$ at time 1, with probabilities recorded in A_1, or move up to $\ell(n+1)$, with probabilities recorded in A_0; from $\ell(n+1)$, the QBD returns to $\ell(n)$ (necessarily before reaching $\ell(n-1)$) with probabilities recorded in G.

This last equation may also be written as

$$\begin{aligned} U &= A_1 + A_0(I - U)^{-1} A_2 \qquad (6.8) \\ &= A_1 + R A_2. \end{aligned}$$

A first important observation is that if any one of the three matrices is known, then we may determine the other two by applying the relevant equation. This we summarize in the next theorem.

Theorem 6.2.9 *If any one of the matrices U, G, or R is known, then we may determine the other two by applying one of the following equations:*

$$\begin{aligned} R &= A_0(I - U)^{-1}, \qquad (6.9) \\ G &= (I - U)^{-1} A_2, \qquad (6.10) \\ U &= A_1 + A_0 G, \qquad (6.11) \\ U &= A_1 + R A_2. \qquad (6.12) \end{aligned}$$

□

A second observation is that (6.8) only involves the unknown matrix U and might serve as a basis from which to actually compute U and then R. That equation is not linear, but it is more parsimonious than our earlier characterization of N from a linear but infinite system. We may also obtain similar equations for R and G; these are given below.

Theorem 6.2.10 *The three matrices U, G, and R, respectively, satisfy the following equations:*

$$U = A_1 + A_0(I - U)^{-1}A_2, \quad (6.13)$$
$$G = A_2 + A_1G + A_0G^2, \quad (6.14)$$
$$R = A_0 + RA_1 + R^2A_2. \quad (6.15)$$

Proof. Starting from (6.9) and postmultiplying both sides by $(I-U)$, we find that $R = A_0 + RU$; using (6.12), we obtain (6.15). Equation (6.14) is similarly proved. □

Remark 6.2.11 These equations are not solely the result of formal manipulations, they also express interesting probabilistic properties.

The case of (6.14) is particularly obvious. The left-hand side records the distribution of the first state visited in $\ell(n-1)$, conditioned on the initial state being in $\ell(n)$; we see in the right-hand side a decomposition according to the first transition: the first term corresponds to the case where the QBD directly moves from $\ell(n)$ to $\ell(n-1)$ in one transition with probabilities recorded in A_2; as for the second term, with probabilities recorded in A_1, the QBD remains in $\ell(n)$, from where it still has to move eventually to $\ell(n-1)$, with probabilities recorded in G; finally, for the last term, note that with probabilities recorded in A_0 the QBD moves up to $\ell(n+1)$, from where it must eventually return to $\ell(n)$, with probabilities recorded in G and then to $\ell(n-1)$, again with probabilities recorded in G. This argument will be used again in Chapter 7.

The interpretation of (6.15) is also simple: recall (see Remark 6.2.8) that R records the expected number of visits to $\ell(n)$, starting from $\ell(n-1)$, before the first return to $\ell(n-1)$. In the right-hand side of (6.15), these visits to $\ell(n)$ are decomposed into three groups according to the level from which a visit to $\ell(n)$ occurs. There has to be a first visit immediately upon starting from $\ell(n-1)$ with probability recorded in A_0, which explains the first term. Each visit to $\ell(n)$—the expected number is recorded in R—may immediately be followed by another visit to $\ell(n)$ with probability recorded in A_1; this explains the second term. Finally, each visit to $\ell(n+1)$—with expected number recorded in R^2—may be followed by a visit to $\ell(n)$ with probability recorded in A_2, which explains the third term.

The interpretation of (6.13) is immediate if we write it as $U = A_1 + \sum_{i\geq 0} A_0 U^i A_2$. Indeed, the term A_1 records the probability of immediately returning to $\ell(n)$; the term $A_0 U^i A_2$ records the probability of returning to $\ell(n)$ after exactly $i+1$ visits to $\ell(n+1)$.

We shall return to these three matrices in Chapter 8, when we address the issue of actually computing the matrix R.

6.3 Boundary Distribution

In order to completely specify the stationary distribution, one needs to determine the subvector $\boldsymbol{\pi}_0$. This we do with the help of Theorem 5.3.1. Consider the restricted process on $T = \ell(0)$. Its transition matrix is readily seen to be given by

$$\begin{aligned} P(T) &= B + A_0(I-U)^{-1}A_2 \\ &= B + A_0 G \qquad \text{by (6.10)} \\ &= B + RA_2 \qquad \text{by (6.9)}. \end{aligned}$$

If the QBD is positive recurrent, then so is the restricted process on $\ell(0)$ and the vector $\boldsymbol{\pi}_0$ is such that $\boldsymbol{\pi}_0 = \boldsymbol{\pi}_0 P(T)$, the normalization factor being determined by the constraint that $\boldsymbol{\pi}\mathbf{1} = 1$. Since

$$\boldsymbol{\pi}\mathbf{1} = \sum_{i\geq 0} \boldsymbol{\pi}_i \mathbf{1} = \boldsymbol{\pi}_0 (I-R)^{-1}\mathbf{1},$$

we have thus proved the following lemma.

Lemma 6.3.1 *The stationary distribution $\boldsymbol{\pi}_0$ of the boundary states in $\ell(0)$ is the unique solution of the system*

$$\begin{aligned} \boldsymbol{\pi}_0(B + A_0 G) &= \boldsymbol{\pi}_0, \\ \boldsymbol{\pi}_0(I-R)^{-1}\mathbf{1} &= 1. \end{aligned}$$

The first equation may also be written as $\boldsymbol{\pi}_0(B + RA_2) = \boldsymbol{\pi}_0$. □

We have seen in section 1.2 an example where $B = A_1 + A_2$. In this case, we may simplify the characterization of the vector $\boldsymbol{\pi}_0$. We need to define the stochastic matrix $A = A_0 + A_1 + A_2$, which will play an important role in the next section.

Lemma 6.3.2 *Assume that the QBD is irreducible, that it is positive recurrent, and that $B = A_1 + A_2$. Then*

$$\boldsymbol{\pi}_n = \boldsymbol{\alpha}(I - R)R^n \qquad \textit{for all } n \geq 0,$$

where R is defined in Theorem 6.2.1 and $\boldsymbol{\alpha}$ is the unique solution of the system $\boldsymbol{\alpha} = \boldsymbol{\alpha}A$, $\boldsymbol{\alpha}\mathbf{1} = 1$, with $A = A_0 + A_1 + A_2$.

Proof. The matrix A is stochastic and defines a Markov chain on the state space $\{1, 2, \ldots, m\}$. It is irreducible under the assumptions that the QBD is irreducible and that $B = A_1 + A_2$: indeed, if it were reducible, then there would exist states which do not communicate in A; without loss of generality, we may assume that these are the states 1 and 2. Then, every subblock in the matrix P of (6.1) would similarly be reducible and there would be no path from any of the states $(n, 1)$ to any of the states $(n', 2)$, in contradiction to our general assumption that the QBD is irreducible.

Now, we have that

$$\begin{aligned}
&\boldsymbol{\pi}_0(I - R)^{-1}A \\
&\quad= \boldsymbol{\pi}_0 \sum_{\nu \geq 0} R^\nu (A_0 + A_1 + A_2) \\
&\quad= \boldsymbol{\pi}_0 \sum_{\nu \geq 0} R^\nu (A_0 + RA_1 + R^2 A_2) + \boldsymbol{\pi}_0(A_1 + A_2 + RA_2) \\
&\quad= \boldsymbol{\pi}_0 \sum_{\nu \geq 1} R^\nu + \boldsymbol{\pi}_0(A_1 + A_2 + RA_2) \qquad \text{by (6.15)} \\
&\quad= \boldsymbol{\pi}_0 \sum_{\nu \geq 1} R^\nu + \boldsymbol{\pi}_0
\end{aligned}$$

since the system which defines $\boldsymbol{\pi}_0$ may be written as

$$\begin{aligned}
\boldsymbol{\pi}_0(A_1 + A_2 + RA_2) &= \boldsymbol{\pi}_0, \\
\boldsymbol{\pi}_0(I - R)^{-1}\mathbf{1} &= 1.
\end{aligned}$$

Thus, if we define $\boldsymbol{\alpha} = \boldsymbol{\pi}_0(I - R)^{-1}$, then $\boldsymbol{\alpha}A = \boldsymbol{\alpha}$, $\boldsymbol{\alpha}\mathbf{1} = 1$. Since A is irreducible, $\boldsymbol{\alpha}$ is unique. □

Remark 6.3.3 In section 6.4 we shall need the following result, which is slightly more general than Theorem 6.2.1. Assume that the transition probabilities from $\ell(0)$ to $\ell(1)$ are different from the transition probabilities from $\ell(1)$ to $\ell(2)$, $\ell(2)$ to $\ell(3)$, and so on, so that the transition matrix P is given by

$$P = \begin{bmatrix} B & B_0 & 0 & 0 & \cdots \\ A_2 & A_1 & A_0 & 0 & \cdots \\ 0 & A_2 & A_1 & A_0 & \cdots \\ 0 & 0 & A_2 & A_1 & \cdots \\ \vdots & \vdots & \vdots & \vdots & \ddots \end{bmatrix},$$

where B_0 may be different from A_0. Then one obtains

$$\boldsymbol{\pi}_1 = \boldsymbol{\pi}_0 B_0 N, \qquad (6.16)$$

$$\boldsymbol{\pi}_{n+1} = \boldsymbol{\pi}_1 R^n \qquad (6.17)$$

instead of (6.2) and the vector $\boldsymbol{\pi}_0$ is such that

$$\boldsymbol{\pi}_0(B + B_0 G) = \boldsymbol{\pi}_0, \qquad (6.18)$$

$$\boldsymbol{\pi}_0\mathbf{1} + \boldsymbol{\pi}_0 B_0 N (I - R)^{-1}\mathbf{1} = 1.$$

We do not give the proof since it is very similar to the proof of Theorem 6.2.1 and Lemma 6.3.1; moreover, we shall consider even more general QBDs in Chapter 12.

6.4 Continuous Time QBDs

A continuous time QBD is a Markov process $\{X(t), t \in \mathbb{R}^+\}$ on the state space $\cup_{n \geq 0}\ell(n)$ with infinitesimal generator Q of the form

$$Q = \begin{bmatrix} B & A_0 & 0 & 0 & \cdots \\ A_2 & A_1 & A_0 & 0 & \cdots \\ 0 & A_2 & A_1 & A_0 & \cdots \\ 0 & 0 & A_2 & A_1 & \cdots \\ \vdots & \vdots & \vdots & \vdots & \ddots \end{bmatrix}, \qquad (6.19)$$

where the matrices A_0 and A_2 are nonnegative and the matrices A_1 and B have nonnegative off-diagonal elements and strictly negative diagonals. The row sums of Q are equal to zero, so that we have $(B + A_0)\mathbf{1} = \mathbf{0}$ and $(A_0 + A_1 + A_2)\mathbf{1} = \mathbf{0}$. We shall assume as before that the process is irreducible and regular.

One may analyze continuous time QBDs by repeating almost verbatim the argument of the three preceding sections. We give the fundamental structural property first.

Theorem 6.4.1 *Assume that the continuous time QBD is positive recurrent. Then, the stationary probability distribution is such that*

$$\boldsymbol{\pi}_n = \boldsymbol{\pi}_0 R^n \qquad \text{for } n \geq 0, \tag{6.20}$$

where the matrix R records the rate of sojourn in the states of $\ell(n+1)$ per unit of the local time of $\ell(n)$.

Furthermore, we have that $R = A_0 N$, where the matrix N records the expected sojourn time in the states of $\ell(n)$, starting from $\ell(n)$, before the first visit to $\ell(n-1)$.

Proof. This is a consequence of Theorem 5.5.1 and Proposition 5.5.4. We do not give the mathematical derivation since it is identical to that in the proof of Theorem 6.2.1. □

As in the discrete time case, we define two matrices U and G as follows:

$$G_{ij} = P[\tau < \infty \;\&\; X(\tau) = (n-1, j) | X(0) = (n, i)],$$

where τ is the first passage time from the level n to the level $n-1$. We have that

$$G = (-A_1)^{-1} A_2 + (-A_1)^{-1} A_0 G^2$$

or

$$0 = A_2 + A_1 G + A_0 G^2.$$

The first equation is justified as follows: the term $(-A_1)^{-1}A_2$ records the probability that, starting from a state in $\ell(n)$, the Markov process $\{X(t)\}$ eventually leaves $\ell(n)$ *and* moves down to $\ell(n-1)$, thereby terminating the first passage time τ. With probabilities recorded in $(-A_1)^{-1}A_0$, the Markov process eventually leaves $\ell(n)$ *and* moves up to $\ell(n+1)$; the factor G^2 records the probability of moving down by two levels, from $\ell(n+1)$ to $\ell(n)$, and then from $\ell(n)$ to $\ell(n-1)$.

The matrix U is defined by

$$U = A_1 + A_0 G$$

and is interpreted as follows: starting from $X(0) \in \ell(n)$, consider the Markov process restricted to $\ell(n)$ until the first visit to $\ell(n-1)$; this is a transient process, and the matrix U is its infinitesimal generator.

In order to see this, we may consider $U_{ij}h$ for some small value h. Clearly, $(A_1)_{ij}h$ is (up to some quantity $o(h)$) equal to the probability that, during an interval of time of length h, the QBD moves from (n, i) to (n, j); while $(A_0 hG)_{ij}$ is the probability that, during an interval of time of length h, the QBD moves from (n, i) to some state in $\ell(n+1)$, from where it eventually returns to $\ell(n)$ in the state (n, j) after an unspecified amount of time.

Since the matrix U is the generator of the Markov process in the local time of $\ell(n)$ before the first visit to $\ell(n-1)$, we see that $[\exp(Uh)]_{ij}$ is the probability of being in (n, j), starting from (n, i), after having spent h units of time in $\ell(n)$ under taboo of $\ell(n-1)$. By the same argument as in Theorem 2.4.3, we find that the expected sojourn time in $\ell(n)$ before the first visit to $\ell(n-1)$ is

$$N = \int_0^\infty \exp(Uh)dh = -U^{-1},$$

so that

$$R = A_0(-U)^{-1}. \tag{6.21}$$

Similarly, $[\exp(Uh)A_2 dh]_{ij}$ is for dh small enough the probability that, starting from (n, i), the QBD spends h units of time in $\ell(n)$ and then moves down to $(n-1, j)$. Thus, we have that

$$G = \int_0^\infty \exp(Uh)A_2 dh = (-U)^{-1}A_2. \tag{6.22}$$

Notice that (6.21) and (6.22) are very similar to (6.9) and (6.10). We finally obtain in the continuous time case a similar set of equations as in discrete time interrelating R, G, and U. These equations are collected below for later reference; we do not give the missing proofs since one may easily obtain them using either the probabilistic interpretation or an algebraic derivation.

Proposition 6.4.2 *The matrices $R, U,$ and G satisfy the following equations:*

$$\begin{aligned} U &= A_1 + A_0 G \\ &= A_1 + R A_2 \\ &= A_1 + A_0(-U)^{-1} A_2, \\ R &= A_0(-U)^{-1}, \\ G &= (-U)^{-1} A_2, \end{aligned}$$

and

$$\begin{aligned} A_2 + A_1 G + A_0 G^2 &= 0, \\ A_0 + R A_1 + R^2 A_2 &= 0. \end{aligned}$$

The boundary distribution is determined as follows.

Lemma 6.4.3 *The stationary probability vector $\boldsymbol{\pi}_0$ of $\ell(0)$ is the unique solution of the system*

$$\boldsymbol{\pi}_0(B + A_0 G) = \mathbf{0}, \tag{6.23}$$

$$\boldsymbol{\pi}_0(I - R)^{-1}\mathbf{1} = 1. \tag{6.24}$$

The first equation may also be written as $\boldsymbol{\pi}_0(B + RA_2) = \mathbf{0}$.

Proof. It immediately results from Theorem 5.5.3 that $\boldsymbol{\pi}_0$ is a solution of (6.23) and the normalizing condition $\boldsymbol{\pi}\mathbf{1} = \mathbf{1}$ is equivalent to (6.24) by (6.20). □

If $m < \infty$, another approach may be followed, whereby one directly translates results on discrete time QBDs into results about continuous time QBDs by considering the jump chain, i.e., the process $\{\widetilde{X}_t, t \in \mathbb{N}\}$ embedded at the epochs of transition for the process $\{X(t)\}$. That approach is of historical interest as it is the one followed by Neuts [79].

The embedded process $\{\widetilde{X}_t\}$ is a discrete time QBD with transition matrix $\widetilde{P}$ given by

$$\widetilde{P} = \begin{bmatrix} \widetilde{B} & \widetilde{B}_0 & 0 & 0 & \cdots \\ \widetilde{A}_2 & \widetilde{A}_1 & \widetilde{A}_0 & 0 & \cdots \\ 0 & \widetilde{A}_2 & \widetilde{A}_1 & \widetilde{A}_0 & \cdots \\ 0 & 0 & \widetilde{A}_2 & \widetilde{A}_1 & \cdots \\ \vdots & \vdots & \vdots & \vdots & \ddots \end{bmatrix}, \tag{6.25}$$

where $\widetilde{B} = I + \Delta_B^{-1}B$, $\widetilde{B}_0 = \Delta_B^{-1}A_0$, $\widetilde{A}_1 = I + \Delta_A^{-1}A_1$, $\widetilde{A}_0 = \Delta_A^{-1}A_0$, and $\widetilde{A}_2 = \Delta_A^{-1}A_2$ with Δ_B and Δ_A diagonal matrices of which the main diagonals are, respectively, the same as those of the matrices B and A_1.

If $\{X(t)\}$ is recurrent, then $\{\widetilde{X}_t\}$ is recurrent and their stationary vectors $\boldsymbol{\pi}$ and $\tilde{\boldsymbol{\pi}}$ are related as follows:

$$\boldsymbol{\pi}_0 = c\tilde{\boldsymbol{\pi}}_0\Delta_B^{-1}, \tag{6.26}$$

$$\boldsymbol{\pi}_n = c\tilde{\boldsymbol{\pi}}_n\Delta_A^{-1} \qquad \text{for } n \geq 1. \tag{6.27}$$

For a general Markov chain, it is possible that one of the series $\sum_{n\geq 0}\boldsymbol{\pi}_n\mathbf{1}$ or $\sum_{n\geq 0}\tilde{\boldsymbol{\pi}}_n\mathbf{1}$ converges but not the other, so that *positive* recurrence of the continuous chain does not imply positive recurrence of the jump chain or vice versa. In the case where $m < \infty$, however, as we assume here, a homogeneous QBD and its jump chain are positive recurrent together and we may apply (6.16), (6.17) to $\widetilde{P}$, thereby obtaining that

$$\begin{aligned}
\tilde{\boldsymbol{\pi}}_1 &= \tilde{\boldsymbol{\pi}}_0\widetilde{B}_0\widetilde{N}, \\
\tilde{\boldsymbol{\pi}}_{n+1} &= \tilde{\boldsymbol{\pi}}_1\widetilde{R}^n \qquad \text{for } n \geq 0,
\end{aligned}$$

where $\widetilde{R} = \widetilde{A}_0\widetilde{N}$. If we define $N = \widetilde{N}\Delta_A^{-1}$, then $R = \Delta_A\widetilde{R}\Delta_A^{-1}$ and it is a simple matter to obtain (6.20) from the equations above, using (6.26), (6.27).

Similarly, it results from (6.18) that

$$\tilde{\boldsymbol{\pi}}_0(\widetilde{B} + \widetilde{B}_0\widetilde{N}\widetilde{A}_2 - I) = \mathbf{0}$$

and (6.23) follows. The last equation merely states that the factor c in (6.26), (6.27) is chosen such that $\boldsymbol{\pi}\mathbf{1} = 1$.

This algebraic proof is short, but hides the probabilistic significance of the result. We recall that $\{\widetilde{X}_t\}$ is the embedded process at the epochs when $\{X(t)\}$ changes state and (see Theorem 6.2.1) that $\widetilde{N}_{ij}$ is the expected *number of visits* to (n, j), starting from the sate (n, i), before the first visit to $\ell(n-1)$.

Moreover, $(\Delta_A^{-1})_{jj}$ is the expected sojourn time in (n, j) for any $n \geq 1$. Thus, we see that N_{ij} is the expected *sojourn time* of $\{X(t)\}$ in (n, j), starting from (n, i), before the first visit to $\ell(n-1)$ for any $n \geq 1$.

Similarly, we see that $R = A_0N$ records the expected rates of visit to the states in $\ell(n+1)$ per unit of the local time of $\ell(n)$ spent in (n, i).

Chapter 7

Stability Condition

Continuing the discussion of irreducible, homogeneous QBDs, we note that the results in the preceding chapter are conditioned on the fact that the process is positive recurrent. We know by Corollary 6.2.4 that $\text{sp}(R) < 1$ is a necessary condition when $m < \infty$. We wish to have a necessary and sufficient condition that is simple and makes it easy to verify directly if a QBD is recurrent or not.

When $m < \infty$, there exists a very simple necessary and sufficient condition which we present in section 7.2. The situation is much more involved, and not satisfactorily settled yet, if $m = \infty$; we show with one example at the end of the chapter some of the difficulties which may arise in that case and briefly return to the question in Chapter 16.

This is another very short chapter. Our intention, in presenting this material separated from the rest, is to emphasize the fact that to determine the matrix-geometric structure on the one hand, and to find stability conditions on the other hand, are two distinct issues, a fact which was somewhat blurred in early presentations of matrix analytic methods.

7.1 First Passage Times

The discussion which follows is based on the analysis of the first passage time δ_n to the level n, starting from the level $n+1$:

$$\delta_n = \inf\{t \geq 0 : X_t \in \ell(n)\}.$$

We define

$$G_{ij}[k] = \mathrm{P}[\delta_n = k, X_{\delta_n} = (n,j)|X_0 = (n+1,i)] \qquad \text{for } k \geq 0$$

and the generating functions

$$G_{ij}(z) = \sum_{k\geq 0} z^k G_{ij}[k] \qquad \text{for } 0 \leq z \leq 1.$$

Observe that the distribution $\{G_{ij}[k], k \geq 0\}$ is independent of n since the one-step transition probabilities are level-independent.

We have that $G[0] = 0$, naturally, and

$$\begin{aligned} G[1] &= A_2, \\ G[2] &= A_1 G[1], \\ G[k] &= A_1 G[k-1] + \sum_{1\leq\nu\leq k-2} A_0 G[\nu] G[k-1-\nu] \qquad \text{for } k \geq 3, \end{aligned}$$

and

$$G(z) = zA_2 + zA_1 G(z) + zA_0 G^2(z), \tag{7.1}$$

where we write $G^2(z)$ for $[G(z)]^2$.

These equations are easily proved using the fact that the first passage time δ_{n+1} from $\ell(n+2)$ to $\ell(n+1)$ and δ_n from $\ell(n+1)$ to $\ell(n)$ have the same distributions. The matrix $G(z)$ evaluated at $z = 1$ is equal to the matrix G defined in section 6.2, so that (6.14) is a special case of (7.1).

In what follows, we say that G is substochastic if $G\mathbf{1} \leq \mathbf{1}$ with at least one strict inequality and we denote by $G'(1)$ the matrix $\partial/\partial_z G(z)$ evaluated at $z = 1$ with

$$(G'(1))_{ij} = \mathrm{E}[\delta_n \mathcal{I}\{X_{\delta_n} = (n,j)\}|X_0 = (n+1,i)], \tag{7.2}$$

where $\mathcal{I}\{\cdot\}$ is the indicator function. We say that $G'(1)$ is finite if all of its elements are finite and that it is infinite if at least one element is infinite.

An important factor, to determine whether a QBD is recurrent or not, is the behavior of the restricted Markov chain on $\ell(0)$. Its transition matrix is

$$P^{\{0\}} = B + A_0 G$$

and is irreducible since the QBD is irreducible.

We define the matrix M as follows: M_{ij} is the expected number of visits to the state $(0,j)$, starting from $(0,i)$, before the first return to $(0,i)$. Since the QBD is irreducible, M is a finite matrix.

Let us fix i and write the matrix $P^{\{0\}}$ as follows, possibly after a permutation of its rows and columns:

$$P^{\{0\}} = \begin{bmatrix} p_{ii}^{\{0\}} & \boldsymbol{p}_{i\bullet}^{\{0\}} \\ \boldsymbol{p}_{\bullet i}^{\{0\}} & D_i \end{bmatrix},$$

where D_i is the matrix obtained by removing from $P^{\{0\}}$ its ith row and its ith column and $\boldsymbol{p}_{i\bullet}^{\{0\}}$ is the vector obtained by removing the ith element from the ith row of $P^{\{0\}}$.

By Lemma 5.1.2, we have that, for all $k, j \neq i$, $[(I-D_i)^{-1}]_{kj}$ is the expected number of visits to $(0,j)$, starting from $(0,k)$, before the first visit to $(0,i)$. Thus, we find that

$$M_{ij} = [\boldsymbol{p}_{i\bullet}^{\{0\}}(I-D_i)^{-1}]_j \qquad \text{for all } j \neq i. \tag{7.3}$$

We also have that

$$M_{ii} = 1. \tag{7.4}$$

Theorem 7.1.1

The QBD is transient if either G is substochastic or if the Markov chain given by $P^{\{0\}} = B + A_0 G$ is transient. Equivalently, the QBD is recurrent if and only if G is stochastic and the Markov chain with transition matrix $P^{\{0\}}$ is recurrent.

The QBD is positive recurrent if and only if G is stochastic, the Markov chain with transition matrix $P^{\{0\}}$ is recurrent, and the vector

$$\boldsymbol{m} = M(\boldsymbol{1} + A_0 G'(1)\boldsymbol{1})$$

is finite, where $G'(1)$ is defined by (7.2) and M is given by (7.3), (7.4).

If m is finite, then the QBD is positive recurrent if and only if G is stochastic and $G'(1)$ is finite.

Proof. Our argument is based on the first return time θ to a specific state in $\ell(0)$, the state $(0,1)$, say (since the QBD is irreducible, any state will do).

If G is substochastic, then there is at least one index j such that $(G\mathbf{1})_j < 1$, so that starting from $(1,j)$, there is a strictly positive probability of never reaching $\ell(0)$. Since the QBD is irreducible, there is a positive probability of reaching $(1,j)$ from $(0,1)$ without returning to $(0,1)$. Hence, $\mathrm{P}[\theta < \infty | X_0 = (0,1)] < 1$ and the QBD is transient. This proves our first claim.

We assume from now on that G is stochastic, so that the transition matrix $P^{\{0\}} = B + A_0 G$ is stochastic, and we demonstrate that the QBD is transient if the restricted Markov chain given by $P^{\{0\}}$ is transient. We take $X_0 = (0,1)$ and define the sequence $\{\tau_i, i \geq 0\}$ of the successive epochs when the QBD returns to $\ell(0)$. Clearly, $\tau_{i+1} \geq \tau_i + 1$, so that $\lim_{i\to\infty} \tau_i = \infty$. Since G is stochastic, $\tau_{i+1} - \tau_i$ is finite a.s.

We further define $K = \inf\{i > 0 : X_{\tau_i} = (0,1)\}$, i.e., K is the first return time to $(0,1)$ in the restricted chain. Clearly, τ_K is the first return time to $(0,1)$ in the QBD and is thus equal to θ. If the Markov chain given by $P^{\{0\}}$ is transient, then $\mathrm{P}[K < \infty | X_0 = (0,1)] < 1$, which implies that $\mathrm{P}[\theta < \infty | X_0 = (0,1)] < 1$, so that the QBD is transient. Conversely, if the restricted Markov chain is recurrent, then K is finite a.s., hence θ is finite a.s. and the QBD is recurrent.

We now turn to the proof of the second statement. We restrict our attention to the case where G is stochastic and $P^{\{0\}}$ is recurrent, and we determine the additional condition under which $\mathrm{E}[\theta]$ is finite, i.e., the condition under which the QBD is *positive* recurrent.

Let N_j denote the number of times the state $(0,j)$ is visited by the QBD in a first return to (0,1), and let $\omega_\nu^{(j)}$ denote the number of steps taken by the QBD between its νth visit to $(0,j)$ and its next visit to $\ell(0)$. We see that θ may also be written as

$$\theta = \sum_{j\geq 1} \sum_{1\leq \nu \leq N_j} \omega_\nu^{(j)}.$$

For any fixed j, the $\omega_\nu^{(j)}$'s are independent and identically distributed and we denote by $\mathrm{E}[\omega^{(j)}]$ their common expected value. By Wald's equation (Ross [107, Page 104]), we find that

$$\begin{aligned} \mathrm{E}[\theta | X_0 = (0,1)] &= \sum_{j\geq 1} \mathrm{E}[N_j | X_0 = (0,1)]\mathrm{E}[\omega^{(j)}] \\ &= \sum_{j\geq 1} M_{1j}\mathrm{E}[\omega^{(j)}]. \end{aligned} \tag{7.5}$$

Now, we have that $\mathrm{E}[\omega^{(j)}] = \mathrm{E}[\tau|X_0 = (0,j)]$, where τ is the first return time to $\ell(0)$. Conditioning on X_1, we easily obtain that

$$\mathrm{E}[\tau|X_0 = (0,j)] = 1 + \sum_{k\geq 1}(A_0 G'(1))_{jk},$$

which, combined with (7.5), shows that the expected return time from $(0,1)$ to $(0,1)$ is equal to the first component of $\boldsymbol{m} = M(\mathbf{1}+A_0G'(1)\mathbf{1})$. It is easy to see by an entirely similar argument that the ith component of $\boldsymbol{m}$ is the expected return time in the QBD to $(0,i)$. Thus, the QBD is positive recurrent if and only if every component of $\boldsymbol{m}$ is finite.

To prove the last statement we note that if m is finite, the condition for the QBD to be positive recurrent becomes simpler to express. If the irreducible Markov chain $P^{\{0\}}$ has finitely many states, then it must be positive recurrent and the matrix M must be finite. Thus, the finiteness of $\boldsymbol{m}$ depends only on the finiteness of the matrix $G'(1)$. □

This theorem has the following consequence.

Corollary 7.1.2 *If m is finite, then the QBD is positive recurrent if and only if* $\mathrm{sp}(R) < 1$.

Proof. We have seen earlier, with Corollary 6.2.4, that $\mathrm{sp}(R) < 1$ is a necessary condition for positive recurrence. To prove that it is sufficient, we observe that $G'(1)$ is finite if and only if $G'(1)\mathbf{1}$ is finite. Recall from Theorem 6.2.1 that the matrix N_{1i} records the expected number of visits to the states in $\ell(i)$, starting from a state in $\ell(1)$, before the first visit to $\ell(0)$. This implies that $G'(1)\mathbf{1} = \mathbf{1} + \sum_{i\geq 1} N_{1i}\mathbf{1}$. By Theorem 6.2.7, $N_{1i} = N_{11}R^{i-1}$, from which $G'(1)\mathbf{1} = \mathbf{1} + N_{11}\sum_{i\geq 0} R^i\mathbf{1}$. If $\mathrm{sp}(R) < 1$, then the series above converges, so that $G'(1)$ is finite, and the QBD is positive recurrent by Theorem 7.1.1. □

7.2 Simple Drift Condition

The major drawback of Theorem 7.1.1 is that it requires the evaluation of the matrix G before one may determine if the QBD is recurrent. If this evaluation is done numerically, and if $G\mathbf{1}$ is close to $\mathbf{1}$, it may be difficult to ascertain whether G is truly stochastic or not. Corollary 7.1.2 does not give a very satisfactory criterion for similar reasons.

It is much preferable to have criteria which may directly be determined from the most basic parameters, i.e., from the matrix P itself.

A very simple necessary and sufficient condition exists in the case where m is finite. It is based on an easily verified property of the matrices A_0, A_1, and A_2 and states that the (suitably) averaged transition probability to lower levels should be greater than the averaged transition probability to higher levels; in other words, there must be a drift toward the level 0. We first prove this under the restriction that the matrix $A = A_0 + A_1 + A_2$ is irreducible.

Before we prove the major Theorem 7.2.3, we need to better know the structure of the matrices G and R. If A is irreducible and m is finite, both G and R have one and only one irreducible class, as we show in the next two theorems (see also Latouche, Pearce, and Taylor [51] and Latouche and Taylor [56]). We say that a matrix is *strictly* lower or upper triangular if it is, respectively, lower or upper triangular with diagonal entries equal to zero.

Theorem 7.2.1 *If the QBD is irreducible, if A is irreducible, and if m is finite, then either G is irreducible or, possibly after a permutation,*

$$G = \begin{bmatrix} G_1 & 0 \\ G_{\bullet 1} & G_\bullet \end{bmatrix}, \qquad (7.6)$$

where G_1 is irreducible and $G_\bullet$ is strictly lower triangular.

Proof. The proof is by contradiction. Assume that the statement of the theorem does not hold. Then we have that either G is strictly lower triangular or $G_\bullet$ is not strictly lower triangular. In the first case, one of the rows of G is identically zero, and this contradicts the assumption that the QBD is irreducible. In the second case, $G_\bullet$ is of the form

$$G_\bullet = \begin{bmatrix} G_2 & 0 \\ G_{32} & G_3 \end{bmatrix}, \qquad (7.7)$$

where G_2 is irreducible and G_3 could be anything (it might even not exist if $G_\bullet$ were irreducible). We partition each level $\ell(n)$ into sublevels $\ell_1(n)$, $\ell_2(n)$, and $\ell_3(n)$ corresponding to the states in G_1, G_2, and G_3, respectively. We prove in three steps that (7.7) is in contradiction with the assumption that A is irreducible.

First, we observe that if X_0 is in $\ell_1(n)$ for some $n \geq 1$, then the first state visited in the bottom level $\ell(0)$ must be in the subset $\ell_1(0)$; indeed, since the matrix G has the form (7.6), G^k has the same structure

$$G^k = \begin{bmatrix} G_1^k & 0 \\ G_{\bullet 1}^{(k)} & G_\bullet^k \end{bmatrix} \qquad \text{for all } k \geq 1.$$

This implies that if the initial state is in $\ell_1(n)$, then the QBD is in $\ell_1(n-1)$, $\ell_1(n-2)$, and so on at each transition to a lower level.

Second, we observe that if the matrix G is of the form (7.6) with $G_\bullet$ of the form (7.7), then for $m, n \geq 1$ there is no path in the QBD from $\ell_1(n)$ to $\ell_2(m)$, except through states in the level $\ell(0)$.

To see this, assume to the contrary that there is a path from $\ell_1(n)$ to $\ell_2(m)$, avoiding the level $\ell(0)$, for some $m \geq 1$. Since G_2 is irreducible, G_2^k is nonzero for all k. This means that there is a positive probability of moving down successively from $\ell_2(m)$ to $\ell_2(m-1)$, $\ell_2(m-2)$, ..., $\ell_2(0)$. Hence, we see that it is possible that, starting from $\ell_1(n)$, the first state visited in $\ell(0)$ is in the subset $\ell_2(0)$. This would be in contradiction with our conclusion in the first step.

Now, since A is irreducible, there exists for all pairs (i, j) a sequence $(i_0 = i, i_1, i_2, \ldots, i_K = j)$ of length $K \leq m$ such that $A_{i_k, i_{k+1}} > 0$ for all k. This means that for n_0 large enough (it suffices that $n_0 \geq K + 1$), there is a sequence $((n_0, i), (n_1, i_1), \ldots, (n_K, j))$ of positive probability for which n_{k+1} is either equal to n_k or $n_k + 1$ or $n_k - 1$ for each k. Now choose i and j such that (n_0, i) is in $\ell_1(n_0)$ and (n_K, j) is in $\ell_2(n_K)$.

We have constructed in this way a path from $\ell_1(n_0)$ to $\ell_2(n_K)$ avoiding level 0. This contradicts step 2; therefore $G_\bullet$ cannot be of the form (7.7) with G_2 irreducible, which completes the proof. □

A family of examples where G has the structure (7.6) with a nonzero $G_\bullet$ is given by the matrices

$$A_2 = \begin{bmatrix} \bullet & 0 & 0 \\ \bullet & 0 & 0 \\ * & \bullet & 0 \end{bmatrix}, \qquad A_1 = \begin{bmatrix} * & * & 0 \\ * & * & 0 \\ * & * & * \end{bmatrix}, \qquad A_0 = \begin{bmatrix} * & * & \bullet \\ * & * & \bullet \\ * & * & \bullet \end{bmatrix},$$

where the symbols $*$ and $\bullet$, respectively, represent nonnegative and strictly positive quantities.

Thus we see that if A is irreducible, then the Markov chain with transition matrix G has a unique irreducible class. Similarly, one proves that R has a unique principal submatrix. The argument is very close to that of Theorem 7.2.1.

Theorem 7.2.2 *If the QBD is irreducible, if A is irreducible, and if m is finite, then either R is irreducible or one has that*

$$R = \begin{bmatrix} R_1 & R_{1\bullet} \\ 0 & R_\bullet \end{bmatrix}, \tag{7.8}$$

possibly after a permutation of rows and columns, where R_1 is irreducible and $R_\bullet$ is strictly upper triangular.

Proof. To prove the theorem, we shall make use of the interpretation given in Remark 6.2.8 for the matrix R and its powers: the quantity $(R^k)_{ij}$ is the expected time spent in the state (k, j), starting from the state $(0, i)$, before the first return to any of the states in the level $\ell(0)$.

The negation of the statement of the theorem is that either R is strictly upper triangular or $R_\bullet$ is not strictly upper triangular.

In the first case, the entire matrix R is strictly upper triangular and so $R^n = 0$ for all $n \geq m$. In view of the interpretation recalled above, this implies that the states beyond the level m cannot be reached starting from the level 0, which is in contradiction with the assumption that the QBD is irreducible.

In the second case, $R_\bullet$ is of the form

$$R_\bullet = \begin{bmatrix} R_2 & R_{23} \\ 0 & R_3 \end{bmatrix}, \tag{7.9}$$

where R_2 is irreducible and R_3 could be anything. Thus, the matrix R and its successive powers have the following structure:

$$R^k = \begin{bmatrix} R_1^k & R_{12}^{(k)} & R_{13}^{(k)} \\ 0 & R_2^k & R_{23}^{(k)} \\ 0 & 0 & R_3^k \end{bmatrix} \quad \text{for all } k \geq 1. \tag{7.10}$$

We partition each level $\ell(n)$ into sublevels $\ell_1(n)$, $\ell_2(n)$, and $\ell_3(n)$ corresponding to the states in R_1, R_2, and R_3, respectively, and we prove in three steps that such a form is in contradiction with the assumption that A is irreducible.

We first observe that with the matrix R of the form (7.10), if the initial state is in $\ell_2(1)$, then the expected time spent in the states of $\ell_1(k)$, before a return to $\ell(1)$ or $\ell(0)$, is equal to zero for all $k \geq 2$. This implies that any path from $\ell_2(1)$ to $\ell_1(k)$ with $k \geq 2$ must visit some other state in the levels 0 or 1.

Second, we conclude that if the matrix R is of the form (7.10), then there is no path in the QBD from $\ell_2(n)$ to $\ell_1(m)$ with $n, m \geq 2$, except through states in the level $\ell(1)$.

To see this, assume the contrary, that there is a path from $\ell_2(n)$ to $\ell_1(m)$, avoiding the level $\ell(1)$, for some $n, m \geq 2$. Since R_2 is irreducible, R_2^k is nonzero for all k. This means that there is a positive probability of moving up successively from $\ell_2(1)$ to $\ell_2(2), \ldots, \ell_2(n)$ and from there eventually to reach the set $\ell_1(m)$, all the while avoiding the level 1. This is in contradiction with our conclusion in the first step.

Finally, since A is irreducible, there exists for all pairs (i, j) a sequence $(i_0 = i, i_1, i_2, \ldots, i_K = j)$ of length $K \leq m$ such that $A_{i_k, i_{k+1}} > 0$ for all k. This means that for n_0 large enough (for example if $n_0 \geq K + 2$), there is a sequence $((n_0, i), (n_1, i_1), \ldots, (n_K, j))$ of positive probability for which n_{k+1} is either equal to n_k, or $n_k + 1$ or $n_k - 1$, and is greater than 1 for each k. Now choose i and j such that (n_0, i) is in $\ell_2(n_0)$ and (n_K, j) is in $\ell_1(n_K)$.

Thus we have constructed a path from $\ell_2(n_0)$ to $\ell_1(n_K)$ avoiding the level 1. This contradicts our second step, therefore $R_\bullet$ cannot be of the form (7.9) with R_2 irreducible, which completes the proof. □

Theorem 7.2.3 *If the QBD is irreducible, if m is finite, and if the stochastic matrix $A = A_0 + A_1 + A_2$ is irreducible, then the QBD is positive recurrent if and only if $\mu = \boldsymbol{\alpha} A_0 \mathbf{1} - \boldsymbol{\alpha} A_2 \mathbf{1} < 0$, where $\boldsymbol{\alpha}$ is the stationary probability vector of A. The QBD is null recurrent if $\mu = 0$, and it is transient if $\mu > 0$.*

Proof. The argument is based on the analysis of a random walk defined on a Markov chain. We generally follow the presentation in Asmussen [3, Pages 235–239], but the details of the proof are slightly different. The argument proceeds in four steps. First, we associate a random walk to the QBD and define its stationary drift μ; then, we successively consider the cases where $\mu > 0$, $\mu < 0$, and $\mu = 0$. The last case is by far the most complex and the bulk of the proof is devoted to it.

Let us consider the Markov chain $\{(\epsilon_t, \varphi_t), t \in \mathbb{N}\}$ on the state space $\{-1, 0, 1\} \times \{1, 2, \ldots, m\}$ with transition probabilities depending only on the second coordinate:

$$\begin{aligned} \mathrm{P}[\epsilon_{t+1} = -1, \varphi_{t+1} = j | \epsilon_t, \varphi_t = i] &= (A_2)_{ij}, \\ \mathrm{P}[\epsilon_{t+1} = 0, \varphi_{t+1} = j | \epsilon_t, \varphi_t = i] &= (A_1)_{ij}, \\ \mathrm{P}[\epsilon_{t+1} = 1, \varphi_{t+1} = j | \epsilon_t, \varphi_t = i] &= (A_0)_{ij}. \end{aligned}$$

Defining $L_t = L_0 + \sum_{1 \le \nu \le t} \epsilon_\nu$, we have a Markov chain $\{(L_t, \varphi_t), t \in \mathbb{N}\}$ on the state space $\mathbb{Z} \times \{1, 2, \ldots, m\}$ and L_t constitutes a random walk on the integers with the increments $\{\epsilon_t\}$ controlled by the Markov chain $\{\varphi_t\}$. Let σ denote the first passage time to the negative half-line: $\sigma = \inf\{t \ge 1 : L_t \le 0\}$.

It is clear that if $X_0 = (L_0, \varphi_0)$ with $L_0 \ge 1$, then the Markov chain $\{(L_t, \varphi_t)\}$ and the QBD $\{X_t\}$ have the same behavior, until the first visit to $\ell(0)$. Moreover, the Markov chain $\{(L_t, \varphi_t)\}$ is completely homogeneous in the levels, so that the conditional distribution of the first passage time to the level $L_0 - 1$ is independent of L_0 and is the same as the distribution of σ when $L_0 = 1$. Therefore, G is stochastic if and only if σ is finite a.s. on $[L_0 = 1]$, and $G'(1)$ is finite if and only if σ has finite expectation on $[L_0 = 1]$.

If we choose φ_0 according to the stationary distribution $\boldsymbol{\alpha}$, then the process ϵ_t is stationary with $\mathrm{E}[\epsilon] = \mu = \boldsymbol{\alpha} A_0 \mathbf{1} - \boldsymbol{\alpha} A_2 \mathbf{1}$ and the process $\{L_t\}$ is a random walk with *stationary* increments, very similar to a traditional random walk with independent increments. In view of this comparison, it is intuitively obvious that the first passage time σ to the negative half-line should be finite when the expected drift μ is negative. The remainder of the proof merely makes that argument tight.

Since the matrix A is irreducible, the Markov chain $\{(\epsilon_t, \varphi_t)\}$ has only one irreducible class of states. It may have ephemeral states, corresponding to those columns of A_0, A_1, or A_2 which are identically zero. Such states can never be visited after time 1, so they are irrelevant and may be removed. Once these states are removed, we find after a little algebra that $\{(\epsilon_t, \varphi_t)\}$ has a stationary distribution given by the nonzero elements of the vector $\boldsymbol{\beta} = (\boldsymbol{\alpha} A_2, \boldsymbol{\alpha} A_1, \boldsymbol{\alpha} A_0)$.

The nonephemeral states of the Markov chain $\{(\epsilon_t, \varphi_t)\}$ being positive recurrent, they form an ergodic process, and therefore $\{\epsilon_t\}$ itself is ergodic. We may then apply the strong law of large numbers for ergodic processes (see, e.g., Karlin and Taylor [36, Theorem 9.5.5]) and conclude that $\lim_{t\to\infty} L_t/t = \mu$ a.s.

If $\mu > 0$, this implies that $\lim_{t\to\infty} L_t = +\infty$, that $\mathrm{P}[\sigma < \infty] < 1$, and therefore that G is substochastic.

If $\mu < 0$, then $\lim_{t\to\infty} L_t = -\infty$ a.s., so that σ is finite a.s. and G is stochastic. To prove that $\mathrm{E}[\sigma]$ itself is finite, we assume that the initial state (ϵ_0, φ_0) is one of the nonephemeral states and define the sequence of renewal epochs $k_0 = 0$, $k_{i+1} = \inf\{t > k_i : \epsilon_t = \epsilon_0, \varphi_t = \varphi_0\}$ when the process $\{(\epsilon_t, \varphi_t)\}$ returns to the initial state. Furthermore, we define $\mathcal{L}_i = L_{k_i}$ as the state of the random walk at the ith return to (ϵ_0, φ_0).

By the strong Markov property, the increments $\{\mathcal{L}_{i+1} - \mathcal{L}_i\}$ are i.i.d. and $\{\mathcal{L}_i\}$ is a regular random walk with independent increments. Since $\mathcal{L}_{i+1} - \mathcal{L}_i = \sum_{k_i+1\le t\le k_{i+1}} \epsilon_t$, we find that $\mathrm{E}[\mathcal{L}_{i+1} - \mathcal{L}_i] = \mu \mathrm{E}[k_1]$ for all initial states (Chung [14, Theorems I.14.5 and I.14.6]) and is strictly negative. Thus, the random walk $\{\mathcal{L}_{i+1} - \mathcal{L}_i\}$ is transient, $\lim_{i\to\infty} \mathcal{L}_i = -\infty$, and $\mathrm{E}[\nu] < \infty$, where $\nu = \inf\{i > 0 : \mathcal{L}_i < 0\}$ (Chung [15, Theorem 8.4.4]). Since $\sigma \le k_\nu$, this proves that $\mathrm{E}[\sigma] \le \mathrm{E}[k_1]\mathrm{E}[\nu]$ is finite when $\mu < 0$.

Finally, if $\mu = 0$, the random walk $\{\mathcal{L}_i\}$ is recurrent and $\underline{\lim}\mathcal{L}_i = -\infty$ a.s. (Theorems 8.3.4 and 8.4.4 in [15]), so that $\underline{\lim} L_t = -\infty$ a.s., since $\{\mathcal{L}_i\}$ is embedded in $\{L_t\}$; this shows that σ is finite a.s. and that G is stochastic.

It only remains to be proved that $\mathrm{E}[\sigma]$ is infinite. We proceed by contradiction and assume that $s_j = \mathrm{E}[\sigma | \varphi_0 = j]$ is finite for all j.

Observe that by Theorem 7.2.1, since A is irreducible, the stochastic matrix G has only one irreducible class of states, but it may have ephemeral states corresponding to the rows and columns of $G_\bullet$. Let $\mathcal{S}$ be that irreducible class, and let j^* be fixed in $\mathcal{S}$. Finally, define $\sigma^* = \inf\{t \ge 1 : L_t < L_0, \varphi_t = j^*\}$.

We necessarily have that $\sigma \le \sigma^*$: equality holds on $[\varphi_\sigma = j^*]$, otherwise,

$$\sigma^* = \sigma + \inf\{t \ge 1 : L'_t < L_0, \varphi'_t = j^*\},$$

where $\{(L'_t, \varphi'_t)\} = \{(L_{t+\sigma}, \varphi_{t+\sigma})\}$ is the process after time σ and has the same law as $\{(L_t, \varphi_t)\}$ by the strong Markov property. Since $L'_0 = L_\sigma = L_0 - 1$,

$$\begin{aligned}\sigma^* &= \sigma + \inf\{t \geq 1 : L'_t < L'_0 + 1, \varphi'_t = j^*\} \\ &\leq \sigma + \inf\{t \geq 1 : L'_t < L'_0, \varphi'_t = j^*\},\end{aligned}$$

so that we eventually find that

$$\mathrm{E}[\sigma^* | \varphi_0 = j] \leq \mathrm{E}[\sigma | \varphi_0 = j] + \sum_{\substack{k \in \mathcal{S} \\ k \neq j^*}} \mathrm{P}[\sigma < \infty, \varphi_\sigma = k | \varphi_0 = j] \mathrm{E}[\sigma^* | \varphi_0 = k].$$

In matrix notations, this is written as

$$\boldsymbol{s}^* \leq \boldsymbol{s} + \widetilde{G}\boldsymbol{s}^*,$$

where the vectors $\boldsymbol{s}^*$ and $\boldsymbol{s}$ have one component for each state in $\mathcal{S}$ and the matrix $\widetilde{G}$ is obtained by replacing the j^*th column of G by a column of 0. The last inequality implies that

$$\boldsymbol{s}^* \leq \sum_{\nu \geq 0} \widetilde{G}^\nu \boldsymbol{s},$$

where the series does converge: we have removed a state from the irreducible class of G, and this makes $\widetilde{G}$ properly substochastic.

By our assumption that s_j is finite for all j, the last inequality implies that s^*_j is finite for all j, which is a contradiction since we know by an earlier argument that at least the component $s^*_{j^*} = \mathrm{E}[\sigma^* | \varphi_0 = j^*]$ is infinite when $\mu = 0$. This completes the proof. □

In continuous time, the same property holds, as we show below, using the same type of argument as we did in the proof of Theorem 6.4.1.

Theorem 7.2.4 *Consider an irreducible, continuous time QBD and assume that m is finite and that the matrix $A = A_0 + A_1 + A_2$ is irreducible. Then, the continuous time QBD is positive recurrent if and only if $\boldsymbol{\alpha} A_2 \mathbf{1} > \boldsymbol{\alpha} A_0 \mathbf{1}$, where $\boldsymbol{\alpha}$ is the unique solution of the system $\boldsymbol{\alpha} A = \mathbf{0}$, $\boldsymbol{\alpha}\mathbf{1} = 1$. It is null recurrent if $\boldsymbol{\alpha} A_2 \mathbf{1} = \boldsymbol{\alpha} A_0 \mathbf{1}$ and transient if $\boldsymbol{\alpha}_2 A_2 \mathbf{1} < \boldsymbol{\alpha} A_0 \mathbf{1}$.*

Proof. We consider the jump chain $\{\widetilde{X}_t : t \in \mathbb{N}\}$ embedded at the epochs of transition of $\{X(t)\}$. This is a discrete time QBD with transition matrix given by (6.25).

Since the matrix A is irreducible, the matrix $\widetilde{A} = \widetilde{A}_0 + \widetilde{A}_1 + \widetilde{A}_2$ which is equal to $I + \mathrm{diag}(A_1)^{-1}A$ is also irreducible, and we may apply Theorem 7.2.3 to $\{\widetilde{X}_t\}$. Since $\{\widetilde{X}_t\}$ is positive recurrent if and only if $\tilde{\boldsymbol{\alpha}}\widetilde{A}_2\mathbf{1} > \tilde{\boldsymbol{\alpha}}\widetilde{A}_0\mathbf{1}$, where $\tilde{\boldsymbol{\alpha}}(\widetilde{A} - I) = \mathbf{0}$, $\tilde{\boldsymbol{\alpha}}\mathbf{1} = 1$, and since $\boldsymbol{\alpha}$ is proportional to $\tilde{\boldsymbol{\alpha}}\mathrm{diag}(A_1)^{-1}$, the theorem is proved. □

7.3 Drift Conditions in General

If the matrix A is reducible, then the situation becomes more complicated, although it remains manageable as long as the number of phases is finite. The matrix A is written, possibly after a permutation of its rows and columns, as

$$A = \begin{bmatrix} C^{(1)} & 0 & \cdots & 0 & 0 \\ 0 & C^{(2)} & \cdots & 0 & 0 \\ \vdots & \vdots & \ddots & \vdots & \vdots \\ 0 & 0 & \cdots & C^{(K)} & 0 \\ D^{(1)} & D^{(2)} & \cdots & D^{(K)} & D^{(0)} \end{bmatrix}, \tag{7.11}$$

where the blocks $C^{(k)}, 1 \le k \le K$ are irreducible and stochastic and $D^{(0)}$ is substochastic. Since the matrices A_0, A_1, and A_2 are nonnegative, they are similarly structured, and we have that

$$A_i = \begin{bmatrix} C_i^{(1)} & 0 & \cdots & 0 & 0 \\ 0 & C_i^{(2)} & \cdots & 0 & 0 \\ \vdots & \vdots & \ddots & \vdots & \vdots \\ 0 & 0 & \cdots & C_i^{(K)} & 0 \\ D_i^{(1)} & D_i^{(2)} & \cdots & D_i^{(K)} & D_i^{(0)} \end{bmatrix}. \tag{7.12}$$

This implies that the state space of the QBD is decomposed into subsets which communicate through the level $\ell(0)$ only. The set $\{1, \ldots, m\}$ is partitioned into subsets $\phi(1), \ldots, \phi(K)$ and $\phi(0)$, and the levels $\ell(1)$, $\ell(2), \ldots$ are partitioned into disjoint subsets $\mathbb{N}_0 \times \phi(1), \ldots, \mathbb{N}_0 \times \phi(K)$

and $\mathbb{N}_0 \times \phi(0)$ with $\mathbb{N}_0 \times \phi(i) = \{(n, j) : n \geq 1, \ j \in \phi(i)\}$. The case when A is irreducible corresponds to $K = 1$, $\phi(0) = \emptyset$.

The dynamics are clear. If $X_0 \in \mathbb{N}_0 \times \phi(1)$ at some level $\ell(n)$ with $n \geq 1$, then the process is forced to remain in $\mathbb{N}_0 \times \phi(1)$ until it reaches $\ell(0)$. Thus, the distribution of the first passage time δ_n from $\ell(n+1)$ to $\ell(n)$ is solely determined by the submatrices $C_0^{(1)}$, $C_1^{(1)}$, and $C_2^{(1)}$. The same argument holds for the other subsets $\mathbb{N}_0 \times \phi(k)$, where $2 \leq k \leq K$.

If $X_0 \in \mathbb{N}_0 \times \phi(0)$, then the situation is different: the QBD may remain in $\mathbb{N}_0 \times \phi(0)$ until it reaches $\ell(0)$, but it may also move to another subset $\mathbb{N}_0 \times \phi(k)$ with $1 \leq k \leq K$, in which case it will remain there for the remainder of the passage time to $\ell(0)$.

The general version of Theorem 7.2.3 is given now.

Theorem 7.3.1 *Assume that the QBD is irreducible and that m is finite. Assume that the stochastic matrix $A = A_0+A_1+A_2$ is partitioned as in (7.11), where $K \geq 1$ and the matrices $C^{(k)}, 1 \leq k \leq K$ are irreducible.*

The QBD is recurrent if and only if $\gamma^{(k)} C_2^{(k)} \mathbf{1} \geq \gamma^{(k)} C_0^{(k)} \mathbf{1}$ for all $k : 1 \leq k \leq K$, where $\gamma^{(k)}$ is the stationary probability vector of $C^{(k)}$.

It is positive recurrent if and only if all the inequalities are strict.

It is transient if and only if there exists at least one $k, 1 \leq k \leq K$, such that $\gamma^{(k)} C_2^{(k)} \mathbf{1} < \gamma^{(k)} C_0^{(k)} \mathbf{1}$.

Proof. We first assume that $X_0 \in \mathbb{N}_0 \times \phi(1)$. Then, as we have observed, δ_n has the same distribution as if the matrices A_0, A_1, A_2, and A were, respectively, given by $C_0^{(1)}$, $C_1^{(1)}$, $C_2^{(1)}$, and $C^{(1)}$. Since $C^{(1)}$ is irreducible, we may use Theorem 7.2.3 to conclude that δ_n is finite a.s. if and only if $\gamma^{(1)} C_0^{(1)} \mathbf{1} \leq \gamma^{(1)} C_2^{(1)} \mathbf{1}$ and has finite expectation if and only if the inequality is strict. The same argument holds if $X_0 \in \mathbb{N}_0 \times \phi(k)$ for $k = 2, \ldots, K$.

If $X_0 \in \mathbb{N}_0 \times \phi(0)$, then we define θ as the first epoch when the QBD either enters the level n or one of the subsets $\mathbb{N}_0 \times \phi(k)$ with $1 \leq k \leq K$. We further define the vectors $\boldsymbol{f}_k^{(t)}$, $t \geq 0$, $k \geq n+1$, where

$$f_{ki}^{(t)} = P[\theta > t | X_0 = (k, i)].$$

We prove by induction that $\boldsymbol{f}_{n+1}^{(t)} \leq (D^{(0)})^t \mathbf{1}$. This obviously holds for $t = 0$. Assuming that it holds for some t, and conditioning on X_1, we have that

$$\begin{aligned} \boldsymbol{f}_k^{(t+1)} &= D_1^{(0)} \boldsymbol{f}_k^{(t)} + D_0^{(0)} \boldsymbol{f}_{k+1}^{(t)} \\ &\le (D_1^{(0)} + D_0^{(0)})(D^{(0)})^t \mathbf{1} \\ &\le (D^{(0)})^{t+1} \mathbf{1}. \end{aligned}$$

Since $D^{(0)}$ is substochastic and finite, the series $\sum_{\nu \ge 0} (D^{(0)})^\nu \mathbf{1}$ converges, which proves that θ is finite a.s. and has finite expectation. Since $\delta_n = \theta + \Delta_n$, where Δ_n is equal to zero if $X_\theta \in \ell(n)$ and is equal to the residual first passage time to $\ell(n)$ otherwise, we find ourselves back in the previously examined case, where the initial state belongs to some $\mathbb{N}_0 \times \phi(k)$ with $k \ge 1$. □

When $m = \infty$ and A is irreducible, the necessary and sufficient condition expressed in Theorem 7.2.3 might not hold. Tweedie [115] has proved in a more general context that if A is irreducible, then $\mu < 0$ is a sufficient condition for the QBD to be positive recurrent, but it is not clear that it is also necessary. We shall briefly return to this question at the end of Chapter 16.

If, in addition to m being infinite, A is reducible, then the situation becomes very messy and one may need to take into account the detailed structure of the matrices A_0, A_1, and A_2. For instance, if there are infinitely many phases in $\phi(0)$, then our proof of Theorem 7.3.1 cannot be valid, since it is possible that the phase never leaves $\phi(0)$.

To illustrate the difficulty, we return to the example of section 1.4 for which we recommended to use the generator (1.12) instead of the more obvious generator (1.11). With (1.12), the matrix A is given by

$$A = \begin{bmatrix} -\lambda_1 & \lambda_1 & \cdot & \cdot & \cdots \\ \mu_1 & -\lambda_1 - \mu_1 & \lambda_1 & \cdot & \cdots \\ \cdot & \mu_1 & -\lambda_1 - \mu_1 & \lambda_1 & \cdots \\ \cdot & \cdot & \mu_1 & -\lambda_1 - \mu_1 & \cdots \\ \vdots & \vdots & \vdots & \vdots & \ddots \end{bmatrix}.$$

This is the irreducible generator of an M/M/1 queue, so that its stationary probability vector $\boldsymbol{\alpha}$ exists if and only if $\mu_1 > \lambda_1$ and is given by $\alpha_i = (1 - \rho_1)\rho_1^i$ with $\rho_1 = \lambda_1/\mu_1$. Since $\boldsymbol{\alpha} A_0 \mathbf{1} - \boldsymbol{\alpha} A_2 \mathbf{1} = \lambda_2 - \mu_2(1 - \rho_1)$, we have $\mu < 0$ (and the QBD is positive recurrent) if $\rho_1 + \rho_2 < 1$, where

$\rho_2 = \lambda_2/\mu_2$. This is the well-known stability condition for the single server queue with two priority classes.

On the other hand, if we use the generator (1.11), then the matrix A is given by

$$A = \begin{bmatrix} -\lambda_2 & \lambda_2 & \cdot & \cdot & \cdots \\ \cdot & -\lambda_2 & \lambda_2 & \cdot & \cdots \\ \cdot & \cdot & -\lambda_2 & \lambda_2 & \cdots \\ \cdot & \cdot & \cdot & -\lambda_2 & \cdots \\ \vdots & \vdots & \vdots & \vdots & \ddots \end{bmatrix},$$

which is the generator of a pure birth process. None of the approaches in this section is then applicable.

Remark 7.3.2 Some authors (Gail and Hantler [23], Mitrani and Mitra [70], Neuts [79], Rama Murthy, Kim, and Coyle [91]) have determined conditions for recurrence or transience which are based on spectral analysis methods under the assumption that m is finite. The most general results along these lines are in [23]: the authors define polynomial matrices $C^{(k)}(z) = C_2^{(k)} + z(C_1^{(k)} - I) + z^2 C_0^{(k)}$ for $1 \le k \le K$ (the same matrices $C_i^{(k)}$ as in (7.12)) and prove that the QBD is recurrent if and only if $d/dz \det C^{(k)}(z)|_{z=1} \le 0$ for all k and is positive recurrent if and only if the inequalities are strict. We do not elaborate on this approach since we feel that Theorem 7.3.1 is simpler to use.

Part IV

Algorithms

Chapter 8

Algorithms for the Rate Matrix

We have described in Chapters 6 and 7 the structure of the stationary probability distribution of a QBD and the conditions for its existence. We now examine the question of its numerical evaluation. In doing so, we are led to the discussion of a number of recurrence equations which are useful in designing iterative algorithms.

The transition matrix of a homogeneous QBD has the form

$$P = \begin{bmatrix} B & A_0 & 0 & 0 & \cdots \\ A_2 & A_1 & A_0 & 0 & \cdots \\ 0 & A_2 & A_1 & A_0 & \cdots \\ 0 & 0 & A_2 & A_1 & \cdots \\ \vdots & \vdots & \vdots & \vdots & \ddots \end{bmatrix}, \tag{8.1}$$

where A_0, A_1, A_2, and B are square matrices of order $m \leq \infty$.

The stationary probability distribution, when it exists, is of the form $\boldsymbol{\pi}_n = \boldsymbol{\pi}_0 R^n$, and a key step is the computation of the so-called rate matrix R, which records the rate of visit to $\ell(n)$ per unit of time spent in $\ell(n-1)$. Two other matrices have also been introduced: the matrix G, which records the probability of moving down from $\ell(n)$ to $\ell(n-1)$, and the matrix U, which records the probability of returning to $\ell(n)$ under taboo of $\ell(n-1)$. We have seen that if any one of the three matrices G, U, or R is known, then the other two can be determined easily.

It is clear that when m is infinite, it is impossible to actually compute any of these matrices in its entirety and that it will be necessary to somehow truncate the set of phases. We do not address this problem, and we assume, when presenting the algorithms and discussing their efficiency, that m is finite. We want to stress, however, that the equations themselves, on which the algorithms are based, are valid for $m = \infty$ as well. For that reason, we do not put any general restriction on m, and we clearly identify those results which only hold when m is finite.

Although we could concentrate on any of the three matrices for the purpose of determining the stationary distribution, we generally favor computing the matrix G, in contrast to most of the literature on this subject, where the focus usually is on directly computing the rate matrix R. Our reason is that the matrix G has the most useful property of being stochastic when the QBD is positive recurrent, which is usually the case of interest, so that it is known a priori that the computed matrix must satisfy the equation $G\mathbf{1} = \mathbf{1}$. By contrast, the matrix U is only known to be substochastic and to satisfy $U\mathbf{1} \leq \mathbf{1}$ and the matrix R is only known to have a maximal eigenvalue strictly less than one. Still, it is the matrix R which is of major interest and this is reflected in the chapter title. Once G is computed, R is easily computed as

$$R = A_0(I - A_1 - A_0G)^{-1} \tag{8.2}$$

by (6.9), (6.11).

We analyze three general-purpose iterative schemes for computing the matrix G based only on the general form (8.1) of the transition matrix and without any further assumption, except that the QBD is irreducible.

The first algorithm is linearly convergent with convergence rate equal to the maximal eigenvalue of the matrix R; the second is also linearly convergent with a convergence rate which is usually much smaller. The third algorithm is more involved but has the advantage of being quadratically convergent.

A nice feature of these algorithms is that they are described in terms of the dynamic behavior of the stochastic process itself and probabilistic arguments are used to prove some of their convergence properties, thereby allowing for an interesting interplay between linear algebra and probability theory.

We have exhibited at the end of sections 6.2 and 6.4 a number of relations which are satisfied by the matrices R, U, and G. We recall them here and show that these equations actually completely characterize the three matrices. In consequence, subject to some minimality condition, each of the equations may serve as the starting point for an algorithm and this has led over time to a variety of methods. To compare all of them is beyond the scope of the present book, and we concentrate on three particular methods, two of which are reasonably efficient and very easy to implement and a third which is among the fastest known today.

In some circumstances, the transition matrix is such that one may exploit specific properties in order to obtain the distribution by more efficient ad hoc approaches. This is illustrated in section 8.5, where we show the major simplifications which occur when A_0 or A_2 is of rank 1.

We conclude this chapter with a brief section on the computation of the boundary probability vector $\boldsymbol{\pi}_0$.

8.1 A Basic Algorithm

Let $\{X_t : t \geq 0\}$ be a QBD with transition matrix (8.1). We begin by defining the first passage times

$$\gamma(i) = \inf\{t \geq 1 : X_t \in \ell(i)\} \qquad \text{for } i \geq 0.$$

In terms of these quantities, the definitions (6.5), (6.6) of the matrices U and G are as follows:

$$G_{ij} = \mathrm{P}[\gamma(0) < \infty \;\&\; X_{\gamma(0)} = (0, j) | X_0 = (1, i)]$$

and

$$U_{ij} = \mathrm{P}[\gamma(1) < \gamma(0) \;\&\; \gamma(1) < \infty \;\&\; X_{\gamma(1)} = (1, j) | X_0 = (1, i)]$$

for $1 \leq i, j \leq m$. In order to simplify the equations, we shall use the short-hand notations

$$G = \mathrm{P}[\gamma(0) < \infty \;\&\; X_{\gamma(0)} | X_0 \in \ell(1)] \tag{8.3}$$

and

$$U = \mathrm{P}[\gamma(1) < \gamma(0) \ \& \ \gamma(1) < \infty \ \& \ X_{\gamma(1)} | X_0 \in \ell(1)]. \qquad (8.4)$$

This slight abuse of notation will help us avoid storms of subscripts and will increase the readability of our equations.

A simple and often efficient algorithm is based on the equations (6.10), (6.11) which we reproduce below:

$$U = A_1 + A_0 G, \qquad (8.5)$$

$$G = (I - U)^{-1} A_2. \qquad (8.6)$$

It is natural to try to solve these equations by successive substitutions: starting somewhat arbitrarily with $G(0) = 0$, we might iteratively compute two sequences $\{U(k), k \geq 1\}$ and $\{G(k), k \geq 1\}$ as follows:

$$G(0) = 0, \qquad (8.7)$$

$$U(k) = A_1 + A_0 G(k-1), \qquad (8.8)$$

$$G(k) = (I - U(k))^{-1} A_2 \qquad \text{for } k \geq 1. \qquad (8.9)$$

Two questions need to be examined. Do these sequences converge? If they do converge, are their limits the matrices U and G defined in (8.3), (8.4) and not some other solution of the system of equations (8.5), (8.6)?

To answer these questions, we show that $G_{ij}(k)$ is the probability that the QBD moves from the state $(1, i)$ at time 0 to the level $\ell(0)$ in a finite amount of time by visiting the specific state $(0, j)$ under the taboo of the states in $\ell(k+1)$ and higher levels. Similarly $U_{ij}(k)$ is shown to be the probability that, starting from $(1, i)$, the QBD returns to the level $\ell(1)$ by visiting the specific state $(1, j)$ under taboo of the states in $\ell(k+1)$ and in $\ell(0)$.

Equivalently, $G(k)$ is the matrix of first passage probabilities from $\ell(n)$ to $\ell(n-1)$ under taboo of $\ell(n+k)$ for any $n \geq 1$, and $U(k)$ is the matrix of first return probabilities to $\ell(n)$ under taboo of $\ell(n-1)$ and $\ell(n+k)$.

Theorem 8.1.1 *The sequences $\{U(k), k \geq 1\}$ and $\{G(k), k \geq 1\}$ defined by*

$$U(k) = \mathrm{P}[\gamma(1) < \gamma(0) \ \& \ \gamma(1) < \gamma(k+1) \& \ X_{\gamma(1)} | X_0 \in \ell(1)] \quad (8.10)$$

and

$$G(k) = \mathrm{P}[\gamma(0) < \gamma(k+1) \;\&\; X_{\gamma(0)} | X_0 \in \ell(1)] \qquad (8.11)$$

for $k \geq 1$ *satisfy* (8.7)–(8.9). *The two sequences are monotonically increasing and respectively converge to the matrices* U *and* G.

Proof. If $X_0 \in \ell(1)$, then necessarily $0 < \gamma(2) < \gamma(3) < \cdots$ since the QBD may not increase by more than one level at a time. For the same reason, we have that $\gamma(k) \geq k-1$, so that $\lim_{k\to\infty} \gamma(k) = \infty$. Thus the last statement is obvious, and we only need to verify that the two sequences satisfy (8.8), (8.9).

We first establish (8.9). Starting from $X_0 \in \ell(1)$, the QBD can make a first passage into $\ell(0)$, avoiding $\ell(k+1)$, in one of two mutually exclusive ways. One way is to visit $\ell(0)$ at the very first step with probabilities recorded in A_2. The other way is for the QBD to return to $\ell(1)$ avoiding $\ell(0)$ and $\ell(k+1)$ with probabilities recorded in $U(k)$ and to start all over again, starting from $X_{\gamma(1)}$. This leads to the equation

$$G(k) = A_2 + U(k)G(k).$$

Since $U(k) \leq U$, $(I - U(k))^{-1}$ exists and we immediately have (8.9).

To prove (8.8) for $k = 1$, we observe that

$$[\gamma(1) < \gamma(0) \;\&\; \gamma(1) < \gamma(2)] \equiv [\gamma(1) = 1] \equiv [X_1 \in \ell(1)],$$

so that $U(1) = A_1$. For $k \geq 2$, we condition on X_1. In order to have $\gamma(1) < \gamma(0)$, it is necessary either that the QBD remains in $\ell(1)$ at the first step (the probabilities of this event are recorded in A_1) or that the QBD moves up to $\ell(2)$ (with probabilities recorded in A_0), from which level the QBD will eventually have to return to $\ell(1)$ avoiding $\ell(k+1)$. Thus, we have that

$$U(k) = A_1 + A_0 \widetilde{G}(k)$$

with

$$\widetilde{G}(k) = \mathrm{P}[\gamma(1) < \gamma(k+1) \;\&\; X_{\gamma(1)} | X_0 \in \ell(2)].$$

Since the QBD is level homogeneous, the matrix $G(k)$ also records for all $n \geq 1$ the probability that, starting from level $\ell(n)$, the QBD eventually visits $\ell(n-1)$ before it visits $\ell(n+k)$. Therefore, $\widetilde{G}(k) = G(k-1)$, which completes the proof. □

Linear Progression Algorithm
$G_\bullet := (I - A_1)^{-1}A_2$;
repeat
  $U_\bullet := A_1 + A_0 G_\bullet$;
  $G_\bullet := (I - U_\bullet)^{-1}A_2$
until $\| \mathbf{1} - G_\bullet \mathbf{1} \|_\infty \leq \epsilon$

Figure 8.1: A basic algorithm for the evaluation of G. The stopping criterion is chosen under the assumption that the QBD is recurrent.

The algorithm[12] in Figure 8.1 uses the recurrence equations (8.7)–(8.9) and iteratively computes an approximation for the matrix G. The array $G_\bullet$ is initially assigned the value of $G(1)$. During the kth iteration through the loop, the arrays $U_\bullet$ and $G_\bullet$ are respectively assigned the values of $U(k+1)$ and $G(k+1)$, so that if we denote by K_{LP} the number of iterations through the loop, the algorithm terminates with the approximation

$$G_\bullet = \mathrm{P}[\gamma(0) < \gamma(K_{LP} + 1) \ \& \ X_{\gamma(0)} | X_0 \in \ell(1)].$$

It is easy to define a stopping criterion when the QBD is recurrent, as we assume in Figure 8.1. Indeed, G is stochastic in that case, so that $(G - G_\bullet)\mathbf{1} = (\mathbf{1} - G_\bullet \mathbf{1})$. Knowing that $G \geq G(k)$ for all k, we have that $\| G - G_\bullet \|_\infty = \| \mathbf{1} - G_\bullet \mathbf{1} \|_\infty$, so that $\| G - G_\bullet \|_\infty \leq \epsilon$ upon exit from the loop.

If the QBD is transient, then there is no really satisfactory stopping criterion. A standard practice is to keep track of two successive matrices in the iteration, G_{old}, say, in addition to $G_\bullet$ and to stop the iterations when $\| G_{\text{old}} - G_\bullet \|_\infty \leq \epsilon$. The problem is that the convergence may be very slow, and it is possible that one has $\| G_{\text{old}} - G_\bullet \|_\infty \leq \epsilon$ while $\| G - G_\bullet \|_\infty$ is actually greater than ϵ.

---

[12] We call this the linear progression algorithm in reference to the boundary $\ell(k+1)$: the algorithm progresses by moving it by one level at each iteration.

Remark 8.1.2 The number K_{LP} of iterations has a probabilistic interpretation. The algorithm proceeds as follows: the matrix $G(k)$ collects the total probability mass of those trajectories which, starting from level 1, return to level 0 without leaving the set of levels 1 to k; all the trajectories that venture into level $k+1$ are discarded. The algorithm collects an increasing set of trajectories by adding one level at a time to the allowed set of levels. Since $\mathbf{1} - G\mathbf{1} \geq \mathbf{0}$, the algorithm stops when $(G\mathbf{1})_i \geq 1 - \epsilon$ for every i, so that it stops when it has collected, for every starting state $(1, i)$, a set of trajectories with a probability mass at least equal to $1 - \epsilon$.

In queueing applications, the event $[\gamma(0) < \gamma(k+1)]$ signifies that the maximum queue size during a busy period is at most equal to k. Thus, we may interpret K_{LP} as the $(1-\epsilon)$-quantile of the maximum queue size distribution during a busy period.

An interesting consequence of Theorem 8.1.1 is that the matrices U and G are completely characterized as minimal nonnegative solutions[13] of (8.5), (8.6); this is proved below. Recall that for a matrix X of finite order, $\sum_{\nu \geq 0} X^\nu = (I - X)^{-1}$ when the series converges (Lemma 5.1.1); also recall that we use the same notation for any substochastic matrix X, even of infinite order.

Theorem 8.1.3 *The matrices U and G are the minimal solutions of the system*

$$X = A_1 + A_0 Y, \qquad Y = \sum_{\nu \geq 0} X^\nu A_2 \tag{8.12}$$

in the set of nonnegative matrices such that the series converges.

Proof. Note first of all that the pair (U, G) satisfies (8.12). Assume that (X^*, Y^*) is another nonnegative solution of (8.12). We show that $U(k) \leq X^*$ and $G(k) \leq Y^*$ for all k, so that $U = \lim_{k\to\infty} U(k) \leq X^*$ and $G = \lim_{k\to\infty} G(k) \leq Y^*$.

We proceed by induction. Since $Y^* \geq 0$, we have that $U(1) = A_1 \leq A_1 + A_0 Y^* = X^*$. Now, let us assume that $U(k) \leq X^*$ for some k.

---

[13]Throughout, we say that a solution X is the *minimal nonnegative solution* of a matrix equation if $X_{ij} \leq Y_{ij}$ for all i and j for any solution $Y \geq 0$.

Then

$$\begin{aligned} G(k) &= (I - U(k))^{-1}A_2 = \sum_{\nu\geq 0} U(k)^\nu A_2 \\ &\leq \sum_{\nu\geq 0}(X^*)^\nu A_2 = Y^* \end{aligned}$$

and

$$U(k+1) = A_1 + A_0 G(k) \leq A_1 + A_0 Y^* = X^*,$$

which proves the induction step. □

A further consequence of Theorem 8.1.1 is that each of the equations at the end of section 6.2 may serve to completely characterize the matrices R, U, and G. This is summarized below.

Theorem 8.1.4

The matrices U and R are the minimal solutions of the system

$$X = A_1 + ZA_2, \qquad Z = A_0 \sum_{\nu\geq 0} X^\nu \tag{8.13}$$

in the set of nonnegative matrices such that the series converges.

The matrix U is the minimal solution of the equation

$$X = A_1 + A_0 \sum_{\nu\geq 0} X^\nu A_2 \tag{8.14}$$

in the set of nonnegative matrices such that the series converges.

The matrix G is the minimal solution of the equation

$$Y = A_2 + A_1 Y + A_0 Y^2 \tag{8.15}$$

in the set of nonnegative matrices.

The matrix R is the minimal solution of the equation

$$Z = A_0 + ZA_1 + Z^2 A_2 \tag{8.16}$$

in the set of nonnegative matrices.

Proof. We have proved in Theorems 6.2.9 and 6.2.10 that the matrices U, G, and R are indeed solutions of the stated equations, so that we only need to prove their minimality property.

To prove the first statement, we define a sequence $\{R(k) : k \geq 1\}$ as follows:

$$R(k) = A_0 \sum_{\nu \geq 0} [U(k)]^\nu. \tag{8.17}$$

Since $\{U(k)\}$ monotonically converges to U, $\{R(k)\}$ monotonically converges to $A_0 \sum_{\nu \geq 0} U^\nu = R$. The proof that the pair (U, R) is the minimal solution of (8.13) can now proceed in a manner exactly similar to that of Theorem 8.1.3. We omit the details which involve an inductive proof that for all k, $R(k) \leq Z^*$ and $U(k) \leq X^*$, where (X^*, Z^*) is any nonnegative solution of (8.13).

To prove the second statement, we assume that X^* is a nonnegative solution of (8.14) and define $Y^* = \sum_{\nu \geq 0} (X^*)^\nu A_2$. Since X^* is nonnegative, $\sum_{\nu \geq 0} (X^*)^\nu \geq 0$ and

$$X^* = A_1 + A_0 \sum_{\nu \geq 0} (X^*)^\nu A_2 \geq A_1 = U(1).$$

Now, we can easily show that $U(k) \leq X^*$, $G(k) \leq Y^*$ for all k (see the proof of Theorem 8.1.3). In particular, this gives $U \leq X^*$.

The proof of the third statement is by contradiction. We assume that there exists a nonnegative solution Y^* of (8.15) such that $Y^* \leq G$, $Y^* \neq G$. We define $X^* = A_1 + A_0 Y^*$. Since Y^* is a solution of (8.15), we have that

$$\begin{aligned} Y^* &= A_2 + (A_1 + A_0 Y^*) Y^* \\ &= A_2 + X^* Y^* = A_2 + X^* A_2 + (X^*)^2 Y^* = \cdots \\ &= \sum_{\nu \geq 0} (X^*)^\nu A_2. \end{aligned} \tag{8.18}$$

The series in (8.18) converges since

$$X^* = A_1 + A_0 Y^* \leq A_1 + A_0 G = U,$$

and the series $\sum_{\nu \geq 0} U^\nu$ converges. We thus find that (X^*, Y^*) is a solution of the system (8.12) and is smaller than (U, G), which is impossible by Theorem 8.1.3.

To prove the last statement of the theorem, we assume that Z^* is a nonnegative solution of (8.16) and we define $X^* = A_1 + Z^*A_2$. We then prove as follows that $U(k) \leq X^*$ and $R(k) \leq Z^*$ for all k.

Since $Z^* \geq 0$,

$$\begin{aligned} Z^* &\geq A_0 + Z^*A_1 \\ &\geq A_0 \sum_{\nu \geq 0} A_1^\nu \\ &= A_0(I - U(1))^{-1} = R(1). \end{aligned}$$

Now, let us assume that $R(k) \leq Z^*$ for some k. Then

$$\begin{aligned} U(k+1) &= A_1 + A_0 G(k) \qquad \text{by (8.8)} \\ &= A_1 + A_0(I - U(k))^{-1} A_2 \qquad \text{by (8.9)} \\ &= A_1 + R(k)A_2 \qquad \text{by (8.17)} \\ &\leq A_1 + Z^*A_2 = X^* \end{aligned}$$

by the induction hypothesis. We also have that

$$\begin{aligned} Z^* &= A_0 + Z^*(A_1 + Z^*A_2) \\ &= A_0 + Z^*R^* \\ &\geq A_0 + Z^*U(k+1) \\ &\geq A_0(I - U(k+1))^{-1} \\ &= R(k+1), \end{aligned}$$

which proves the induction step. Thus, $R = \lim_{k\to\infty} R(k) \leq Z^*$. □

Equations (8.13)–(8.16) also lend themselves to the definition of iterative sequences. Wallace [116] and Neuts [76, 79] originally used

$$R_0 = 0, \tag{8.19}$$

$$R_{n+1} = A_0 + R_n A_1 + R_n^2 A_2, \tag{8.20}$$

which is derived from (8.16). Few of these algorithms have been examined by probabilistic means, the reason being that the successive sets of sample paths have a more complex description than those in Theorem 8.1.1. We refer to Lucantoni [60, Page 26] for a probabilistic proof of the algorithm based on (8.15) (see also Neuts [79, Page 98]).

A detailed comparison of the relative merits of these algorithms remains to be done. One finds in Latouche [48] that the linear progression algorithm (which is called algorithm U there) requires the smallest number of iterations; the actual computation times for a few examples are also reported. This, however, is sketchy. The effect of using parallel architectures is not measured, and the comparisons are made for matrices which are filled. In many applications (and especially so for continuous time models), the matrices A_0, A_1, and A_2 are sparse, although the matrices R, U, and G are filled. For a survey of recent work not reported here, we refer to Bini and Meini [7] and Favati and Meini [19] who exploit FFT and functional relaxation methods.

8.2 Analysis of the Linear Algorithm

We shall assume at several places in this section that the number m of phases is finite. It is customary to measure the efficiency of numerical algorithms by the number of arithmetic operations (also called flops) executed on two floating-point numbers. The sum and the product of two matrices of finite order m cost m^2 and $2m^3 - m^2$ flops, respectively; an expression of the form $G_\bullet = (I - U_\bullet)^{-1}A_2$ is computed by solving the linear system $(I - U_\bullet)X = A_2$, at a cost of $\frac{8}{3}m^3 - \frac{11}{6}m^2 + \frac{1}{6}m$ flops (see Isaacson and Keller [33]). This somewhat depends on the specific implementation.

In summary, the complexity of the linear progression algorithm is approximately $\frac{14}{3}m^3 K_{LP}$, where the number K_{LP} of iterations is the factor of greater interest. It can be shown to be asymptotically equal to $|\log \epsilon|/|\log \eta|$, where ϵ is the precision required in the stopping criterion and η is the maximal eigenvalue of R. Clearly, K_{LP} will be very large in those cases where η is close to 1. The proof given below is based on Latouche and Ramaswami [53]. A purely algebraic proof is given in Meini [66].

In view of (8.3), (8.11), we have that

$$G - G(k) = \mathrm{P}[\gamma(0) < \infty \;\&\; \gamma(0) > \gamma(k+1) \;\&\; X_{\gamma(0)} | X_0 \in \ell(1)],$$

i.e., $G - G(k)$ records the probability of moving down to $\ell(0)$ after moving up to $\ell(k+1)$. We may factor that probability matrix as

$$G - G(k) = F(1, k+1)G^{k+1}, \tag{8.21}$$

where $F(k, k')$ records the probability, starting from $\ell(k)$, of reaching $\ell(k')$ before $\ell(0)$:

$$F(k, k') = \mathrm{P}[\gamma(0) > \gamma(k') \ \& \ X_{\gamma(k')} | X_0 \in \ell(k)]. \tag{8.22}$$

One easily shows by conditioning on $X_{\gamma(k+1)}$ that for $k' > k$,

$$\begin{aligned} F(k, k') &= F(k, k+1)F(k+1, k') \\ &= \prod_{0 \leq i \leq k'-k-1} F(k+i, k+i+1) \\ &= \prod_{k \leq i \leq k'-1} F(i), \end{aligned} \tag{8.23}$$

where $F(i) = F(i, i+1)$ is the first passage probability matrix from $\ell(i)$ to $\ell(i+1)$ under taboo of $\ell(0)$. In particular,

$$F(1, k+1) = \prod_{1 \leq i \leq k} F(i). \tag{8.24}$$

In order to study this product, it is useful to consider another QBD, denoted by $\{\hat{X}_t, t \geq 0\}$, called the level-reversed process. Its transition matrix $\hat{P}$ is given by

$$\hat{P} = \begin{bmatrix} \hat{B}_1 & A_2 & 0 & 0 & \cdots \\ A_0 & A_1 & A_2 & 0 & \cdots \\ 0 & A_0 & A_1 & A_2 & \cdots \\ 0 & 0 & A_0 & A_1 & \cdots \\ \vdots & \vdots & \vdots & \vdots & \ddots \end{bmatrix}, \tag{8.25}$$

so that the events which lead to an increase of level for the original process lead here to a decrease and vice versa. The two processes are intimately related, and we prove below that the sequence $\{F(k)\}$ converges to $\hat{G}$, the matrix G for the level-reversed process.

Lemma 8.2.1 *The matrices* $\{F(k) : k \geq 1\}$ *satisfy the recurrence equations*

$$\begin{aligned} F(1) &= (I - A_1)^{-1} A_0, & (8.26)\\ F(k+1) &= [I - A_1 - A_2 F(k)]^{-1} A_0 \qquad \text{for } k \geq 1. & (8.27) \end{aligned}$$

The sequence is monotone increasing and converges to the matrix $\hat{G}$ *of first passage probabilities from* $\ell(n)$ *to* $\ell(n-1)$ *for the QBD defined by* (8.25). *If the original QBD defined by* (8.1) *is positive recurrent, then the matrix* $\hat{G}$ *is substochastic; if that QBD is null recurrent or transient, then the matrix* $\hat{G}$ *is stochastic. Finally, the matrix* $\hat{G}$ *is the minimal nonnegative solution of the equation*

$$\hat{G} = A_0 + A_1 \hat{G} + A_2 \hat{G}^2 \qquad (8.28)$$

and the spectral radius of $\hat{G}$ *is strictly positive.*

Proof. It is clear that $F(1) = (I - A_1)^{-1} A_0$ since, starting from $\ell(1)$, we have $\gamma(0) > \gamma(2)$ if and only if the first state visited outside of $\ell(1)$ is in $\ell(2)$. For $k \geq 1$, we condition on X_1 and find that

$$\begin{aligned} F(k+1) &= A_0 + A_1 F(k+1) + A_2 F(k, k+2)\\ &= A_0 + A_1 F(k+1) + A_2 F(k) F(k+1) \end{aligned}$$

by (8.23), so that

$$F(k+1) = A_0 + (A_1 + A_2 F(k)) F(k+1).$$

The matrix $A_1 + A_2 F(k)$ is substochastic, since it records the probability of returning to $\ell(k+1)$, starting from $\ell(k+1)$, under taboo of both $\ell(0)$ and $\ell(k+2)$; therefore (8.27) holds.

If we apply Theorem 8.1.1 to the level-reversed process and compute the matrices $\{\hat{G}(k)\}$ which converge to $\hat{G}$, we find after some algebraic manipulations that $F(k) = \hat{G}(k)$ for all $k \geq 1$. Thus the sequence $\{F(k)\}$ monotonically converges to $\hat{G}$, and the remainder of the lemma follows. □

Theorem 8.2.2 *If the number* m *of phases is finite, we have that* $G - G(k) = O(\xi^k \gamma^k)$ *asymptotically as* $k \to \infty$, *where* $\xi = \mathrm{sp}(\hat{G}) \leq 1$ *and* $\gamma = \mathrm{sp}(G) \leq 1$. *If the QBD is positive recurrent or transient, then* $\xi\gamma < 1$.

Proof. The proof is immediate: from (8.21), (8.24) we get

$$\begin{aligned} 0 \le G - G(k) &= \prod_{1 \le i \le k} F(i) G^{k+1} \\ &\le \hat{G}^k G^{k+1} \end{aligned}$$

by Lemma 8.2.1, which shows that $G - G(k) = O(\xi^k \gamma^k)$. The matrices G and $\hat{G}$ being (sub)stochastic, one has that $\xi,\ \gamma \le 1$. They are both stochastic, and $\gamma = \xi = 1$, if and only if the QBD is null recurrent; otherwise, one is substochastic while the other is stochastic and $\xi\gamma$ is strictly less than one. □

We have thus proved that when the QBD is ergodic, the rate of convergence of the linear progression algorithm is governed by the maximal eigenvalue ξ of $\hat{G}$. The following theorem shows that ξ is also equal to the largest eigenvalue of the matrix R, which we denoted by η earlier. We shall show in section 9.1 that η (which has also been called the *caudal characteristic*) is an important indicator of the tail behavior of the QBD. It appears here that the caudal characteristic is related, via the speed of convergence of the linear progression algorithm, to the distribution of the maximum level reached during a first passage time from $\ell(1)$ to $\ell(0)$ (see Remark 8.1.2).

Theorem 8.2.3 *Assume that m is finite and that $A = A_0 + A_1 + A_2$ is irreducible. Then the maximal eigenvalues of $\hat{G}$ and R are the same.*

Proof. If the QBD is transient or null recurrent, then both $\hat{G}$ and R have a maximal eigenvalue equal to one (see Corollary 7.1.2 and Lemma 8.2.1). Therefore, we only need to consider the ergodic case, when $\xi < 1$ and $\eta < 1$. Let us denote, respectively, by $\boldsymbol{x}$ and $\boldsymbol{y}$ two nonnegative vectors such that

$$\hat{G}\boldsymbol{x} = \xi \boldsymbol{x}, \qquad \boldsymbol{y}R = \eta \boldsymbol{y}.$$

We postmultiply (8.28) by $\boldsymbol{x}$, premultiply (6.15) by $\boldsymbol{y}$, and obtain that

$$(A_0 + \xi A_1 + \xi^2 A_2)\boldsymbol{x} = \xi \boldsymbol{x}, \qquad \boldsymbol{y}(A_0 + \eta A_1 + \eta^2 A_2) = \eta \boldsymbol{y}.$$

The matrix $A(x) = A_2 + xA_1 + x^2 A_2$ is irreducible and nonnegative for $x > 0$. By the theorem of Perron–Frobenius, its spectral radius, which we denote by $\chi(x)$, is an eigenvalue, $A(x)$ has a strictly positive left eigenvector $\boldsymbol{u}(x)$, and every other left eigenvector has components of mixed signs (see [25, Remark 3, Section 13.2.5]). Now, η is strictly positive; otherwise, we would have that $R^m = 0$, which would imply that $\boldsymbol{\pi}_n = \mathbf{0}$ for n large enough, in contradiction with our assumption that the QBD is irreducible. Therefore, the equation $\boldsymbol{y}A(\eta) = \eta\boldsymbol{y}$, $\boldsymbol{y} \geq \mathbf{0}$ implies that $\eta = \chi(\eta)$, and η is a solution in (0,1) of the equation $x = \chi(x)$.

By repeating that argument for the equation $A(\xi)\boldsymbol{x} = \xi\boldsymbol{x}$, we find that both ξ and η are solutions in $(0, 1)$ of the equation $z = \chi(z)$. By Neuts [79, Section 1.3], $z = \chi(z)$ has a unique solution in $(0, 1)$, hence $\xi = \eta$. □

Remark 8.2.4 We also define for future reference the matrix $\hat{R}$, the matrix R of the level-reversed process. By Theorem 8.2.3 above, the maximal eigenvalues of G and $\hat{R}$ are equal. Therefore, if $m < \infty$ and the QBD is positive recurrent, then $\mathrm{sp}(R) < 1 = \mathrm{sp}(\hat{R})$; if it is null recurrent, then $\mathrm{sp}(R) = 1 = \mathrm{sp}(\hat{R})$; and if it is transient, then $\mathrm{sp}(\hat{R}) < 1 = \mathrm{sp}(R)$.

8.3 An Improved Algorithm

Instead of starting with $G(0) = 0$, one could try to make the process converge faster by choosing a better initial value for G. In the case where the QBD is recurrent, knowing that G is stochastic, one might choose a stochastic $G(0)$. This indeed speeds up the convergence of the procedure. In the theorem below, we use $G(0) = I$; we shall comment later on this choice.

Theorem 8.3.1 *If m is finite and the QBD is recurrent, then the sequences $\{\bar{U}(k)\}$ and $\{\bar{G}(k)\}$, where*

$$\bar{G}(0) = I, \tag{8.29}$$
$$\bar{U}(k) = A_1 + A_0\bar{G}(k-1), \tag{8.30}$$
$$\bar{G}(k) = (I - \bar{U}(k))^{-1}A_2 \qquad \text{for } k \geq 1, \tag{8.31}$$

are such that $\bar{U}(k)$ is substochastic and $\bar{G}(k)$ is stochastic for all k and $\lim_{k\to\infty} \bar{U}(k) = U$, $\lim_{k\to\infty} \bar{G}(k) = G$. We further have that $\bar{G}(k) \geq G(k)$, $\bar{U}(k) \geq U(k)$ for all k.

Proof. The proof is rather involved, mostly because we need to show that the matrices $\{\bar{G}(k)\}$ are well defined and stochastic for all k.

We consider a *finite* QBD $\{\bar{X}_t, t \geq 0\}$ on the levels $\ell(0) \cup \ell(1) \cup \cdots \cup \ell(k)$ with the transition matrix

$$\bar{P}(k) = \begin{bmatrix} B_1 & A_0 & 0 & \cdots & 0 & 0 \\ A_2 & A_1 & A_0 & \cdots & 0 & 0 \\ 0 & A_2 & A_1 & \cdots & 0 & 0 \\ \vdots & \vdots & \vdots & \ddots & \vdots & \vdots \\ 0 & 0 & 0 & \cdots & A_1 & A_0 \\ 0 & 0 & 0 & \cdots & A_2 & A_1 + A_0 \end{bmatrix}. \tag{8.32}$$

We define $\gamma(0)$ and $\gamma(1)$, respectively, as the first epoch of visit to any state in $\ell(0)$ and as the first epoch of return to $\ell(1)$, and we define the matrices

$$\bar{G}(k) = \mathrm{P}[\gamma(0) < \infty \;\&\; \bar{X}_{\gamma(0)} | \bar{X}_0 \in \ell(1)]$$

and

$$\bar{U}(k) = \mathrm{P}[\gamma(1) < \gamma(0) \;\&\; \bar{X}_{\gamma(1)} | \bar{X}_0 \in \ell(1)]$$

for $k \geq 1$.

If we can prove that $\bar{P}(k)$ is irreducible for all k, then it is a simple matter to prove by induction that $I - \bar{U}(k)$ is nonsingular for all k, that $\bar{G}(k)$ is stochastic, and that $\{\bar{U}(k)\}$ and $\{\bar{G}(k)\}$ satisfy equations (8.29)–(8.31).

Let us assume that k is arbitrary but fixed. In order to prove that $\bar{P}(k)$ is irreducible, we shall show that for every n $(1 \leq n \leq k)$ and i, there exists a path from (n, i) to the level $\ell(n-1)$ immediately below in the Markov chain with the matrix (8.32).

Since the infinite QBD is irreducible, there exists at least one path $s_0 = (n, i)$, $s_1 = (x_1, \varphi_1)$, ..., $s_T = (x_T, \varphi_T)$ of successive states such that $s_T \in \ell(n-1)$. We consider two cases. In the first case, $x_t \leq k$ for all $t \leq T$: the trajectory is wholly included in $\ell(n-1) \cup \cdots \cup \ell(k)$. Then the same path exists in the Markov chain with transition matrix $\bar{P}(k)$, and the claim is proved.

In the second case, there is some t_0 such that $x_{t_0} = k+1$, i.e., the infinite QBD moves to the level $k+1$. Our objective is to construct a new path which is feasible for the Markov process (8.32) beyond t_0.

Naturally, the path $s_0, s_1, \ldots, s_T$ will eventually move down after t_0 and there is a first instant $t_1 \geq t_0$ where $x_{t_1+1} = x_{t_1} - 1$. We define below a time $T_1 < T$ and a new trajectory $s_0^{(1)}, \ldots, s_{T_1}^{(1)}$ so that the phase of $s_t^{(1)}$ is the same as the phase of s_t for all t; only in the level will there be differences:

- For $t \leq t_0 - 1$, we set $x_t^{(1)} = x_t$: the two paths coincide as long as $\{s_t\}$ remains below the level $k+1$.
- For $t_0 \leq t \leq t_1$, we set $x_t^{(1)} = k$: during that interval of time, the infinite QBD moves steadily upward in the levels beyond $\ell(k)$ and the new path is forced to remain in the level k, which is possible with the transition matrix $\bar{P}(k)$.
- For $t_1 + 1 \leq t \leq T$, we set $x_t^{(1)} = x_t - x_{t_1} + k$: this means that we allow for the new trajectory the same up and down movements as for the original path.

Clearly, $x_t^{(1)} < x_t$ for $t \geq t_1 + 1$ and there is a smallest index $T_1 < T$ such that $x_{T_1}^{(1)} = n - 1$. The path $s_0^{(1)}, s_1^{(1)}, \ldots, s_{T_1}^{(1)}$ with $s_t^{(1)} = (x_t^{(1)}, \varphi_t)$ is feasible for (8.32) up to time $t_1 + 1$ and is feasible for the infinite QBD after $t_1 + 1$ and up to T_1.

If $x_t^{(1)} \leq k$ between $t_1 + 1$ and T_1, then the new trajectory is wholly feasible for the truncated process and we are done. Otherwise, we repeat the procedure and define new trajectories $\{s_t^{(2)}\}$, $\{s_t^{(3)}\}$, etc. Since we strictly increase at each iteration the length of the path until the first visit to $\ell(k+1)$ and reduce the total length of the path, this may not be pursued ad infinitum and we eventually obtain a feasible path for (8.32).

We finally prove that the sequence $\{\bar{G}(k)\}$ converges to G, which will also prove that the sequence $\{\bar{U}(k)\}$ converges to U. Since $\bar{G}(k)\mathbf{1} = \mathbf{1}$ for all k, the matrices remain in the compact set of stochastic matrices and the sequence has at least one accumulation point. We denote by G^* one such point and choose a subset of indices $\{k_1, k_2, \ldots\}$

Modified Boundary Algorithm
$G_\bullet := (I - A_1 - A_0)^{-1} A_2$;
repeat
 $G_{\text{old}} := G_\bullet$;
 $U_\bullet := A_1 + A_0 G_\bullet$;
 $G_\bullet := (I - U_\bullet)^{-1} A_2$
until $\| G_\bullet - G_{\text{old}} \|_\infty \leq \epsilon$

Figure 8.2: An alternative, faster algorithm for the evaluation of G when the QBD is recurrent.

---

such that $\lim_{i\to\infty} \bar{G}(k_i) = G^*$. By Theorem 8.1.1, we also have that $\lim_{i\to\infty} G(k_i) = G$ and we necessarily have that $G^*\mathbf{1} = G\mathbf{1} = \mathbf{1}$.

We note that $\bar{G}(0) \geq G(0)$, and we prove by induction that $\bar{G}(k) \geq G(k)$ for all k. Indeed, if $\bar{G}(k) \geq G(k)$ for some k, then

$$\bar{U}(k+1) = A_1 + A_0\bar{G}(k) \geq A_1 + A_0 G(k) = U(k+1)$$

and

$$\begin{aligned}\bar{G}(k+1) &= (I - \bar{U}(k+1))^{-1} A_2 = \sum_{\nu\geq 0} (\bar{U}(k+1))^\nu A_2 \\ &\geq \sum_{\nu \geq 0} (U(k+1))^\nu A_2 = G(k+1),\end{aligned}$$

so that $G^* \geq G$. Since $G \geq 0$ and $G^*\mathbf{1} = G\mathbf{1} = \mathbf{1}$, $G^* = G$. Thus, the sequence $\{\bar{G}(k)\}$ may have only one accumulation point, equal to G, which proves the theorem. □

We now have a second algorithm to compute the matrix G, when it is stochastic, based on equations (8.29)–(8.31). That algorithm[14] is given in Figure 8.2.

---

[14]We call this the modified boundary algorithm in reference to the probabilistic interpretation given in Remark 8.3.2: the transition probabilities are modified at the upper boundary.

The array $G_\bullet$ is initially assigned the value $\bar{G}(1)$, and during the kth iteration through the loop, the arrays $U_\bullet$, $G_\bullet$, and G_{old} are assigned the values $\bar{U}(k+1)$, $\bar{G}(k+1)$, and $\bar{G}(k)$, respectively. The stopping criterion, based on the comparison of two successive approximations, is not very good, as we have discussed earlier. We may not, unfortunately, use the same criterion as in the linear progression algorithm, since here every computed approximation of G is stochastic.

Remark 8.3.2 Since the transition matrices $\bar{P}(k)$ converge to the matrix P of (8.1) as k tends to infinity, one does expect that $\{\bar{G}(k)\}$ converges to G. This is akin to the procedure known as augmented truncation, used to compute an approximation of the stationary vector $\boldsymbol{\pi}$ for an infinite Markov chain: one isolates the upper-left quadrant $P_{[N]}$, of size N, of the infinite matrix, augments some elements of $P_{[N]}$ in order to obtain a stochastic matrix $\bar{P}_{[N]}$, solves the linear system $\boldsymbol{\pi}_{[N]}\bar{P}_{[N]} = \boldsymbol{\pi}_{[N]}, \boldsymbol{\pi}_{[N]}\mathbf{1} = 1$, and hopes that, as N tends to infinity, the vectors $\boldsymbol{\pi}_{[N]}$ converge to $\boldsymbol{\pi}$. This, however, is not always true (see Gibson and Seneta [27]), the fundamental reason being that the processes have different transition structures for different values of N. Hence our obligation in Theorem 8.3.1 to prove the convergence of the sequence $\{\bar{G}(k)\}$ to G.

In contrast, the sequence $\{G(k)\}$ of (8.7)–(8.9) results from an application of truncation *without* augmentation, since $G(k)$ records the first passage probabilities from $\ell(1)$ to $\ell(0)$ for the substochastic matrix

$$P_{[k]} = \begin{bmatrix} B_1 & B_0 & 0 & \cdots & 0 & 0 \\ A_2 & A_1 & A_0 & \cdots & 0 & 0 \\ 0 & A_2 & A_1 & \cdots & 0 & 0 \\ \vdots & \vdots & \vdots & \ddots & \vdots & \vdots \\ 0 & 0 & 0 & \cdots & A_1 & A_0 \\ 0 & 0 & 0 & \cdots & A_2 & A_1 \end{bmatrix}.$$

One may select different ways of augmenting the matrix $P_{[N]}$, thereby obtaining different matrices $\bar{P}_{[N]}$; here, $\bar{G}(0) = I$ is an arbitrary choice. If we have reasons to believe that G has some particular structural property, we might impose this constraint on $\bar{G}(0)$. The only precaution that one should take is that the truncated/augmented transition matrix should not be reducible.

In order to show that some precaution is necessary, we consider one example with $m = 2$,

$$\begin{aligned} A_0 &= \begin{bmatrix} a_{0,1} & a_{0,2} \\ a_{0,3} & a_{0,4} \end{bmatrix}, \\ A_1 &= \begin{bmatrix} a_{1,1} & 0 \\ a_{1,3} & a_{1,4} \end{bmatrix}, \\ A_2 &= \begin{bmatrix} 0 & 0 \\ a_{2,3} & a_{2,4} \end{bmatrix}. \end{aligned}$$

If

$$\bar{G}(0) = \begin{bmatrix} 1 & 0 \\ 1 & 0 \end{bmatrix},$$

then

$$\bar{U}(1) = \begin{bmatrix} 1 & 0 \\ \bar{U}_3(1) & \bar{U}_4(1) \end{bmatrix},$$

so that $(I - \bar{U}(1))$ is singular.

Remark 8.3.3 The modified boundary algorithm is called the algorithm U* in Latouche [48] and Latouche and Ramaswami [53]. The algorithms described in Grassmann and Heyman [29] and Kao [35] are essentially equivalent to it.

Remark 8.3.4 Instead of using the probabilistic interpretation given above, we may characterize the sequence $\{\bar{G}(k)\}$ as follows: each matrix in the sequence has the same Perron–Frobenius eigenvalue and right eigenvector as the matrix G. As mentioned above, one may define sequences $\{R(k)\}$ which converge to R; since it is possible to evaluate the Perron–Frobenius eigenvalue and left eigenvector of R (see section 9.1), one may also define sequences $\{\bar{R}(k)\}$ which share these eigencharacteristics. The interested reader is referred to Neuts [79, 81] or Latouche [48].

The complexity of the modified boundary algorithm is approximately equal to $\frac{14}{3} K_{MB} m^3$ flops, where K_{MB} denotes the number of

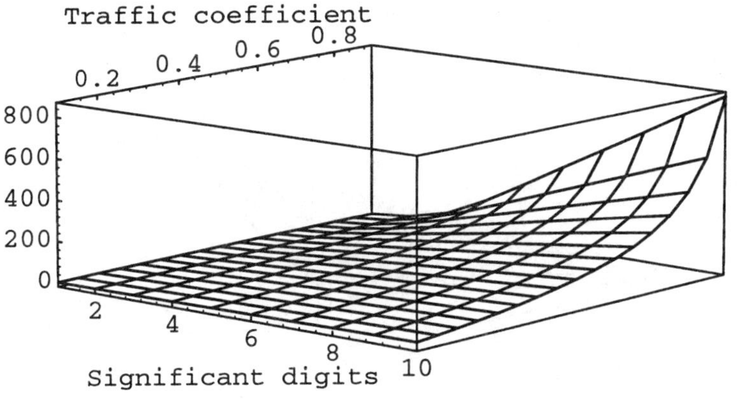

Figure 8.3: Number K_{LP} of iterations of the linear progression algorithm for various values of the traffic coefficient ρ and the required precision ϵ. The number of significant digits is defined as $|\log_{10}\epsilon|$. The QBD is an M/M/1 queue in a random environment.

iterations. Numerical experiments reported in the literature indicate that K_{MB} is much smaller than the number K_{LP} of iterations of the linear progression algorithm, so that the modified boundary algorithm is to be preferred (see Neuts [81], Latouche [48], and Latouche and Ramaswami [53]). Meini [66] shows that the asymptotic rate of convergence is equal to the second largest eigenvalue of the matrix $G \otimes R$, of which the dominant eigenvalue is η. Thus, unless η is a multiple eigenvalue of $G \otimes R$, we see on theoretical grounds that the modified boundary algorithm is asymptotically faster than the linear progression algorithm.

In order to illustrate the behavior of the two algorithms, we present a few experimental results in Figures 8.3–8.5. We have applied the two algorithms to an M/M/1 queue in a random environment; that queue will be defined and discussed in greater detail in section 9.1 (case 1: constant service rates); suffice it to say that η is very close to 1, even for moderate values of the traffic coefficient. We give in Figures 8.3 and 8.4 the numbers K_{LP} and K_{MB} of iterations for various values of ϵ and of the traffic coefficient ρ. Since η is quite high, we expect that

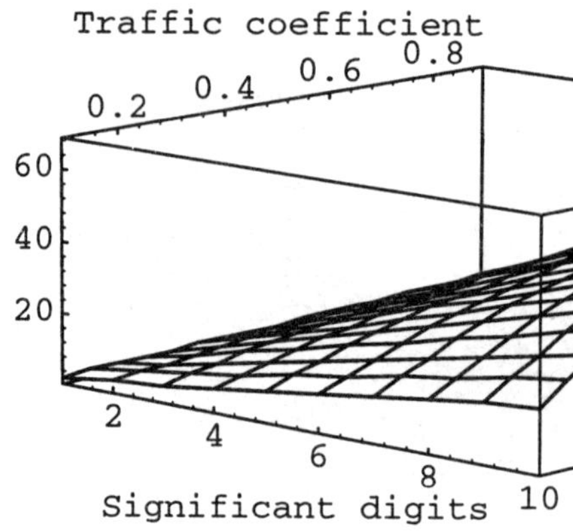

Figure 8.4: Number K_{MB} of iterations of the modified boundary algorithm for various values of the traffic coefficient ρ and the required precision ϵ. The number of significant digits is defined as $|\log_{10}\epsilon|$. The QBD is an M/M/1 queue in a random environment.

K_{LP} will be very large when ρ is close to 1, this is clearly confirmed by Figure 8.3. In contrast, we observe that K_{MB} increases much more slowly as ρ approaches 1 or ϵ becomes very small.

In order to measure the accuracy of the modified boundary algorithm, we have compared ϵ and the actual error on the approximation $\bar{G}$. The error is estimated by $\| G^* - \bar{G} \|_\infty$, where G^* is an approximation of G, computed by the algorithm of the next section, and such that $\| G - G^* \|_\infty \leq 10^{-12}$, so that we expect to have $\| G^* - \bar{G} \|_\infty$ approximately equal to $\| G - \bar{G} \|_\infty$. We show in Figure 8.5 the ratio $r = \| G^* - \bar{G} \|_\infty / \epsilon$. As we suspected, $\| G - \bar{G} \|_\infty$ is not always less than ϵ. The main factor appears to be the traffic coefficient: when the QBD is more nearly *null* recurrent, convergence to G is not quite ensured by the stopping criterion.

The numerical stability of these algorithms has hardly been investigated, if at all. One expects them to be stable over a large range of QBDs since only the computation of $(I - U_\bullet)^{-1}A_2$ might require some care if $I - U_\bullet$ should happen to be nearly singular. This, however, would indicate that the QBD has a tendency to remain in the same level

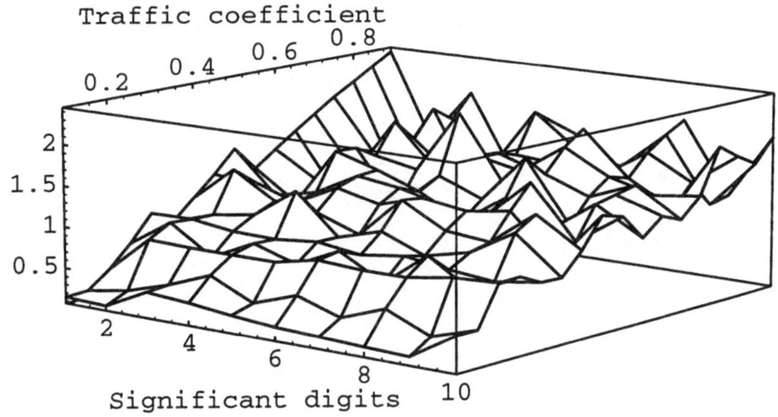

Figure 8.5: Precision attained by the modified boundary algorithm, measured by the ratio $||G - \bar{G}||_\infty/\epsilon$, for various values of the traffic coefficient ρ and the required precision ϵ. The number of significant digits is defined as $|\log_{10}\epsilon|$. The QBD is an M/M/1 queue in a random environment.

---

for extraordinarily long periods of time and that is not likely to occur very often in practice. The other operations do not involve negative quantities so that there is no risk of catastrophic cancelations.

8.4 Quadratically Convergent Algorithms

In order to improve the rate of convergence, it is a standard practice to consider Newton's sequence. Indeed, this is shown in Latouche [49] to be quadratically and monotonically convergent with a very small number of iterations needed to reach $\| G - \tilde{G}(k) \|_\infty \leq \epsilon$ (see also Neuts [75] and Ramaswami [93]). Unfortunately, the complexity of each iteration is $O(m^6)$ instead of $O(m^3)$, so that Newton's sequence is actually slower than the two algorithms discussed above (some computational experience is reported in [49]).

Another quadratically convergent algorithm with $O(m^3)$ complexity at each iteration is based on a direct representation of the matrix G.

We start from the equation $G = A_0 + A_1 G + A_2 G^2$, which we transform as

$$(I - A_1)G = A_2 + A_0 G^2$$

or

$$G = (I - A_1)^{-1} A_2 + (I - A_1)^{-1} A_0 G^2,$$

and which we write as

$$G = L + HG^2 \tag{8.33}$$

with

$$L = (I - A_1)^{-1} A_2, \qquad H = (I - A_1)^{-1} A_0. \tag{8.34}$$

This equation[15] has the following interpretation.

We denote by $\{Y_n : n \geq 0\}$ the restricted process obtained by observing the QBD $\{X_n : n \geq 0\}$ only at those epochs where a change of level occurs. The new process is a QBD also, and its transition matrix is given by

$$\widetilde{P} = \begin{bmatrix} 0 & H_0 & 0 & 0 & \cdots \\ L & 0 & H & 0 & \cdots \\ 0 & L & 0 & H & \cdots \\ 0 & 0 & L & 0 & \cdots \\ \vdots & \vdots & \vdots & \vdots & \ddots \end{bmatrix},$$

where H and L are defined in (8.34) and $H_0 = (I - B)^{-1} A_0$. The equation (8.33) states that the matrix G is the same for the two processes.

The first term in the right-hand side of (8.33) is

$$L = \mathrm{P}[\gamma(0) < \gamma(2) \;\&\; X_{\gamma(0)} | X_0 \in \ell(1)];$$

it records the probability that the QBD moves down to $\ell(0)$ immediately upon the first exit from $\ell(1)$. The second term is

$$\begin{aligned} HG^2 &= \mathrm{P}[\gamma(2) < \gamma(0) \;\&\; X_{\gamma(2)} | X_0 \in \ell(1)] \\ &\quad \mathrm{P}[\gamma(0) < \infty \;\&\; X_{\gamma(0)} | X_0 \in \ell(2)] \\ &= \mathrm{P}[\gamma(2) < \gamma(0) < \infty \;\&\; X_{\gamma(0)} | X_0 \in \ell(1)] \end{aligned}$$

[15] Incidentally, it can be proved that the matrix G is the minimal nonnegative solution of this equation also.

by the strong Markov property; i.e., it records the probability that the QBD reaches up to the level 2 before eventually moving down to $\ell(0)$.

We might attempt to directly compute G^2, so that the matrix G would be explicitly given by (8.33). Since the matrix G^2 records the first passage probabilities across *two* levels, we are led to define the censored process $\{X_n^{(1)} : n \geq 0\}$, which is $\{Y_n : n \geq 0\}$ observed when it visits one of the even-numbered levels. This is also a QBD, and its transition matrix is readily seen to be

$$P^{(1)} = \begin{bmatrix} B_1^{(1)} & B_0^{(1)} & 0 & 0 & \cdots \\ A_2^{(1)} & A_1^{(1)} & A_0^{(1)} & 0 & \cdots \\ 0 & A_2^{(1)} & A_1^{(1)} & A_0 & \cdots \\ 0 & 0 & A_2^{(1)} & A_1^{(1)} & \cdots \\ \vdots & \vdots & \vdots & \vdots & \ddots \end{bmatrix}$$

with

$$A_2^{(1)} = L^2, \qquad A_1^{(1)} = HL + LH, \qquad A_0^{(1)} = H^2,$$

and

$$B_1^{(1)} = H_0 L, \qquad B_0^{(1)} = H_0 H.$$

If we denote by $G^{(1)}$ the matrix of first passage probabilities to lower levels for the process $\{X_n^{(1)} : n \geq 0\}$, we find after repeating the argument at the beginning of this section that

$$G^{(1)} = L^{(1)} + H^{(1)}[G^{(1)}]^2, \tag{8.35}$$

where

$$L^{(1)} = (I - A_1^{(1)})^{-1} A_2^{(1)}, \qquad H^{(1)} = (I - A_1^{(1)})^{-1} A_0^{(1)}.$$

Since $G^{(1)} = G^2$, it results from (8.33), (8.35) that

$$G = L + HL^{(1)} + HH^{(1)} G^4. \tag{8.36}$$

The first term, as we saw earlier, is

$$L = \mathrm{P}[\gamma(0) < \gamma(2) \ \& \ X_{\gamma(0)} | X_0 \in \ell(1)]$$

and records the probability of reaching $\ell(0)$ before $\ell(2)$. The second term is

$$\begin{aligned} HL^{(1)} &= \mathrm{P}[\gamma(2) < \gamma(0) \ \& \ X_{\gamma(2)} | X_0 \in \ell(1)] \\ &\quad\ \mathrm{P}[\gamma(0) < \gamma(4) \ \& \ X_{\gamma(0)} | X_0 \in \ell(2)] \\ &= \mathrm{P}[\gamma(2) < \gamma(0) < \gamma(4) \ \& \ X_{\gamma(0)} | X_0 \in \ell(1)] \end{aligned}$$

and records the probability of moving to $\ell(0)$ *after* reaching $\ell(2)$ but *before* reaching $\ell(4)$. The last term is

$$\begin{aligned} HH^{(1)}G^4 &= \mathrm{P}[\gamma(2) < \gamma(0) \ \& \ X_{\gamma(2)} | X_0 \in \ell(1)] \\ &\quad\ \mathrm{P}[\gamma(4) < \gamma(0) \ \& \ X_{\gamma(4)} | X_0 \in \ell(2)] \\ &\quad\ \mathrm{P}[\gamma(0) < \infty \ \& \ X_{\gamma(0)} | X_0 \in \ell(4)] \\ &= \mathrm{P}[\gamma(4) < \gamma(0) < \infty \ \& \ X_{\gamma(0)} | X_0 \in \ell(1)] \end{aligned}$$

and records the probability of moving down to $\ell(0)$ after having moved up to $\ell(4)$.

This motivates us to examine the censored process restricted to the levels $\ell(0)$, $\ell(4)$, $\ell(8)$, ... Repeating the procedure ad infinitum, we eventually obtain that

$$G = \sum_{k \geq 0} \mathrm{P}[\delta(k) < \gamma(0) < \delta(k+1) \ \& \ X_{\gamma(0)} | X_0 \in \ell(1)], \tag{8.37}$$

where $\{\delta(k) : k \geq 0\}$ is the sequence of first passage times to the levels which are powers of 2:

$$\delta(k) = \gamma(2^k) = \inf\{t \geq 0 : X_t \in \ell(2^k)\}. \tag{8.38}$$

To show why (8.37) holds, we note that $\delta(0) = 0 < \delta(1) < \delta(2), \ldots$, that $\delta(k) < \infty$ a.s. for all k, and that $\lim_{k \to \infty} \delta(k) = \infty$. Thus $\gamma(0)$ is finite if and only if it is in one of the intervals $(\delta(i), \delta(i+1))$.

We show in the next theorem how to determine recursively the successive terms in the right-hand side of (8.37). For $k = 0$, the empty product in (8.39) is, by convention, equal to the identity matrix.

Theorem 8.4.1 *The matrix G is given by*

$$G = \sum_{k \geq 0} \left(\prod_{0 \leq i \leq k-1} H^{[i]} \right) L^{[k]}, \tag{8.39}$$

where

$$\begin{aligned} H^{[0]} &= (I - A_1)^{-1} A_0, && (8.40) \\ L^{[0]} &= (I - A_1)^{-1} A_2, && (8.41) \\ H^{[k+1]} &= (I - U^{[k]})^{-1} (H^{[k]})^2, && (8.42) \\ L^{[k+1]} &= (I - U^{[k]})^{-1} (L^{[k]})^2 && (8.43) \end{aligned}$$

for $k \geq 0$ *with*

$$U^{[k]} = H^{[k]} L^{[k]} + L^{[k]} H^{[k]}. \tag{8.44}$$

Proof. We first show that

$$H^{[k]} = \mathrm{P}[\delta(k+1) < \gamma(0) \ \& \ X_{\delta(k+1)} | X_0 \in \ell(2^k)] \tag{8.45}$$

and that

$$L^{[k]} = \mathrm{P}[\gamma(0) < \delta(k+1) \ \& \ X_{\gamma(0)} | X_0 \in \ell(2^k)]. \tag{8.46}$$

These equations clearly hold for $k = 0$, and we make the induction assumption that they hold for some $k \geq 0$. We observe that, since the transition probabilities are level-independent, we also have

$$H^{[k]} = \mathrm{P}[\gamma(n + 2^k) < \gamma(n - 2^k) \ \& \ X_{\gamma(n+2^k)} | X_0 \in \ell(n)]$$

and

$$L^{[k]} = \mathrm{P}[\gamma(n - 2^k) < \gamma(n + 2^k) \ \& \ X_{\gamma(n-2^k)} | X_0 \in \ell(n)]$$

for all $n > 2^k$. Now, if we define the matrix M as

$$M = \mathrm{P}[\gamma(0) < \gamma(2^{k+2}) \ \& \ X_{\gamma(0)} | X_0 \in \ell(2^{k+1})]$$

and if we consider the state of the QBD after level displacements of amplitude 2^k, we then find by a simple argument that

$$\begin{aligned} M &= L^{[k]} L^{[k]} + L^{[k]} H^{[k]} M + H^{[k]} L^{[k]} M \\ &= (L^{[k]})^2 + U^{[k]} M, \end{aligned}$$

where $U^{[k]}$ is defined by (8.44) and records the probability that, starting from $\ell(2^{k+1})$, the QBD returns to that level after visiting either $\ell(2^k)$ or $\ell(2^{k+1} + 2^k)$ but before visiting either $\ell(0)$ or $\ell(2^{k+2})$.

Since the QBD is irreducible, $U^{[k]}$ is substochastic and $(I - U^{[k]})$ is nonsingular, so that $M = (I - U^{[k]})^{-1}(L^{[k]})^2 = L^{[k+1]}$ by (8.43), which proves (8.46). The proof of (8.45) is similar.

The second step in our proof is to show that

$$\prod_{0\le i\le k} H^{[i]} = \mathrm{P}[\delta(k+1) < \gamma(0) \;\&\; X_{\delta(k+1)} | X_0 \in \ell(1)] \tag{8.47}$$

and that

$$\prod_{0\le i\le k-1} H^{[i]} L^{[k]} = \mathrm{P}[\delta(k) < \gamma(0) < \delta(k+1) \;\&\; X_{\gamma(0)} | X_0 \in \ell(1)], \tag{8.48}$$

so that (8.39) will result immediately from (8.37). These equations clearly hold for $k = 0$, since $\prod_{0\le i\le -1} H^{[i]} = I$ by convention. Assuming that they hold for some $k \ge 0$ and conditioning on the state at time $\delta(k+1)$, we find that

$$\begin{aligned}
&\mathrm{P}[\delta(k+1) < \gamma(0) < \delta(k+2) \;\&\; X_{\gamma(0)} | X_0 \in \ell(1)] \\
&\quad = \prod_{0\le i\le k} H^{[i]} \mathrm{P}[\gamma(0) < \delta(k+2) \;\&\; X_{\gamma(0)} | \delta(k+1) < \gamma(0) \;\&\; X_{\delta(k+1)}] \\
&\quad = \prod_{0\le i\le k} H^{[i]} \mathrm{P}[\gamma(0) < \delta(k+2) \;\&\; X_{\gamma(0)} | X_0 \in \ell(2^{k+1})]
\end{aligned}$$

by the strong Markov property. From this last equation and (8.46), we immediately obtain (8.48). Equation (8.47) is proved by a similar argument. □

The series (8.39) may serve to define yet another converging sequence $\{G^*(k), k \ge 0\}$, where

$$G^*(k) = \sum_{0\le j\le k} \left(\prod_{0\le i\le j-1} H^{[i]} \right) L^{[j]}, \tag{8.49}$$

from which, when $m < \infty$, one can derive iterative algorithms such as the one[16] given in Figure 8.6. The array $G_\bullet$ is initially assigned the value of $G^*(0)$. One can easily verify that during the kth iteration of

[16] We call this the logarithmic reduction algorithm because its number of iterations is equal to the logarithm of the number of iterations of the linear progression algorithm.

Logarithmic Reduction Algorithm
$H^\bullet := (I - A_1)^{-1}A_0$;
$L^\bullet := (I - A_1)^{-1}A_2$;
$G_\bullet := L^\bullet$;
$T := H^\bullet$;
repeat
  $U^\bullet := H^\bullet L^\bullet + L^\bullet H^\bullet$;
  $M := (H^\bullet)^2$;
  $H^\bullet := (I - U^\bullet)^{-1}M$;
  $M := (L^\bullet)^2$;
  $L^\bullet := (I - U^\bullet)^{-1}M$;
  $G_\bullet := G_\bullet + TL^\bullet$;
  $T := TH^\bullet$
until $\| \mathbf{1} - G_\bullet \mathbf{1} \|_\infty \leq \epsilon$

Figure 8.6: A quadratically convergent algorithm for the evaluation of G. The stopping criterion is chosen under the assumption that the QBD is recurrent.

---

the loop, the arrays $U^\bullet$, $H^\bullet$, $L^\bullet$, $G_\bullet$, and T are, respectively, assigned the values $U^{[k]}$, $H^{[k]}$, $L^{[k]}$, $G^*(k)$, and $\prod_{0 \leq i \leq k} H^{[i]}$.

One readily observes that if the QBD is either positive recurrent or transient, then the sequence $\{G^*(k), k \geq 0\}$ is *quadratically* convergent to G. Indeed, we have that $G^*(k) = G(2^{k+1}-1)$ by (8.11), (8.38), (8.48), (8.49). By Theorems 8.2.2 and 8.2.3, we may write that $G - G^*(k) = O((\eta\gamma)^{2^{n+1}})$, where η and γ are, respectively, the maximal eigenvalues of R and G and $\eta\gamma < 1$.

A point worth mentioning is that one may expect the logarithmic reduction algorithm to be numerically well behaved. In particular, we show below that the matrices $U^{[k]}$ converge to zero, as k increases to infinity, so that there is no problem in solving (8.42), (8.43).

Theorem 8.4.2 *If the QBD is ergodic, then*

$$\lim_{k\to\infty} H^{[k]} = 0, \qquad \lim_{k\to\infty} L^{[k]} = \mathbf{1}\cdot \boldsymbol{g},$$

where $\boldsymbol{g}$ *is the stationary probability vector of* G.

If the QBD is transient, then

$$\lim_{k\to\infty} H^{[k]} = \mathbf{1}\cdot \hat{\boldsymbol{g}}, \qquad \lim_{k\to\infty} L^{[k]} = 0,$$

where $\hat{\boldsymbol{g}}$ *is the stationary probability vector of* $\hat{G}$.

In both cases, we have that

$$\lim_{k\to\infty} U^{[k]} = \lim_{k\to\infty} (H^{[k]}L^{[k]} + L^{[k]}H^{[k]}) = 0.$$

Proof. First we consider the ergodic case. We may rewrite (8.45) as follows:

$$\begin{aligned} H^{[k]} &= \mathrm{P}[\gamma(2^{k+1}) < \gamma(0) \ \& \ X_{\gamma(2^{k+1})} | X_0 \in l(2^k)] \\ &= F(2^k, 2^{k+1}) \qquad \text{by (8.22)} \\ &= \prod_{i=1}^{2^k} F(2^k + i - 1) \qquad \text{by (8.23)} \\ &\leq \hat{G}^{2^k} \end{aligned}$$

by Lemma 8.2.1. If the original QBD is ergodic, then $\hat{G}$ is substochastic and we have that

$$0 \leq \lim_{k\to\infty} H^{[k]} \leq \lim_{k\to\infty} \hat{G}^k = 0,$$

which proves the first statement.

We have seen earlier that $H^{[k]} + L^{[k]}$ is stochastic for all k. This implies that $\{L^{[k]} : k \geq 1\}$ is a bounded sequence and therefore it has accumulation points. Let us denote by L^* one such accumulation point. The second statement of the theorem will be proved once we show that $L^* = \mathbf{1}\cdot \boldsymbol{g}$ is the only possibility.

By an argument similar to the one we used above, we conclude that

$$\begin{aligned}
L^{[k]} &= \mathrm{P}[\gamma(0) < \gamma(2^{k+1}) \ \&\ X_{\gamma(0)} | X_0 \in l(2^k)] \\
&= \prod_{i=1}^{2^k-1} \mathrm{P}[\gamma(2^k - i) < \gamma(2^{k+1}) \ \&\ X_{\gamma(2^k-i)} | X_0 \in l(2^k - i + 1)] \\
&\quad \times \mathrm{P}[\gamma(0) < \gamma(2^{k+1}) \ \&\ X_{\gamma(0)} | X_0 \in l(1)] \\
&= \prod_{i=1}^{2^k-1} \mathrm{P}[\gamma(0) < \gamma(2^{k+1} - 2^k + i) \ \&\ X_{\gamma(0)} | X_0 \in l(1)] \\
&\quad \times \mathrm{P}[\gamma(0) < \gamma(2^{k+1}) \ \&\ X_{\gamma(0)} | X_0 \in l(1)]
\end{aligned}$$

by the spatial homogeneity of the transition matrix; hence,

$$\begin{aligned}
L^{[k]} &= \prod_{i=1}^{2^k} G(2^k + i) \\
&\leq G^{2^k}
\end{aligned}$$

by Theorem 8.1.1.

Since the QBD is ergodic, G is stochastic and $\lim_{k\to\infty} G^k = \mathbf{1} \cdot \boldsymbol{g}$. We then conclude that $L^{[k]} < (1 + \epsilon)\mathbf{1} \cdot \boldsymbol{g}$ for k large enough and for any $\epsilon > 0$. This implies that $L^* \leq (1 + \epsilon)\mathbf{1} \cdot \boldsymbol{g}$ for any $\epsilon > 0$ and hence that $L^* \leq \mathbf{1} \cdot \boldsymbol{g}$.

Furthermore, L^* is stochastic since $H^{[k]} + L^{[k]}$ is stochastic for all k and $\lim_{k\to\infty} H^{[k]} = 0$. Therefore, L^* is necessarily equal to $\mathbf{1} \cdot \boldsymbol{g}$ and the second statement is proved.

The case when the original QBD is transient is treated by using the relation between the original and the level-reversed QBD. We first define the first passage times $\hat{\gamma(0)} = \inf\{t \geq 1 : \hat{X}_t \in l(0)\}$ and $\hat{\delta}_k = \inf\{t \geq 0 : \hat{X}_t \in l(2^k)\}$.

We then define the matrices

$$\begin{aligned}
\hat{H}^{[k]} &= \mathrm{P}[\hat{\delta}_{k+1} < \hat{\gamma(0)} \ \&\ \hat{X}_{\hat{\delta}_{k+1}} | \hat{X}_0 \in l(2^k)], \\
\hat{L}^{[k]} &= \mathrm{P}[\hat{\gamma(0)} < \hat{\delta}_{k+1} \ \&\ \hat{X}_{\hat{\gamma(0)}} | \hat{X}_0 \in l(2^k)].
\end{aligned}$$

Since the original QBD is transient, the reverse process is ergodic and we may repeat the preceding argument, obtaining that

$$\lim_{k\to\infty} \hat{H}^{[k]} = 0, \qquad \lim_{k\to\infty} \hat{L}^{[k]} = \mathbf{1} \cdot \hat{\boldsymbol{g}}.$$

Simple algebraic manipulations suffice to prove that $H^{[k]} = \hat{L}^{[k]}$ and that $L^{[k]} = \hat{H}^{[k]}$, which concludes the proof. □

The complexity of the logarithmic reduction algorithm (when m is finite) is given by $\frac{50}{3}\lfloor \log_2 K_{LP} \rfloor m^3$, where K_{LP} is the number of iterations of the linear reduction algorithm. In order to see this, we consider a fixed value of ϵ and we denote by K_{LR} the number of iterations of the logarithmic reduction algorithm. Then, recalling that the sequence $\{G(k), k \geq 1\}$ is monotone (see Theorem 8.1.1) and that $G^*(k) = G(2^{k+1})$, we write that

$$G(K_{LP} - 1)\mathbf{1} < 1 - \epsilon \leq G(K_{LP})\mathbf{1}$$

and

$$G(2^{K_{LR}} - 1)\mathbf{1} < 1 - \epsilon \leq G(2^{K_{LR}+1} - 1)\mathbf{1}$$

from which we conclude that $K_{LR} \leq \lfloor \log_2 K_{LP} \rfloor$.

Remark 8.4.3 Recall from Theorems 8.2.2 and 8.2.3 that, in the case of a positive recurrent QBD, the difference $G - G^*(k)$ is of the order of $\eta^{2^{k+1}}$, where $\eta < 1$ is the maximal eigenvalue of R. Therefore, K_{LR} is approximately equal to $\log_2 |\log \epsilon| - \log_2 |\log \eta|$. In floating-point arithmetic, ϵ and η are, respectively, bounded away from zero and from one, so that the number of iterations K_{LR} is effectively bounded for all practical purposes.

The logarithmic reduction algorithm is clearly superior to the linear progression algorithm. It is asymptotically faster than the modified boundary algorithm since the latter is linearly convergent. There is, however, no clear relationship between the numbers K_{MB} and K_{LR} of iterations. A few comparisons made in Latouche and Ramaswami [53] indicate that $K_{LR} < K_{MB}$, although the actual execution times may be close due to the larger constant $\frac{50}{3}$.

We may also write for U and R expressions like (8.39) for G. We give below one for R, to which we shall refer in Chapter 15. The proof is in Latouche and Ramaswami [53].

Theorem 8.4.4 *The matrix R is given by*

$$R = \sum_{k \geq 0} C_0^{[k]} \left(\prod_{0 \leq i \leq k-1} C_2^{[k-1-i]} \right),$$

where

$$\begin{aligned} C_0^{[0]} &= A_0(I - A_1)^{-1}, \\ C_2^{[0]} &= A_2(I - A_1)^{-1}, \\ C_i^{[k+1]} &= (C_i^{[k]})^2(I - C_0^{[k]}C_2^{[k]} - C_2^{[k]}C_0^{[k]})^{-1} \end{aligned}$$

for $k \geq 0,\ i = 1, 2.$

8.5 Special Structure

The algorithms described in the preceding sections are quite general, but there exist models for which they may be considerably improved upon because it is possible to use some special property of the blocks A_0, A_1, and A_2. The most striking cases are when A_0 or A_2 is of rank 1, these are discussed below. Other interesting cases are when it is known a priori that the matrix G has a number of columns which are identically zero (see Chandramouli, Neuts, and Ramaswami [13], for instance) or when it has some repetitive structure which may be exploited, as in Ramaswami and Lucantoni [99] and Gail, Hantler, and Taylor [24].

For instance, if the matrix A_2 is the product $A_2 = \boldsymbol{c} \cdot \boldsymbol{r}$ of a column vector $\boldsymbol{c}$ by a row vector $\boldsymbol{r}$, as for the M/PH/1 queue and for the queue with priority classes (sections 1.3 and 1.4), then the following property holds.

Theorem 8.5.1 *Assume that the QBD is recurrent and that* $A_2 = \boldsymbol{c} \cdot \boldsymbol{r}$, *where* $\boldsymbol{c}$ *is a column vector and* $\boldsymbol{r}$ *is a row vector normalized by* $\boldsymbol{r}\mathbf{1} = 1$. *Then the matrix* G *is given by* $G = \mathbf{1} \cdot \boldsymbol{r}$.

Proof. We obtain from (8.6) that G is the product $G = \boldsymbol{c}' \cdot \boldsymbol{r}$ of some column vector $\boldsymbol{c}'$ by $\boldsymbol{r}$. Since G is stochastic, we have that $\mathbf{1} = G\mathbf{1} = \boldsymbol{c}'\boldsymbol{r}\mathbf{1} = \boldsymbol{c}'$. □

In such a case, the matrix R is explicitly given by

$$R = A_0(I - A_1 - A_0\mathbf{1} \cdot \boldsymbol{r})^{-1}.$$

In the symmetrical case where the matrix A_0 is the product of a column vector by a row vector (see, e.g., the PH/M/1 queue in section 1.3), a similar property holds.

Theorem 8.5.2 *Assume that the QBD is positive recurrent and that* $A_0 = \boldsymbol{c} \cdot \boldsymbol{r}$, *where* $\boldsymbol{c}$ *is a column vector and* $\boldsymbol{r}$ *is a row vector normalized by* $\boldsymbol{r}\mathbf{1} = 1$. *Then the matrix* R *is given by* $R = \boldsymbol{c} \cdot \boldsymbol{\xi}$, *where*

$$\boldsymbol{\xi} = \boldsymbol{r}(I - A_1 - \eta A_2)^{-1}$$

and $\eta = \boldsymbol{\xi c}$ *with* $\eta = \mathrm{sp}(R)$ *when* $m < \infty$.

Proof. By (8.2), R is the product $R = \boldsymbol{c} \cdot \boldsymbol{\xi}$ of $\boldsymbol{c}$ by some row vector $\boldsymbol{\xi}$. If we define $\eta = \boldsymbol{\xi c}$, we have that $R^n = \eta^{n-1}\boldsymbol{c} \cdot \boldsymbol{\xi}$ for all $n \geq 1$ and $\eta < 1$ (see Corollary 6.2.4 and the succeeding remark). Thus, we may write (8.16) for R as

$$\boldsymbol{c} \cdot \boldsymbol{\xi} = \boldsymbol{c} \cdot \boldsymbol{r} + \boldsymbol{c} \cdot \boldsymbol{\xi} A_1 + \eta \boldsymbol{c} \cdot \boldsymbol{\xi} A_2,$$

so that

$$\boldsymbol{\xi}(I - A_1 - \eta A_2) = \boldsymbol{r},$$

from which the theorem follows. In order to verify that the matrix $(I - A_1 - \eta A_2)$ is nonsingular, it is simple to define a new QBD with transition matrix given by

$$\begin{bmatrix} B & A_0 & 0 & \cdots \\ (1-\eta)A_2 & A_1 + \eta A_2 & A_0 & \cdots \\ 0 & (1-\eta)A_2 & A_1 + \eta A_2 & \cdots \\ \vdots & \vdots & \vdots & \ddots \end{bmatrix}.$$

If the original QBD is irreducible, then the new QBD is also irreducible, the expected time spent in any level before moving to another level is finite, and therefore $(I - A_1 - \eta A_2)^{-1}$ exists. □

This theorem does not provide a thoroughly explicit expression for the matrix R, since it depends on η. The quantity η, however, may be directly evaluated, as we shall discuss at the end of section 9.1. Alternatively, one may solve the equation $z = \boldsymbol{r}(I - A_1 - zA_2)^{-1}\boldsymbol{c}$ for z in $(0, 1)$, as discussed in Ramaswami and Latouche [98], so that the algorithmic problem reduces to solving a *scalar* equation in this case.

8.6 Boundary Probabilities

As we saw in section 6.3, the boundary probability vector $\boldsymbol{\pi}_0$ is determined by solving the system $\boldsymbol{\pi}_0 = \boldsymbol{\pi}_0 P(T)$, where $P(T) = B + A_0 G$ is the transition probability matrix for the process restricted on the level $\ell(0)$, together with the normalizing equation $\boldsymbol{\pi}_0 (I - R)^{-1}\mathbf{1} = 1$.

In certain cases, the matrix $P(T)$ has a structure which may be exploited in the numerical evaluation of $\boldsymbol{\pi}_0$. For instance, in the case of the multiprogramming queues defined in section 1.6, the process restricted at the level 0 is itself a QBD, albeit on a finite state space, and one may use the results presented in Chapter 10.

If $P(T)$ does not have a useful special structure, one may proceed numerically and solve the system $\boldsymbol{v}(P(T) - I) = \mathbf{0}$, normalized by $\boldsymbol{v}\mathbf{1} = 1$; then $\boldsymbol{\pi}_0 = [\boldsymbol{v}(I - R)^{-1}\mathbf{1}]^{-1}\boldsymbol{v}$. If m is small enough, the numerical procedure of choice is Gaussian elimination in one of its variants as discussed in section 5.4. For details on the method and its numerical properties, we refer to Isaacson and Keller [33] and Ralston and Rabinowitz [90]. If the order of the linear system is large, then one should turn to iterative methods. A discussion of these is clearly outside the scope of this book and the reader is directed to Stewart [113].

8.7 The Continuous Time Case

At this stage, it should be obvious to the reader that the algorithms need very little change in order to become applicable to continuous time QBDs. The corresponding version of the linear reduction and modified boundary algorithms are based, instead of on (8.5), (8.6), on the continuous time equations

$$\begin{aligned} U &= A_1 + A_2 G, \\ G &= (-U)^{-1} A_2 \end{aligned}$$

(see Proposition 6.4.2). In continuous time, Theorem 8.1.1 becomes the following.

Theorem 8.7.1 *The sequences $\{U(k), k \geq 1\}$ and $\{G(k), k \geq 1\}$ defined by*

$$\begin{aligned} U(1) &= A_1, \\ G(k) &= (-U(k))^{-1}A_2, \\ U(k+1) &= A_1 + A_0G(k) \end{aligned}$$

for $k \geq 1$ converge to the matrices U and G, respectively. The sequence $\{G(k), k \geq 1\}$ is monotone. □

We do not give the proof of this theorem since it is nearly identical to that of Theorem 8.1.1: here, too, $G(k)$ is the matrix of first passage probabilities from $\ell(n)$ to $\ell(n-1)$ under taboo of $\ell(n+k)$ for any $n \geq 1$, while $U(k)$ is the generator matrix for the terminating Markov process of sojourn in $\ell(n)$ under taboo of $\ell(n-1)$ and $\ell(n+k)$.

For finite m, we also have a faster converging sequence as stated below (the proof is the same as that of Theorem 8.3.1).

Theorem 8.7.2 *If m is finite and the QBD is recurrent, then the sequences $\{\bar{U}(k)\}$ and $\{\bar{G}(k)\}$, where*

$$\begin{aligned} \bar{U}(1) &= A_1 + A_0, \\ \bar{G}(k) &= (-\bar{U}(k))^{-1}A_2, \\ \bar{U}(k+1) &= A_1 + A_0\bar{G}(k) \end{aligned}$$

for $k \geq 1$, are such that $\bar{G}(k)$ is stochastic for all k and converge to U and G, respectively. □

Finally, we may proceed as in section 8.4 and observe the continuous time QBD at discrete epochs where changes of level occur. This gives the following result, for which the proof is identical to that of Theorem 8.4.1.

Theorem 8.7.3 *The matrix G is given by*

$$G = \sum_{k\geq 0} \left(\prod_{0\leq i\leq k-1} H^{[i]} \right) L^{[k]},$$

where

$$\begin{aligned} H^{[0]} &= (-A_1)^{-1}A_0, \\ L^{[0]} &= (-A_1)^{-1}A_2, \\ H^{[k+1]} &= (I - U^{[k]})^{-1}(H^{[k]})^2, \\ L^{[k+1]} &= (I - U^{[k]})^{-1}(L^{[k]})^2 \end{aligned}$$

for $k \geq 0$ with

$$U^{[k]} = H^{[k]}L^{[k]} + L^{[k]}H^{[k]}.$$

□

For further details, the reader is referred to, among others, Neuts [79], Latouche [45], and Latouche and Ramaswami [53].

It is a simple matter, using these theorems, to adapt to the continuous time case those algorithms which we have given in the previous sections.

Chapter 9

Spectral Analysis

The maximal eigenvalue η of R plays a very special role. We already saw in section 8.2 that it determines the speed of convergence of the linear progression algorithm. We show in this chapter that it also provides us with useful information about the dynamic behavior of the stochastic process itself. We give two illustrative examples in section 9.2.

It is interesting to note that one may in principle construct the matrices R and G by computing from scratch their Jordan canonical form. We show in the second part of this chapter how this can be done for the matrix G (like in Chapter 8, we shall take the view that this matrix suffices to determine the whole distribution). We show in section 9.3 how the eigenvalues of G may be determined by using the polynomial matrix $x^2 A_0 + x(A_1 - I) + A_2$. The eigenvectors and generalized eigenvectors are characterized in section 9.4, and we show in section 9.5 how this might be used to actually compute the Jordan normal form of G. An entirely similar approach is directly applicable to the matrix R. The practical aspects are mentioned briefly at the end of the chapter.

As in the rest of the book, in this chapter also, we concentrate mainly on QBDs, although the spectral approach applies to more general models. For instance, a thorough analysis of the spectral properties of M/G/1-type Markov chains may be found in Gail, Hantler, and Taylor [23].

9.1 The Caudal Characteristic

In the case where m is finite, we have given in Chapter 7 three equivalent conditions under which the QBD is positive recurrent. Two of these conditions bear a clear relation to some aspect of the dynamic behavior of the Markov process: the condition in Theorem 7.1.1 states that, starting from any level n, the QBD must reach the level $n-1$ in an amount of time which is finite a.s. and has finite expectation; the condition in Theorem 7.2.3 states that suitably averaged one-step transitions must show a drift towards lower levels. Unlike these, however, the condition in Corollary 7.1.2 is about the spectrum of R, and it is not obvious that this is directly related to the Markov process.

The numerical value of the matrix R is rarely, if ever, of interest by itself. Its major use is in providing, together with the vector $\boldsymbol{\pi}_0$, a compact representation of the full stationary vector $\boldsymbol{\pi}$. This representation is useful in obtaining various marginal distributions and moments, as we have seen in section 6.2. The numerical value of sp(R), however, is of interest as we show in this section. We assume throughout that the QBD is positive recurrent.

We restrict ourselves to the case where $m < \infty$ and A is irreducible. Then by Theorem 7.2.2, we have that R is either irreducible or has the structure (7.8). If R is irreducible, then we know by the Perron–Frobenius theorem (Gantmacher [25, Section 13.3, Theorem 3]) that there exists a real eigenvalue $\eta > 0$ such that $\eta = \text{sp}(R)$; i.e., $|\eta'| \leq \eta$ for each eigenvalue η' of R. The corresponding left and right eigenvectors $\boldsymbol{u}$ and $\boldsymbol{v}$ may be chosen to be strictly positive and normalized by $\boldsymbol{u}\mathbf{1} = 1$ and $\boldsymbol{u}\boldsymbol{v} = 1$.

If R has the form (7.8), then its eigenvalues different from zero are those of R_1. Since R_1 is irreducible, we have again that there exists a real eigenvalue η which dominates all the others and corresponding nonnegative eigenvectors $\boldsymbol{u}$ and $\boldsymbol{v}$ such that $\boldsymbol{u}\mathbf{1} = 1$ and $\boldsymbol{u}\boldsymbol{v} = 1$.

In both cases, one has that $R^k = O(\eta^k)$, as k tends to infinity, and if R is nonperiodic, then (Seneta [108, Theorem 1.2])

$$R^k = \eta^k \boldsymbol{v} \cdot \boldsymbol{u} + o(\eta^k) \qquad \text{as } k \to \infty.$$

If the QBD is positive recurrent, then $\eta < 1$ by Corollary 7.1.2 and it results from Theorem 6.2.1 that $\boldsymbol{\pi}_n = O(\eta^n)$ as $n \to \infty$. Thus, η

is indicative of the tail behavior of the stationary distribution; it has been called the *caudal characteristic* by Neuts [81].

The caudal characteristic is also indicative of the speed of convergence of the iterative algorithms, as we saw in section 8.2. We show here that η is related to the dynamic behavior of the QBD when it is away from the boundary. In order to simplify the presentation, we assume that R is nonperiodic.

If we denote by $\{p_n, n \geq 0\}$ the marginal stationary distribution of the level, then we have $p_n = \boldsymbol{\pi}_0 R^n \mathbf{1}$ by (6.4), so that

$$p_n = \kappa\eta^n + o(\eta^n) \qquad \text{as } n \to \infty$$

with $\kappa = \boldsymbol{\pi}_0 \boldsymbol{v}$. Similarly, $\sum_{k \geq n} p_k = \boldsymbol{\pi}_0 R^n (I - R)^{-1}\mathbf{1}$, so that

$$\sum_{k \geq n} p_k = \kappa\eta^n (1-\eta)^{-1} + o(\eta^n) \qquad \text{as } n \to \infty.$$

Finally, we obtain that

$$\frac{1}{p_n} \sum_{k \geq n} p_k = O((1-\eta)^{-1}) \qquad \text{as } n \to \infty,$$

which shows that the ratio of the expected time spent in $\cup_{k \geq 1} \ell(n+k)$ to the expected time spent in the level $\ell(n)$ is asymptotically $(1-\eta)^{-1}$ as n gets large. The consequence of this is as follows.

In absolute terms, the probability of being in a high level is principally determined by the ratio $\rho = \boldsymbol{\alpha} A_0 \mathbf{1} / \boldsymbol{\alpha} A_2 \mathbf{1}$ of the (average) probability of moving up by one level to that of moving down by one level. If the traffic coefficient ρ is very small, then the stationary probability mass is concentrated on the low levels. Even then, the QBD occasionally does visit a level $\ell(n)$ for a high value of n. If the caudal characteristic η is much lower than 1, then the QBD will rapidly be forced back to lower levels. If, on the contrary, η is close to 1, then the QBD will spend a long time in higher levels before being forced back towards the boundary. In summary, a high value of η indicates that the occasional excursions far away from the boundary are long-lived.

Thus, it appears that the traffic coefficient ρ and the caudal characteristic η provide complementary information about the behavior of the

QBD. The coefficient ρ indicates how strong the overall trend towards the lower levels is; the characteristic η indicates how strong that trend is under those circumstances when the QBD has reached a high level.

It is worth noting that one may evaluate η even if the matrix R is not available. This would be particularly useful in circumstances when m is so large that the matrix R cannot be computed and stored. Then, the caudal characteristic might be the only quantitative information which would be easily obtainable.

In order to show this in the discrete time case, we repeat the argument of Theorem 8.2.3. Let us assume that the matrix A is irreducible; then the matrix $A(x) = A_2 + xA_1 + x^2 A_2$ is an irreducible nonnegative matrix for $x > 0$, its spectral radius, which we denote by $\chi(x)$, is an eigenvalue, and η is the unique solution in (0,1) of the equation $x = \chi(x)$.

Thus, in order to determine η when R is not available, we only need to solve the equation $x = \chi(x)$ in $(0, 1)$ (by bisection or otherwise). The maximal eigenvalue $\chi(x)$ of $A(x)$ may be computed easily (see, e.g., Ralston and Rabinowitz [90, Chapter 18]) provided that m is reasonably small or when the matrices A_0, A_1, and A_2 are sparse, which is often the case.

9.2 Examples

Example 9.2.1 PH/M/1 queues. It is often informative to examine η as a function of ρ or some other parameter of the QBD. We display in Figure 9.1 the function $\eta(\rho)$ for two PH/M/1 queues (see the definition in section 1.3). The first is the E_4/M/1 queue for which the interarrival intervals have the Erlang distribution with four phases. We assume that the phases have parameter 4, so that the expected interarrival interval is equal to 1. With this choice of time unit, the traffic coefficient ρ is equal to $1/\mu$, where μ is the rate of the service time distribution. The curve $\eta(\rho)$ for this queue is marked with the label (1) in Figure 9.1.

It appears that $\eta(\rho) < \rho$ always; this comparison is particularly interesting for small values of ρ, where we see that η is very small. The queue being positive recurrent, it is certain that congestion will

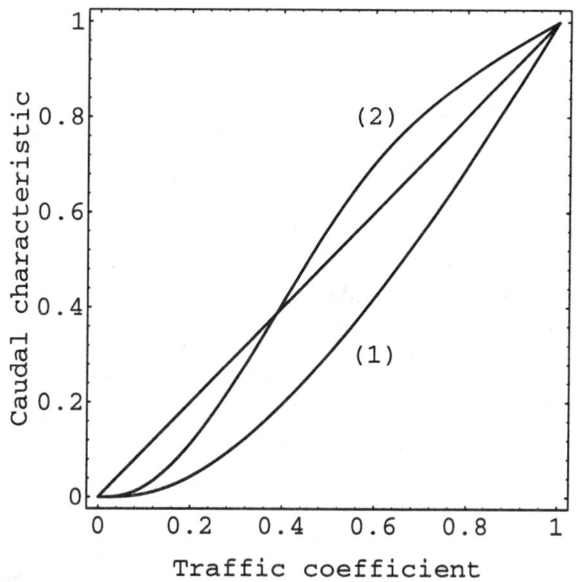

Figure 9.1: Caudal characteristic of two PH/M/1 queues. The expected interarrival times are equal to 1. In case (1), the intervals between arrivals have the Erlang $F_{4\nu}(\cdot)$ distribution; in case (2), the distribution is the mixture $pF_{50\nu}(\cdot) + (1-p)F_{4\nu}(\cdot)$ with $p = 0.05$.

occasionally occur; the fact that η is small indicates that this condition will be of very short duration; the queue will soon be forced back to lower levels. This is due to the highly regular arrival process and the fact that services are fast; the set of circumstances that leads to a build-up of the queue does not persist over long intervals of time.

For the second queue, the interarrival time distribution is the mixture $pF_{m,\nu}(\cdot) + (1-p)F_{k,\nu}(\cdot)$ of two Erlang distributions with $m = 50$, $k = 4$, $\nu = 6.3$, and $p = 0.05$. Again, the expected interval of time between two arrivals is equal to 1. However, most of the intervals are shorter, with mean equal to 0.635; occasionally an interval is much longer. It is as if the arrivals consisted of platoons of regularly spaced customers, the platoons being separated by intervals of mean duration equal to $m/\nu = 7.937$. The curve $\eta(\rho)$ for this queue is marked with the label (2).

For small values of ρ, the second curve has the same shape as for the $E_4/M/1$ queue. This is because services are very fast and there is little opportunity for the queue to build up and, when it does, the regular spacing of customers within a platoon becomes the dominant factor. For larger values of ρ, however, we observe a different behavior. When the queue does build up, it may be because of an exceptionally long service for the first few customers in a platoon, or exceptionally short intervals between two platoons, or an exceptionally large number of customers within a platoon. Whatever the reason for the first build-up, since ρ is large, service durations are large and that perpetuates the build-up as it is more likely that the queue will not become empty before the next platoon begins to arrive.

For this specific model, the change of regimen occurs around $\rho = 0.4$.

Example 9.2.2 M/M/1 queues in a random environment. As a further example, we consider various queues in a random environment (see section 1.2). There are 8 environmental phases, and the generator S of the environmental process is given by

$$S = \begin{bmatrix} -1 & 1 & \cdot & \cdots & \cdot & \cdot \\ \cdot & -1 & 1 & \cdot & \cdots & \cdot \\ \cdot & \cdot & -1 & \cdots & \cdot & \cdot \\ \vdots & \vdots & \vdots & \ddots & \vdots & \vdots \\ \cdot & \cdot & \cdot & \cdots & -1 & 1 \\ 1 & \cdot & \cdot & \cdots & \cdot & -1 \end{bmatrix}.$$

The environmental process cycles through the phases 1–8 and back to 1, remaining in each phase for an interval with exponential distribution and expectation equal to one unit. The vector of arrival rates is given by

$$\boldsymbol{\lambda} = (0.2\rho, 0.2\rho, 0.2\rho, 0.2\rho, 13\rho, \rho, \rho, 0.2\rho).$$

This may model a situation where the arrival process occasionally experiences a sharp increase during a short period. The stationary probability vector $\boldsymbol{\alpha}$ ($\boldsymbol{\alpha} S = \mathbf{0}$, $\boldsymbol{\alpha}\mathbf{1} = 1$) for the environmental process is uniform, and we have that the overall arrival rate is $\boldsymbol{\alpha\lambda} = 2\rho$ customers per unit of time.

We consider four sets of values for the service rate, which are given in the table below.

Case	$\boldsymbol{\mu}$							
1	2	2	2	2	2	2	2	2
2	1	1	1	1	5	5	1	1
3	0.4	0.4	0.4	0.4	10	2	2	0.4
4	0.2	0.2	0.2	0.2	13	1	1	0.2

In all four cases, the overall service rate is $\boldsymbol{\alpha\mu} = 2$. In case (1), the service rate is constant; in case (4), the service rate perfectly matches the arrival rates, so that $\lambda_i/\mu_i = \rho$ for all i. In cases (2) and (3), the service rates may be thought of as having been selected in order to match the arrival rates pattern, but the match is imperfect with case (3) being very close to the perfect match of case (4). The generator Q is given by (1.2) so that if we apply Theorem 7.2.3, we find that the four queues are positive recurrent if and only if $\boldsymbol{\alpha\lambda} < \boldsymbol{\alpha\mu}$ or $\rho < 1$.

The functions $\eta(\rho)$ are displayed in Figure 9.2. In case (4), the marginal distribution of the queue length is that of a simple M/M/1 queue (see Neuts [79, Remark f, Page 259]), so that $\eta(\rho) = \rho$. In the other three cases, we have that $\eta(\rho) > \rho$ always, the most extreme being case (1).

For case (1), we have that the stationary probability p_0 that the server is idle is given by $1 - \rho$ (see [79, Remark e, Page 259]), so that the server is on the whole not more utilized than in the simple M/M/1 case. The dynamic behavior of the queue, however, is very different, and excursions to high queue levels tend to persist for a long time, when they occur, even for low values of ρ. This is clearly due to the fact that the queue is unstable during the 5th environmental phase even for values of ρ as low as 0.16; during that phase, the queue may build up so much that the server will need a long time to clear the build-up when the arrival rate decreases.

In order to visualize this, we present in Figure 9.3 the conditional stationary expected queue size m_i given that the environmental process is at the end of a sojourn in phase i. It results from [79, Corollary 6.2.1] that these are given by

$$m_i = \alpha_i^{-1}(\boldsymbol{\alpha} R(I-R)^{-2})_i \qquad \text{for } 1 \leq i \leq 8.$$

The expected queue size respectively reaches its minimum and its maximum immediately before and at the end of phase 5. The minimum

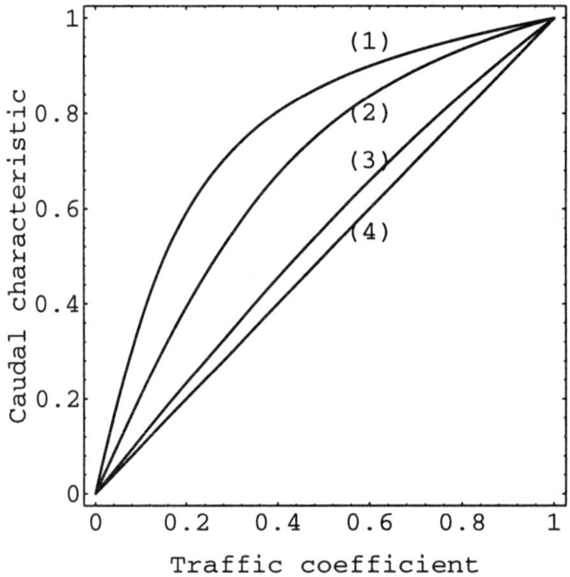

Figure 9.2: Caudal characteristic of the M/M/1 queue in a cyclic random environment. The arrival rates are proportional to (0.2, 0.2, 0.2, 0.2, 13, 1, 1, 0.2). In case (1), the service rate is constant and equal to 2; in case (3), the service rates are (1, 1, 1, 1, 5, 5, 1, 1); in case (3), the service rates are (0.4, 0.4, 0.4, 0.4, 10, 2, 2, 0.4); in case (4), the service rates are proportional to the arrival rates.

---

ranges from $m_4 = 0.012$ for $\rho = 0.1$ to 40 for $\rho = 0.9$, and the maximum ranges from $m_5 = 0.58$ to 50. This may be compared to the expected queue size of the M/M/1 queue, which increases from 0.11 to 9 only. Clearly, the effect of the large arrival rate in phase 5 is propagated throughout the environmental cycle.

The situation of case (1) is rather extreme, of course, but we see with the curve for case (2) that persisting excursions happen even if the service rate is somewhat adapted to the arrival rate. It is only for a service rate pattern such as that of case (3) that the caudal characteristic is substantially reduced. Further investigations show that case (3) represents one set of circumstances where it is possible to have one's

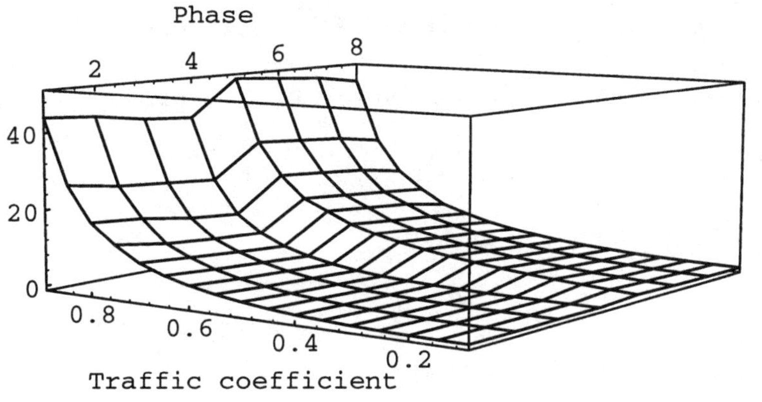

Figure 9.3: Conditional expectation of the queue size for an M/M/1 queue in a random environment given the environmental phase. The service rate is constant and equal to 2; the arrival rates are proportional to (0.2, 0.2, 0.2, 0.2, 13, 1, 1, 0.2).

---

cake and eat it too. Numerical results (not reported here) would show that we have $p_0 > 1-\rho$ with differences up to 20% for cases (2) and (3); we only give one typical set of values for η, p_0, and the overall expected queue length $\mathrm{E}[L]$ in the following table for $\rho = 0.5$.

Case	1	2	3	4
η	0.86	0.77	0.56	0.50
p_0	0.5	0.60	0.62	0.5
$\mathrm{E}[L]$	3.4	1.5	0.76	1.0

Compared to the M/M/1 queue of case (4), we see that case (2) is favorable to the server since $p_0 > 0.5$, but detrimental to the customers since $\mathrm{E}[L] > 1.0$ and $\eta > 0.5$; case (3) is favorable to *both* the server (since $p_0 > 0.5$) *and* the customers (since $\mathrm{E}[L] < 1.0$), while η is not much larger than 0.5.

The interested reader may find in Neuts [79, 81] and Latouche [44] other examples and a more detailed discussion of the caudal characteristic.

9.3 The Eigenvalues of the Matrix G

Inspired by the fact that G is a solution of the equation

$$A_0G^2 + (A_1 - I)G + A_2 = 0 \tag{9.1}$$

(see (6.14)), we define the matrix

$$\Gamma(x) = x^2A_0 + x(A_1 - I) + A_2 \tag{9.2}$$

and its determinant $\gamma(x) = \det(\Gamma(x))$.

Theorem 9.3.1 *Assume that the QBD is positive recurrent. The polynomial $\gamma(x)$ has exactly m roots in the disk $|z| \le 1$, which are the eigenvalues of the matrix G, and at most m roots with $|z| > 1$, which are the reciprocals of the nonzero eigenvalues of the matrix R.*

Proof. The proof is in three parts.

First, we prove by direct verification that the matrix $\Gamma(x)$ may be factored as

$$\Gamma(x) = \Phi_G(x)(xI - G), \tag{9.3}$$

where

$$\Phi_G(x) = xA_0 + A_1 - I + A_0G.$$

Since $\gamma(x) = \det(\Gamma(x)) = \det(\Phi_G(x))\det(xI - G)$, all the eigenvalues of G are roots of $\gamma(x)$ with at least the same multiplicity. Thus, $\gamma(x)$ has at least m roots in $|z| \le 1$. It remains to be proved that there are exactly m such roots.

For the second step, we consider the equation

$$R = A_0 + RA_1 + R^2A_2$$

and the matrix

$$\Xi(x) = x^2A_2 + x(A_1 - I) + A_0$$

with determinant $\xi(x)$. We prove by direct verification that

$$\Xi(x) = (xI - R)\Phi_R(x),$$

where

$$\Phi_R(x) = xA_2 + A_1 - I + RA_2.$$

Then, by a similar argument as above, we know that the eigenvalues of R are roots of $\xi(x)$, so that the polynomial $\xi(x)$ has at least m roots in $|z| < 1$.

Since $\Xi(x) = x^2\Gamma(x^{-1})$, the determinants $\xi(x)$ and $\gamma(x)$ satisfy the equation $\xi(x) = x^{2m}\gamma(x^{-1})$ and we see that, if R has the eigenvalue 0 with multiplicity r, then $\gamma(x)$ has degree $2m - r$ at most. Also, the reciprocals of the $m - r$ nonzero eigenvalues of R are roots of $\gamma(x)$ in $|z| > 1$.

We thus find that $\gamma(x)$ has exactly m roots in the unit disk, and the theorem follows. □

9.4 The Jordan Normal Form

We briefly recall (Faddeev and Faddeeva [18]) that one may associate with a square matrix A its Jordan normal form, i.e., a matrix J such that J is similar to A: it holds that $A = TJT^{-1}$ for some nonsingular matrix T; J is block diagonal:

$$J = \begin{bmatrix} J_1 & 0 & \dots & 0 \\ 0 & J_2 & \dots & 0 \\ \vdots & \vdots & \ddots & \vdots \\ 0 & 0 & \dots & J_r \end{bmatrix};$$

and each diagonal block has the form

$$J_i = \begin{bmatrix} \lambda_i & 1 & 0 & \dots & 0 \\ 0 & \lambda_i & 1 & \ddots & \vdots \\ \vdots & 0 & \ddots & \ddots & 0 \\ \vdots & & \ddots & \lambda_i & 1 \\ 0 & \dots & \dots & 0 & \lambda_i \end{bmatrix},$$

where λ_i is an eigenvalue of A.

When writing $A = TJT^{-1}$, the Jordan normal form J is given up to a permutation of the blocks; the similarity matrix T is, however, not unique since we may also write that $A = (UT)J(UT)^{-1}$, where U is any nonsingular matrix commuting with A.

From Theorem 9.3.1, it results that we may directly determine the eigenvalues of G from the matrix $\Gamma(x)$. We now show that the whole Jordan structure of G may also be obtained from that matrix. This is adapted from Rama Murthy, Kim, and Coyle [91], who consider the matrix R.

Theorem 9.4.1 *Let λ be an eigenvalue of G or, equivalently, a root of $\gamma(x)$ in $|z| \le 1$. The column vector $\boldsymbol{v}$ is an eigenvector of G associated with λ if and only if it is a solution of $\Gamma(\lambda)\boldsymbol{v} = \mathbf{0}$.*

Proof. In the proof of Theorem 9.3.1, we saw that

$$\gamma(x) = \det(\Phi_G(x))\det(xI - G)$$

and that $\gamma(x)$ has exactly the same m roots as $\det(xI - G)$ in $|z| \le 1$. We conclude, therefore, that $\Phi_G(x)$ is nonsingular in $|x| \le 1$. In particular, $\Phi_G(\lambda)$ is nonsingular for any eigenvalue of G. Thus, $\Gamma(\lambda)\boldsymbol{v} = \Phi_G(\lambda)(\lambda I - G)\boldsymbol{v} = \mathbf{0}$ if and only if $(\lambda I - G)\boldsymbol{v} = \mathbf{0}$. □

Following the terminology in [18], we shall say that $\boldsymbol{w}$ is a root vector for G relative to λ if $(\lambda I - G)^k\boldsymbol{w} = \mathbf{0}$ for some $k \ge 1$. These are also called "generalized eigenvectors." A root vector $\boldsymbol{w}$ is at the height k if $(\lambda I - G)^k\boldsymbol{w} = \mathbf{0}$ and $(\lambda I - G)^{k-1}\boldsymbol{w} \ne \mathbf{0}$. Thus, the eigenvectors are root vectors at the height 1.

Assume that $\boldsymbol{w}$ is a root vector at height k and that we define the vectors $\boldsymbol{v}_k = \boldsymbol{w}$, $\boldsymbol{v}_{k-1} = (\lambda I - G)\boldsymbol{v}_k, \ldots, \boldsymbol{v}_1 = (\lambda I - G)\boldsymbol{v}_2$. Assume furthermore that there is no vector $\boldsymbol{y}$ such that $\boldsymbol{w} = (\lambda I - G)\boldsymbol{y}$. Then the sequence $(\boldsymbol{v}_k, \boldsymbol{v}_{k-1}, \ldots, \boldsymbol{v}_1)$ is called a tower[17]; the vector $\boldsymbol{v}_i$ is a root vector at the height i; in particular, the vector $\boldsymbol{v}_1$ at the bottom of the tower is an eigenvector for the eigenvalue λ. A tower is maximal in the sense that there is no possibility of increasing its length at the right ($\boldsymbol{v}_0 = (\lambda I - G)\boldsymbol{v}_1 = \mathbf{0}$) or at the left (there is no $\boldsymbol{v}_{k+1}$ such that $\boldsymbol{v}_k = (\lambda I - G)\boldsymbol{v}_{k+1}$) and the vectors $\boldsymbol{v}_1, \boldsymbol{v}_2, \ldots, \boldsymbol{v}_k$ are linearly independent. We shall say that $\boldsymbol{v}_p, \boldsymbol{v}_{p-1}, \ldots, \boldsymbol{v}_1$ are *in* a tower if $\boldsymbol{v}_1$ is an eigenvector and $\boldsymbol{v}_i = (\lambda I - G)\boldsymbol{v}_{i+1}$ for $1 \le i \le p - 1$.

[17]Towers are also called Jordan chains because of their role in the Jordan normal form of a matrix.

Theorem 9.4.2 *The vectors $\boldsymbol{v}_p, \boldsymbol{v}_{p-1}, \ldots, \boldsymbol{v}_1$ are in a tower for the matrix G relative to λ if and only if the following holds:*

$$\begin{aligned} \Gamma(\lambda)\boldsymbol{v}_1 &= \mathbf{0}, \\ \Gamma(\lambda)\boldsymbol{v}_2 &= \Gamma^{(1)}(\lambda)\boldsymbol{v}_1, \\ \Gamma(\lambda)\boldsymbol{v}_{i+1} &= \Gamma^{(1)}(\lambda)\boldsymbol{v}_i - A_0\boldsymbol{v}_{i-1}, \qquad 2 \le i \le p-1, \end{aligned}$$

with $\Gamma^{(1)}(x) = 2xA_0 + A_1 - I$.

Proof. We shall refer to (9.3) and to the definition of a tower, so that we need to prove that the following holds: we have

$$\begin{aligned} (\lambda I - G)\boldsymbol{v}_1 &= \mathbf{0}, & (9.4) \\ (\lambda I - G)\boldsymbol{v}_{i+1} &= \boldsymbol{v}_i, \qquad 1 \le i \le p-1, & (9.5) \end{aligned}$$

if and only if

$$\begin{aligned} \Phi_G(\lambda)(\lambda I - G)\boldsymbol{v}_1 &= \mathbf{0}, & (9.6) \\ \Phi_G(\lambda)(\lambda I - G)\boldsymbol{v}_2 &= A_0(\lambda I - G)\boldsymbol{v}_1 + \Phi_G(\lambda)\boldsymbol{v}_1, & (9.7) \\ \Phi_G(\lambda)(\lambda I - G)\boldsymbol{v}_{i+1} &= A_0(\lambda I - G)\boldsymbol{v}_i + \Phi_G(\lambda)\boldsymbol{v}_i - A_0\boldsymbol{v}_{i-1}, \\ & \qquad 2 \le i \le p-1. & (9.8) \end{aligned}$$

We proceed as follows. Since $\Phi_G(\lambda)$ is nonsingular, (9.4) and (9.6) are equivalent. Similarly, if (9.4) holds, then (9.7) is equivalent to

$$\Phi_G(\lambda)(\lambda I - G)\boldsymbol{v}_2 = \Phi_G(\lambda)\boldsymbol{v}_1,$$

which is equivalent to (9.5) with $i = 1$.

We now make the induction assumption that the two sets of equations have been shown to be equivalent for $i \le j-1$ for some j, and we prove that they are equivalent for $i = j$. For $i = j$, (9.8) is written as

$$\begin{aligned} \Phi_G(\lambda)(\lambda I - G)\boldsymbol{v}_{j+1} &= \Phi_G^{(1)}(\lambda)[(\lambda I - G)\boldsymbol{v}_j - \boldsymbol{v}_{j-1}] + \Phi_G(\lambda)\boldsymbol{v}_j \\ &= \Phi_G(\lambda)\boldsymbol{v}_j \end{aligned}$$

since (9.5) holds for $i = j-1$; this is equivalent to (9.5) with $i = j$. To complete the proof, we observe that

$$A_0(\lambda I - G) + \Phi_G(\lambda) = 2\lambda A_0 + A_1 - I.$$

□

Towers are fundamental in expressing the Jordan canonical form of a matrix: in $G = MJM^{-1}$, where J is the Jordan normal form for G, the columns of M are formed by disjoint towers. The number of Jordan blocks is equal to the number of towers; the size of a Jordan block is equal to the number of vectors in the corresponding tower. Clearly, one way of computing the matrix G is by construction of a complete set of towers, using the theorems in this section.

9.5 Construction of the Towers

We show in this last section that it is indeed possible, in principle, to compute the matrix G by constructing a complete set of towers. Such a set, for a given eigenvalue λ, may be represented in visual form as follows:

$v_r^{(r,1)}$	$\dots$	$v_r^{(r,k_r)}$								
$\vdots$		$\vdots$	$\vdots$							
$v_2^{(r,1)}$	$\dots$	$v_2^{(r,k_r)}$	$\dots$	$v_2^{(2,1)}$	$\dots$	$v_2^{(2,k_2)}$				
$v_1^{(r,1)}$	$\dots$	$v_1^{(r,k_r)}$	$\dots$	$v_1^{(2,1)}$	$\dots$	$v_1^{(2,k_2)}$	$v_1^{(r,1)}$	$\dots$	$v_1^{(r,k_1)}$	.

Every column represents a sequence of vectors which form a tower. In the case represented here, there will be $k_1 + k_2 + \cdots + k_r$ Jordan blocks for the eigenvalue λ, respectively, of size 1, 2, $\dots, r$. The total multiplicity of the eigenvalue λ is $k_1 + 2k_2 + \cdots + rk_r$.

In order to compute the Jordan canonical form, we may proceed as follows. We shall use visual representations such as the one above, in a more compact form. Thus, we wish to compute for every eigenvalue the structure

$V_r^{(r)}$					
$V_{r-1}^{(r)}$	$V_{r-1}^{(r-1)}$				
$\vdots$	$\vdots$	$\vdots$			,
$V_2^{(r)}$	$V_2^{(r-1)}$	$\dots$	$V_2^{(2)}$		
$V_1^{(r)}$	$V_1^{(r-1)}$	$\dots$	$V_1^{(2)}$	$V_1^{(1)}$	

where

$$V_i^{(j)} = [\boldsymbol{v}_i^{(j,1)} \cdots \boldsymbol{v}_i^{(j,k_j)}]$$

is a matrix of root vectors at height i, which belong to towers of maximal height j. We shall denote by span(M) the vector space spanned by the columns of the matrix M.

Step 0. Determine the eigenvalues of G, and their multiplicity, by finding the m roots of $\gamma(x)$ in the disk $|z| \le 1$ (see Theorem 9.3.1).

Step 1. Determine the eigenvectors. For each eigenvalue λ, find a matrix $W_{1,1}(\lambda)$ such that span$(W_{1,1}(\lambda))$ is the space of all the solutions of the equation $\Gamma(\lambda)\boldsymbol{x} = \mathbf{0}$. By Theorem 9.4.1, these are all the eigenvectors of G for the eigenvalue λ.

Let $p_1(\lambda)$ be the number of columns of $W_{1,1}(\lambda)$. If $p_1(\lambda)$ is equal to the multiplicity of λ, then set $V_1^{(1)}(\lambda) = W_{1,1}(\lambda)$; all the towers for λ have been identified. Otherwise, we only have that span$(W_{1,1}(\lambda)) =$ span$(V_1^{(r)}(\lambda), V_1^{(r-1)}(\lambda), \ldots, V_1^{(2)}(\lambda), V_1^{(1)}(\lambda))$ and more work needs to be done. Visually, one has established the ground level of the pyramid drawn earlier, but nothing more:

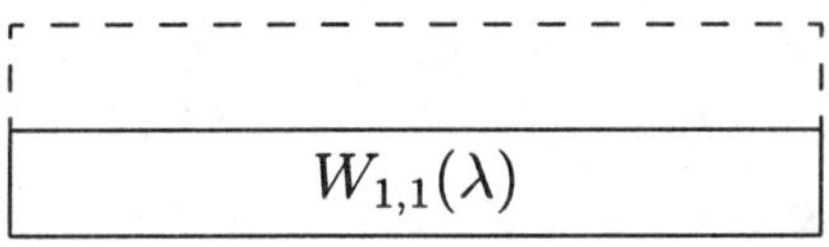

.

Step 2. Determine the second level. Let M_2 be a matrix such that $[M_2, W_{1,1}(\lambda_1), \ldots, W_{1,1}(\lambda_k)]$ spans the whole m-dimensional vector space, where $W_{1,1}(\lambda_i)$ are the matrices determined in step 1. For each eigenvalue λ such that its towers remain to be identified, determine all the solutions of the system

$$\Gamma(\lambda)\boldsymbol{v}_2 - \Gamma^{(1)}(\lambda)\boldsymbol{v}_1 = \mathbf{0},$$

where $\boldsymbol{v}_2 \in$ span(M_2) and $\boldsymbol{v}_1 \in$ span$(W_{1,1}(\lambda))$. By Theorem 9.4.2, the vectors $\boldsymbol{v}_2$ and $\boldsymbol{v}_1$ will be in a tower. Equivalently, we may solve the system

$$[\Gamma(\lambda)M_2 - \Gamma^{(1)}(\lambda)W_{1,1}(\lambda)] \begin{bmatrix} \boldsymbol{y} \\ \boldsymbol{x} \end{bmatrix} = \mathbf{0}, \tag{9.9}$$

where $\boldsymbol{y}$ and $\boldsymbol{x}$, respectively, have the same dimensions as span(M_2) and span$(W_{1,1}(\lambda))$. Thus, we need to find here a matrix $U_2(\lambda)$ such that $\text{span}(U_2(\lambda))$ is the space of all the solutions to (9.9) and which we write as follows:

$$U_2(\lambda) = \begin{bmatrix} Y_2(\lambda) \\ X_2(\lambda) \end{bmatrix}.$$

We then define the matrices $W_{2,2}(\lambda)$ and $V_1^{(1)}(\lambda)$ as follows:

$$\begin{aligned} W_{2,2}(\lambda) &= M_2 Y_2(\lambda), \\ W_{2,1}(\lambda) &= W_{1,1}(\lambda) X_2(\lambda), \end{aligned}$$

and $V_1^{(1)}(\lambda)$ is such that $\text{span}(W_{2,1}(\lambda), V_1^{(1)}(1)) = \text{span}(W_{1,1}(\lambda))$. Visually, we have the following pyramid:

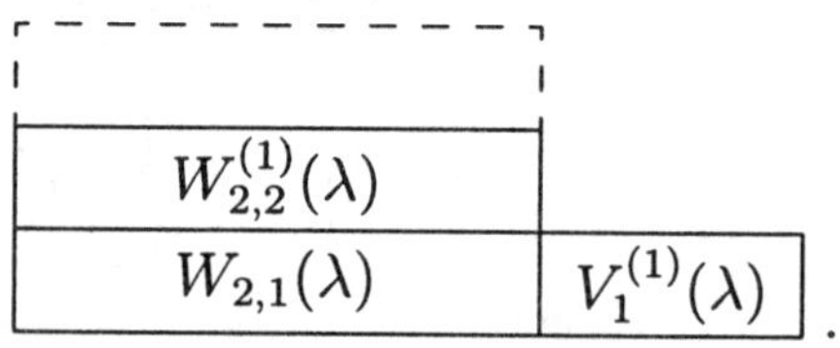

By Theorem 9.4.2, a column of $W_{2,2}(\lambda)$ and the corresponding column of $W_{2,1}(\lambda)$ are two vectors in the same tower. Consequently, the columns of $V_1^{(1)}(\lambda)$ are eigenvectors which cannot be extended into a tower; they are eigenvectors corresponding to a Jordan block of size 1.

Let $p_2(\lambda)$ be the number of columns of $W_{2,2}(\lambda)$ (and $W_{2,1}(\lambda)$). If $p_1(\lambda)+p_2(\lambda)$ is equal to the multiplicity of λ, then set $V_2^{(2)}(\lambda) = W_{2,2}(\lambda)$ and $V_1^{(2)}(\lambda) = W_{2,1}(\lambda)$: all the towers have been identified for that particular eigenvalue.

Step 3. This is the general step, which is repeated until all the eigenvalues have been completely treated. At some stage, the towers have been identified up to some maximal height $i-1 \geq 2$: graphically, we have that

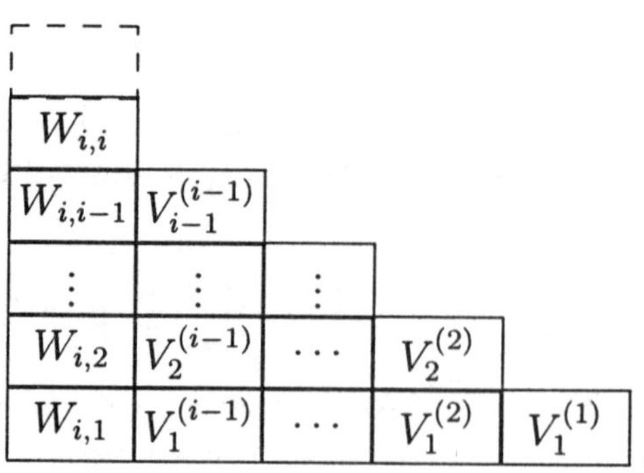

One then determines the towers of height i. In order to do this, one identifies the root vectors at the height $i+1$.

Let M_{i+1} be a matrix such that its columns, together with the columns in the union over all eigenvalues λ of $\cup_{1\le j\le i}\cup_{j\le k\le i-1} V_j^{(k)}(\lambda)$ and $\cup_{1\le j\le i} W_{ij}(\lambda)$, span the whole m-dimensional vector space. Then, for each eigenvalue λ such that some of its towers remain to be identified, determine all the solutions of the system

$$\Gamma(\lambda)\boldsymbol{v}_{i+1} - \Gamma^{(1)}\boldsymbol{v}_i + A_0\boldsymbol{v}_{i-1} = \boldsymbol{0},$$

where $\boldsymbol{v}_{i+1} \in \operatorname{span}(M_{i+1})$, $\boldsymbol{v}_i \in \operatorname{span}(W_{i,i}(\lambda))$, and $\boldsymbol{v}_{i-1}$ is the corresponding vector in $\operatorname{span}(W_{i,i-1}(\lambda))$. Equivalently, we solve the system

$$\Gamma(\lambda)M_{i+1}\boldsymbol{y} - [\Gamma^{(1)}(\lambda)W_{i,i}(\lambda) - A_0W_{i,i-1}(\lambda)]\boldsymbol{x} = \boldsymbol{0}.$$

We denote by $U_{i+1}(\lambda)$ a matrix which spans the space of all solutions $(\boldsymbol{y},\boldsymbol{x})$, and we write

$$U_{i+1}(\lambda) = \begin{bmatrix} Y_{i+1}(\lambda) \\ X_{i+1}(\lambda) \end{bmatrix}.$$

Then, we define the matrices

$$\begin{aligned} W_{i+1,i+1}(\lambda) &= M_{i+1}Y_{i+1}(\lambda), \\ W_{i+1,j}(\lambda) &= W_{i,j}(\lambda)X_{i+1}(\lambda), \qquad 1\le j\le i, \\ V_j^{(i)}(\lambda) &= W_{i,j}(\lambda)Z_{i+1}(\lambda), \qquad 1\le j\le i, \end{aligned}$$

where $Z_{i+1}(\lambda)$ is a matrix such that $\operatorname{span}(X_{i+1}(\lambda), Z_{i+1}(\lambda)) = \operatorname{span}(I)$, I being the identity matrix of order $p_i(\lambda)$.

Let $p_{i+1}(\lambda)$ be the number of columns of $W_{i+1,i+1}(\lambda)$. If the number $p_1(\lambda)+\cdots+p_{i+1}(\lambda)$ is equal to the multiplicity of λ, then the eigenvalue λ needs no further treatment and we set $V_j^{(i+1)}(\lambda) = W_{i+1,j}(\lambda)$.

The last step is repeated until all the eigenvalues are completely treated. In short, this mimics a procedure to determine the Jordan canonical form of the matrix G, that unknown matrix being replaced by the polynomial matrix $\Gamma(x)$.

The determination of the Jordan canonical form of a matrix is known to be difficult and arduous (Faddeev and Faddeeva [18, Page 71],

Moler and Van Loan [71]) and the procedure described above seems to be less efficient than the algorithms in Chapter 8, so that we do not recommend it as a general procedure. Akar and Sohraby [1] propose a quadratically convergent algorithm based on another approach to the computation of the Jordan form of R.

One may obtain useful properties in special cases. For instance, Mitra and Mitrani [69] consider a queueing system modeled as a QBD, for which they prove that the eigenvalues of R are distinct. In cases such as this, there is no need to embark upon the determination of towers, since there is a full complement of eigenvectors.

We do not elaborate further on approaches based on the spectral decomposition of R since they are outside the realm of probability altogether and not much practical experience has been reported so far.

Chapter 10

Finite QBDs

In many applications, the level is the number of customers in a queue. In the preceding chapters, we have systematically assumed that the queue is unbounded. Here, we shall assume that buffers are bounded, that the state space is restricted to $\cup_{0 \le n \le M} \ell(n)$ where M is finite, and that the transition matrix has, in the discrete time case, the form below:

$$P = \begin{bmatrix} P_{00} & P_{01} & 0 & \cdots & 0 & 0 & 0 \\ P_{10} & A_1 & A_0 & \cdots & 0 & 0 & 0 \\ 0 & A_2 & A_1 & \cdots & 0 & 0 & 0 \\ \vdots & \vdots & \vdots & \ddots & \vdots & \vdots & \vdots \\ 0 & 0 & 0 & \cdots & A_1 & A_0 & 0 \\ 0 & 0 & 0 & \cdots & A_2 & A_1 & P_{M-1M} \\ 0 & 0 & 0 & \cdots & 0 & P_{MM-1} & P_{MM} \end{bmatrix}. \tag{10.1}$$

In order to determine the stationary distribution, many approaches may be followed; most of them are algorithmic in nature. We shall give several, showing how we may use special structural properties in order to improve algorithmic efficiency.

The first method is inspired from Gaussian elimination; it is applicable in all cases where P is a band matrix. The next two methods are specifically tailored to the case of a *homogeneous* finite QBD, such as we have here, where the transition probabilities do not depend on the level, except near the level 0 and the level M. The last will only be applicable under very restrictive assumptions about the matrices A_0, A_1, and A_2 themselves.

In order to simplify the presentation in this chapter, we shall assume that $P_{01} = P_{M-1M} = A_0$ and $P_{10} = P_{MM-1} = A_2$, so that the matrix P is given by

$$P = \begin{bmatrix} P_{00} & A_0 & 0 & \cdots & 0 & 0 & 0 \\ A_2 & A_1 & A_0 & \cdots & 0 & 0 & 0 \\ 0 & A_2 & A_1 & \cdots & 0 & 0 & 0 \\ \vdots & \vdots & \vdots & \ddots & \vdots & \vdots & \vdots \\ 0 & 0 & 0 & \cdots & A_1 & A_0 & 0 \\ 0 & 0 & 0 & \cdots & A_2 & A_1 & A_0 \\ 0 & 0 & 0 & \cdots & 0 & A_2 & P_{MM} \end{bmatrix}. \tag{10.2}$$

At this stage, it should be obvious that the results in this chapter will all have very similar counterparts if the QBD process evolves in continuous time. We leave it to the reader to determine these, or to look up the referenced papers, where the continuous time case is the one often treated.

10.1 Linear Level Reduction

The first method is adapted from Gaver, Jacobs, and Latouche [26] and proceeds in two phases. During the first phase, we progressively reduce the state space by removing one level at each step, until we are left with a Markov chain on the level M only. Once that Markov chain is solved, we construct the stationary vector in the second phase by adding back one level at each step.

Starting from the complete state space $S_0 = \cup_{0 \le n \le M} \ell(n)$, we restrict the range of the QBD successively to the sets $S_1, S_2, \ldots, S_M$, where $S_i = \cup_{i \le n \le M} \ell(n)$.

Lemma 10.1.1 *The transition matrix of the process restricted on S_i is given by*

$$P_{(i)} = \begin{bmatrix} C_i & A_0 & & & \\ A_2 & A_1 & \ddots & & \\ & \ddots & \ddots & & \\ & & & A_1 & A_0 \\ & & & A_2 & P_{M,M} \end{bmatrix} \qquad \textit{for } 0 \le i \le M-1$$

and $P_{(M)} = C_M$*, where the matrices* $C_i, 0 \le i \le M$*, are recursively defined as follows:*

$$\begin{aligned} C_0 &= P_{00}, \\ C_i &= A_1 + A_2(I - C_{i-1})^{-1}A_0, \qquad 1 \le i \le M-1, \\ C_M &= P_{MM} + A_2(I - C_{M-1})^{-1}A_0. \end{aligned} \tag{10.3}$$

For $0 \le i \le M-1$*, the matrix* C_i *records the probability, starting from the level* i*, of returning to the level* i *before reaching the level* $i+1$*; the matrix* $(I - C_i)^{-1}A_0$ *records the first passage probabilities from* $\ell(i)$ *to* $\ell(i+1)$.

Proof. The property obviously holds for $i = 0$. We make the induction assumption that it holds for i and prove that it holds for $i+1$ by using Theorem 5.3.1: we set $E = S_{i+1}$, $E^c = \ell(i)$, then we have that $(I - P_{E^c})^{-1} = (I - C_i)^{-1}$ and simple algebraic manipulations show that $P^E = P_{(i+1)}$ with C_{i+1} given by (10.3). □

The next theorem indicates how one may determine the stationary probability vector.

Theorem 10.1.2 *The stationary vector* $\boldsymbol{\pi} = (\boldsymbol{\pi}_0, \boldsymbol{\pi}_1, \ldots, \boldsymbol{\pi}_M)$ *is determined by the equations*

$$\boldsymbol{\pi}_M = \boldsymbol{\pi}_M C_M, \tag{10.4}$$

$$\boldsymbol{\pi}_i = \boldsymbol{\pi}_{i+1} A_2 (I - C_i)^{-1} \tag{10.5}$$

for $0 \le i \le M-1$,

$$\sum_{0 \le i \le M} \boldsymbol{\pi}_i \mathbf{1} = 1. \tag{10.6}$$

Proof. By Theorem 5.3.1, the stationary probability vector of C_M is proportional to $\boldsymbol{\pi}_M$, so that the first equation holds. For $0 \le i \le M-1$, the stationary probability vector of the process restricted on S_i is proportional to $(\boldsymbol{\pi}_i, \boldsymbol{\pi}_{i+1}, \ldots, \boldsymbol{\pi}_M)$: we use Theorem 5.2.3 with $E = S_{i+1}$ and $E^c = \ell(i)$. Then, we find that (5.6) reduces to (10.5).

Equations (10.4), (10.5) only characterize $\boldsymbol{\pi}$ up to a multiplicative constant. That constant is determined by the normalizing equation (10.6). □

Linear Level Reduction Algorithm
$C_0 := P_{00}$;
for $i := 1$ **to** $M-1$ **do** $C_i := A_1 + A_2(I - C_{i-1})^{-1}A_0$ **od**;
$C_M := P_{MM} + A_2(I - C_{M-1})^{-1}A_0$;
Solve $\boldsymbol{p}_M(I - C_M) = 0,\ \boldsymbol{p}_M\mathbf{1} = 1$;
for $i := M-1$ **to** 0 **do** $\boldsymbol{p}_i := \boldsymbol{p}_{i+1}A_2(I - C_i)^{-1}$ **od**;
$\boldsymbol{\pi} := (\boldsymbol{p}\mathbf{1})^{-1}\boldsymbol{p}$

Figure 10.1: Computation of the stationary distribution by linear level reduction.

These two lemmas clearly indicate that the stationary probability vector may be numerically determined by the procedure described in Figure 10.1. We proceed there by determining the stationary probability vector $\boldsymbol{p}_M$ of the stochastic matrix C_M and by normalizing the vector $\boldsymbol{\pi}$ once the whole vector $\boldsymbol{p}$ is computed. This might lead to problems of overflow if the actual value of $\boldsymbol{\pi}_M\mathbf{1}$ is much smaller than 1. To overcome such problems, one might renormalize the vector $\boldsymbol{p}$ each time a new subvector is computed.

Instead of cutting the levels off, starting from the level 0 and moving up to the level M, one might also proceed in the reverse direction and obtain the following form for the stationary vector.

Theorem 10.1.3 *The stationary vector* $\boldsymbol{\pi}$ *is given by*

$$\begin{aligned} \boldsymbol{\pi}_0 &= \boldsymbol{\pi}_0 U^{(0)}, \\ \boldsymbol{\pi}_i &= \boldsymbol{\pi}_{i-1} R^{(i)} \qquad \text{for } 1 \le i \le M, \end{aligned} \tag{10.7}$$

where

$$\begin{aligned} R^{(i)} &= A_0(I - U^{(i)})^{-1}, && 1 \le i \le M, \\ U^{(M)} &= P_{MM}, && (10.8) \\ U^{(i)} &= A_1 + A_0(I - U^{(i+1)})^{-1}A_2, && 1 \le i \le M-1, \quad (10.9) \\ U^{(0)} &= P_{00} + A_0(I - U^{(1)})^{-1}A_2. \end{aligned}$$

For $1 \le i \le M$, the matrix $U^{(i)}$ records the transition probability from $\ell(i)$ to $\ell(i)$ before a visit to $\ell(i-1)$ and the matrix $(I - U^{(i)})^{-1}A_2$ records the first passage probabilities from $\ell(i)$ to $\ell(i-1)$. □

We omit the proof since it merely consists of a repetition of the argument above. The equations are presented in a form which emphasizes the similarity with the infinite QBD, for which we have that

$$\begin{aligned} \boldsymbol{\pi}_n &= \boldsymbol{\pi}_{n-1}R, \\ R &= A_0(I-U)^{-1}, \\ U &= A_1 + A_0(I-U)^{-1}A_2. \end{aligned}$$

In the finite case, naturally, the matrices U_i do depend on the levels, since there is a finite upper boundary M.

Remark 10.1.4 There is a link between the matrix-geometric property of Theorem 6.2.1 and the matrix-product form of Theorem 10.1.3. If $P_{M,M-1} = A_2$ and $P_{M,M} = U$ in (10.1), then the matrices $R^{(i)}$ are independent of i, so that (10.7) becomes $\boldsymbol{\pi}_i = \boldsymbol{\pi}_0 R^i$ for $0 \le i \le M$. In a way, one may interpret the matrix U as providing the right augmentation if one wants to truncate the state space of an infinite QBD while keeping the same stationary probabilities (up to a normalizing constant).

Remark 10.1.5 The computational complexity of these algorithms is $2\frac{1}{3}Mm^3 + O(Mm^2)$. Indeed, the major effort is spent on solving the systems $X(I - C_i) = A_2$ and multiplying the resulting matrix X by A_0 in order to obtain $A_2(I - C_i)^{-1}A_0$ for $1 \le i \le M$. The rest consists of vector operations, provided one stores the matrices $A_2(I - C_i)^{-1}$ at a total storage cost of Mm^2 floating-point numbers.

Remark 10.1.6 We observe that the argument given in this section, which leads us to the algorithm given in Figure 10.1, in fact furnishes a probabilistic justification for block-Gaussian elimination. Indeed, let us write $(I - P) = LU$, where L and U are block-partitioned, L is lower triangular and U is upper triangular, and we assume that all the diagonal blocks of U are the identity matrix. One proves by direct verification that if P is given by (10.2), then $I - P = LU$ with

$$L = \begin{bmatrix} I - C_0 & 0 & 0 & \cdots & 0 & 0 \\ -A_2 & I - C_1 & 0 & \cdots & 0 & 0 \\ 0 & -A_2 & I - C_2 & \cdots & 0 & 0 \\ \vdots & \vdots & \vdots & \ddots & \vdots & \vdots \\ 0 & 0 & 0 & \cdots & I - C_{M-1} & 0 \\ 0 & 0 & 0 & \cdots & -A_2 & I - C_M \end{bmatrix}$$

and

$$U = \begin{bmatrix} I & -(I - C_0)^{-1}A_0 & 0 & \cdots & 0 & 0 \\ 0 & I & -(I - C_1)^{-1}A_0 & \cdots & 0 & 0 \\ 0 & 0 & I & \cdots & 0 & 0 \\ \vdots & \vdots & \vdots & \ddots & \vdots & \vdots \\ 0 & 0 & 0 & \cdots & I & -(I - C_{M-1})^{-1}A_0 \\ 0 & 0 & 0 & \cdots & 0 & I \end{bmatrix}.$$

In Gauss's algorithm, the forward elimination phase serves to replace the system $\boldsymbol{\pi}(I - P) = \mathbf{0}$ by the equivalent system $\boldsymbol{\pi}L = \mathbf{0}$ (notice that U is nonsingular). The triangular system $\boldsymbol{\pi}L = \mathbf{0}$ is then solved recursively during the backward-substitution phase. It is obvious that $\boldsymbol{\pi}L = \mathbf{0}$ is equivalent to the system (10.4), (10.5).

Remark 10.1.7 Le Boudec [58] observes that if one is only interested in computing a few moments, and not the whole stationary distribution $\boldsymbol{\pi}$, then it is not necessary to store the matrices C_0, C_1, etc. Instead, one performs the backward-substitution phase on the fly, so to speak, using the fact that (10.5) may also be written as

$$\boldsymbol{\pi}_i = \boldsymbol{\pi}_M \prod_{j=M-1}^{i} A_2(I - C_j)^{-1},$$

where in the product above, the matrix $(I - C_i)^{-1}$ appears as the right-most factor.

We show in Figure 10.2 how to compute in this manner the expression $\sum_{0 \le i \le M} f_i \boldsymbol{\pi}_i \mathbf{1}$. One easily verifies that at the end of each iteration

On the Fly Algorithm
$C_0 := P_{00}$;
$\boldsymbol{v} := \mathbf{0}$;
$\boldsymbol{s} := \mathbf{0}$;
for $i := 1$ **to** $M-1$ **do**
    $C_i := A_1 + A_2(I - C_{i-1})^{-1}A_0$;
    $\boldsymbol{v} := A_2(I - C_{i-1})^{-1}(\boldsymbol{v} + f_{i-1}\mathbf{1})$;
    $\boldsymbol{s} := A_2(I - C_{i-1})^{-1}(\boldsymbol{s} + \mathbf{1})$
od;
$C_M := P_{MM} + A_2(I - C_{M-1})^{-1}$;
$\boldsymbol{v} := A_2(I - C_{M-1})^{-1}(\boldsymbol{v} + f_{M-1}\mathbf{1}) + f_M\mathbf{1}$;
$\boldsymbol{s} := A_2(I - C_{M-1})^{-1}(\boldsymbol{s} + \mathbf{1}) + \mathbf{1}$;
Solve $\boldsymbol{p}_M(I - C_M) = \mathbf{0},\ \boldsymbol{p}_M\mathbf{1} = 1$;
$sum := \boldsymbol{p}_m\boldsymbol{s}$;
$e := \boldsymbol{p}_m\boldsymbol{v}/sum$

Figure 10.2: On the fly computation of $e = \sum_{0 \le i \le M} f_i\boldsymbol{\pi}_i\mathbf{1}$.

---

of the loop, the vectors $\boldsymbol{v}$ and $\boldsymbol{s}$, respectively, are equal to

$$\boldsymbol{v} = \sum_{0 \le k \le i-1} \prod_{j=i-1}^{k} A_2(I - C_j)^{-1} f_k\mathbf{1}$$

and

$$\boldsymbol{s} = \sum_{0 \le k \le i-1} \prod_{j=i-1}^{k} A_2(I - C_j)^{-1}\mathbf{1},$$

so that at the end of the algorithm, *sum* is equal to $\boldsymbol{p}\mathbf{1}$, and we have that

$$e = (\boldsymbol{p}\mathbf{1})^{-1} \sum_{0 \le i \le M} f_i\boldsymbol{p}_i\mathbf{1} = \sum_{0 \le i \le M} f_i\boldsymbol{\pi}_i\mathbf{1}.$$

Remark 10.1.8 Albright [2] is an early paper where methods based on block-Gaussian elimination are presented for the resolution of Markov chains. A general discussion may be found in Stewart [112]; that author also discusses block Gauss–Seidel iterative methods and approaches which may be used when A_0 or A_2 is nonsingular (see Wong, Giffin, and Disney [117] for an example where A_2 is the identity matrix); Stewart signals that the inversion of such subdiagonal blocks must be carefully monitored, since it is prone to numerical instability.

O'Leary and Wu [88] propose a block version of the Grassman–Taksar–Heyman method which was described in section 5.4. Their purpose is to make use of modern computer architectures in the resolution of arbitrary Markov chains. Their method may be used, naturally, in the present circumstances in the evaluation of $A_2(I - C_i)^{-1}$ for $i = 1, \ldots, M-1$.

10.2 The Method of Folding

Ye and Li [118] propose a very efficient method, which they call the *folding algorithm* and which is a form of odd–even reduction.

Let us temporarily assume that the maximum level M is even with $M = 2K$. We partition the state space into two subsets E and E^c such that E comprises the even-numbered levels and E^c comprises the odd-numbered levels: $E = \cup_{0 \le i \le K} \ell(2i)$ and $E^c = \cup_{1 \le i \le K} \ell(2i-1)$.

Permuting the rows and columns of P so that the levels in E are enumerated first, we find that

$$P = \left[\begin{array}{ccccc|cccc}
P_{00} & & & & & A_0 & & & \\
 & A_1 & & & & A_2 & A_0 & & \\
 & & \ddots & & & & A_2 & \ddots & \\
 & & & A_1 & & & & \ddots & A_0 \\
 & & & & P_{MM} & & & & A_2 \\
\hline
A_2 & A_0 & & & & A_1 & & & \\
 & A_2 & \ddots & & & & A_1 & & \\
 & & \ddots & A_0 & & & & \ddots & \\
 & & & A_2 & A_0 & & & & A_1
\end{array}\right] \tag{10.10}$$

and that (5.6) becomes

$$\pi_{2i+1} = (\pi_{2i} A_0 + \pi_{2i+2} A_2)(I - A_1)^{-1} \tag{10.11}$$

for $0 \le i \le K-1$, so that the stationary probability of the odd-numbered levels is very easily determined once the stationary probabilities are determined for the even-numbered levels. By Theorem 5.3.1, the vector $(\boldsymbol{\pi}_0, \boldsymbol{\pi}_2, \ldots, \boldsymbol{\pi}_{2K})$ is proportional to the stationary probability vector $(\boldsymbol{\pi}^*_0, \boldsymbol{\pi}^*_1, \ldots, \boldsymbol{\pi}^*_K)$ of the filtered chain on the even-numbered levels. That filtered chain is easily seen to form a QBD with transition matrix P^* given by

$$P^* = \begin{bmatrix} P^*_{00} & A^*_0 & 0 & \cdots & 0 & 0 \\ A^*_2 & A^*_1 & A^*_0 & \cdots & 0 & 0 \\ 0 & A^*_2 & A^*_1 & \cdots & 0 & 0 \\ \vdots & \vdots & \vdots & \ddots & \vdots & \vdots \\ 0 & 0 & 0 & \cdots & A^*_1 & A^*_0 \\ 0 & 0 & 0 & \cdots & A^*_2 & P^*_{MM} \end{bmatrix} \tag{10.12}$$

with

$$\begin{aligned} P^*_{00} &= P_{00} + A_0^{\{0\}} A_2, \\ A^*_0 &= A_0^{\{0\}} A_0, \qquad A^*_2 = A_2^{\{0\}} A_2, \\ A^*_1 &= A_1 + A_2^{\{0\}} A_0 + A_0^{\{0\}} A_2, &(10.13)\\ P^*_{MM} &= P_{MM} + A_2^{\{0\}} A_0, &(10.14) \end{aligned}$$

where

$$A_i^{\{0\}} = A_i (I - A_1)^{-1} \qquad \text{for } i = 0 \text{ and } 2.$$

Similarly, if M is an odd number, $M = 2K+1$, say, then we find that (10.11) must be completed by the equation

$$\boldsymbol{\pi}_{2K+1} = \boldsymbol{\pi}_{2K} A_0 (I - P_{MM})^{-1}, \tag{10.15}$$

and that the matrix P^*_{MM} is given by

$$P^*_{MM} = A_1 + A_0 (I - P_{MM})^{-1} A_2 \tag{10.16}$$

instead of (10.14).

In summary, in order to solve a QBD on the levels 0 to M, it suffices to solve a smaller QBD on the levels 0 to $\lfloor M/2 \rfloor$ and then to apply the elementary equations (10.11) and possibly (10.15). Naturally, the QBD with transition matrix P^* may itself be solved by the same method, so that we may recursively solve smaller and smaller systems until we have to solve a QBD with only two levels, which may be done, e.g., by Gaussian elimination. The recursive procedure is given in Figure 10.3. The vector $\boldsymbol{\pi}$ is not properly normalized by that procedure, so the complete algorithm is

$$\text{solve-by-folding}(\boldsymbol{\pi}, 1, M, P_{00}, A_0, A_1, A_2, P_{MM}),$$
$$\text{renormalize}$$

Remark 10.2.1 There is some similarity between the method of folding and the quadratic algorithm of section 8.4: one skips every second level in both cases. The greatest difference is that in section 8.4 we compute the matrix G and are therefore only interested in the state at epochs when the QBD makes a change of level, while one computes here the stationary distribution and one therefore keeps track of the QBD whenever it enters some level, either directly (the contribution in (10.13) is through the term A_1) or coming from the neighboring level (the two other terms in (10.13)).

Remark 10.2.2 The complexity of this algorithm is approximately $8\lfloor log_2 M\rfloor m^3$. Indeed, the number of recursive descents is equal to $\lfloor log_2 M\rfloor$, and the major computational burden is carried in the preparation of the recursive call. The remainder consists of vector operations. The storage cost is $4\lfloor log_2 M\rfloor m^2$ floating-point numbers. This is clearly superior to the linear level reduction described earlier (see Remark 10.1.5) even for moderate values of M.

Remark 10.2.3 Recursive algorithms are generally less efficient than equivalent, iterative algorithms. Ye and Li [118] give an iterative algorithm using external files as storage media.

10.3 Matrix-Geometric Combination

Hajek [32] observes that the stationary probability vector $\boldsymbol{\pi}$ of the matrix P in (10.2) may be expressed as a linear combination of two matrix-geometric vectors (see also Naoumov, Krieger, and Wagner [72]). This result is obtained in three steps.

Folding Algorithm

Procedure solve-by-folding($\boldsymbol{\pi}, \delta, M, P_{00}, A_0, A_1, A_2, P_{MM}$)

    if $M \geq 2$ **then**

        $A_0^{\{0\}} := A_0(I - A_1)^{-1}$;

        $A_2^{\{0\}} := A_2(I - A_1)^{-1}$;

        $P_{00}^* := P_{00} + A_0^{\{0\}} A_2$;

        $A_0^* := A_0^{\{0\}} A_0$;

        $A_1^* := A_1 + A_0^{\{0\}} A_2 + A_2^{\{0\}} A_0$;

        $A_2^* := A_2^{\{0\}} A_2$;

        if $even(M)$ **then**

            $P_{MM}^* := P_{MM} + A_2^{\{0\}} A_0$

        else

            $P_{MM}^* := A_1 + A_0(I - P_{MM})^{-1} A_2$

        fi;

        solve-by-folding($\boldsymbol{\pi}, 2\delta, \lfloor M/2 \rfloor, P_{00}^*, A_0^*, A_1^*, A_2^*, P_{MM}^*$);

        for $i := 0$ **to** $\lfloor M/2 \rfloor - 1$ **do**

            $\boldsymbol{\pi}_{(2i+1)\delta} := \boldsymbol{\pi}_{2i\delta} A_0^{\{0\}} + \boldsymbol{\pi}_{2(i+1)\delta} A_2^{\{0\}}$

        od;

        if odd(M) **then**

            $\boldsymbol{\pi}_{M\delta} := \boldsymbol{\pi}_{(M-1)\delta} A_0 (I - P_{MM})^{-1}$

        fi

    else

        solve the system

$$(\boldsymbol{\pi}_0, \boldsymbol{\pi}_{M\delta}) \begin{bmatrix} I - P_{00} & -A_0 \\ -A_2 & I - P_{MM} \end{bmatrix} = \mathbf{0},$$

        $\boldsymbol{\pi}_0 \mathbf{1} + \boldsymbol{\pi}_{M\delta} \mathbf{1} = 1$

    fi

Figure 10.3: The procedure solve-by-folding recursively computes the subvector $\pi_0, \pi_\delta, \pi_{2\delta}, \ldots, \pi_{M\delta}$; the coefficient blocks in the transition matrix (10.2) are P_{00}, A_0, A_1, A_2, and P_{MM}.

First, we define the matrices $N_{ij}^{(M)}$ for $1 \le i, j \le M-1$, which record the expected time spent in level j, starting from level i, before the first visit to either the level 0 or to the level M. We use the superscript (M) to emphasize the difference from the matrices N_{ij} defined in section 6.2.

We also define $\hat{R}$, the rate matrix for the level-reversed process defined in section 8.2. Thus, $\hat{R}$ is the minimal nonnegative solution of the equation

$$\hat{R} = A_2 + \hat{R}A_1 + \hat{R}^2 A_0.$$

We finally define $\hat{U} = A_1 + \hat{R}A_0 = A_1 + A_2\hat{G}$.

Lemma 10.3.1 *The matrices $N_{1j}^{(M)}$ and $N_{M-1,j}^{(M)}$, $1 \le j \le M-1$ satisfy the following system of equations:*

$$(I-U)N_{1j}^{(M)} + R^{M-1}A_2N_{M-1j}^{(M)} = R^{j-1}, \qquad (10.17)$$

$$(I-\hat{U})N_{M-1j}^{(M)} + \hat{R}^{M-1}A_0N_{1j}^{(M)} = \hat{R}^{M-j-1} \qquad (10.18)$$

for $1 \le j \le M-1$.

Proof. In the QBD with infinitely many levels, the matrix N_{1j} which records the expected number of visits to the level j, starting from the level 1, before the first visit to the level 0, is given by

$$N_{1j} = N_{11}R^{j-1},$$

as proved by Theorem 6.2.7 and (6.9). We may subdivide the visits to $\ell(j)$ into several groups: the visits which occur before the first passage to the level M, the visits which occur between the first and second visits to $\ell(M)$, those between the second and third visits, etc.

The expected number of visits in the first group is $N_{1j}^{(M)}$ by definition. The expected number of visits in each of the succeeding groups is given by $A_2N_{M-1j}^{(M)}$: with probability in A_2, the infinite QBD moves from $\ell(M)$ to $\ell(M-1)$ and the expected number of visits to $\ell(j)$ before the first visit to either $\ell(0)$ or $\ell(M)$ is $N_{M-1j}^{(M)}$ by definition.

Since the expected number of visits to $\ell(M)$, starting from the level 1, before returning to the level 0, for the infinite QBD is $N_{1M} = N_{11}R^{M-1}$, we find by adding the expected number of visits in each group that

$$N_{1j} = N_{1j}^{(M)} + N_{11}R^{M-1}A_2N_{M-1j}^{(M)},$$

so that

$$N_{11}R^{j-1} = N_{1j}^{(M)} + N_{11}R^{M-1}A_2N_{M-1j}^{(M)},$$

from which (10.17) results since $N_{11} = (I - U)^{-1}$; see (6.7).

Equation (10.18) is proved in a similar manner. □

The proof of the property below is purely algebraic.

Theorem 10.3.2 *If $\boldsymbol{\alpha}A_0\mathbf{1} \neq \boldsymbol{\alpha}A_2\mathbf{1}$, then the stationary distribution of the QBD with transition matrix* (10.2) *is given by*

$$\boldsymbol{\pi}_i = \boldsymbol{x}_0R^i + \boldsymbol{x}_M\hat{R}^{M-i}, \qquad 0 \le i \le M, \tag{10.19}$$

where $(\boldsymbol{x}_0, \boldsymbol{x}_M)$ is the solution of the system

$$(\boldsymbol{x}_0, \boldsymbol{x}_M)\begin{bmatrix} P_{00} + RA_2 - I & R^M(P_{MM} - I) + R^{M-1}A_0 \\ \hat{R}^M(P_{00} - I) + \hat{R}^{M-1}A_2 & P_{MM} + \hat{R}A_0 - I \end{bmatrix} = \mathbf{0} \tag{10.20}$$

and

$$\boldsymbol{x}_0 \sum_{0 \le i \le M} R^i\mathbf{1} + \boldsymbol{x}_M \sum_{0 \le i \le M} \hat{R}^i\mathbf{1} = 1. \tag{10.21}$$

Proof. We first write (10.17), (10.18) in matrix form:

$$\begin{bmatrix} I - U & R^{M-1}A_2 \\ \hat{R}^{M-1}A_0 & I - \hat{U} \end{bmatrix}\begin{bmatrix} N_{1j}^{(M)} \\ N_{M-1j}^{(M)} \end{bmatrix} = \begin{bmatrix} R^{j-1} \\ \hat{R}^{M-j-1} \end{bmatrix}.$$

We then premultiply both sides of this equation by the matrix

$$\begin{bmatrix} R & 0 \\ 0 & \hat{R} \end{bmatrix},$$

thereby obtaining that

$$\begin{bmatrix} I & R^M \\ \hat{R}^M & I \end{bmatrix}\begin{bmatrix} A_0 & 0 \\ 0 & A_2 \end{bmatrix}\begin{bmatrix} N_{1j}^{(M)} \\ N_{M-1j}^{(M)} \end{bmatrix} = \begin{bmatrix} R^j \\ \hat{R}^{M-j} \end{bmatrix}.$$

We have observed earlier (Remark 8.2.4) that either $sp(R) < 1 = sp(\hat{R})$ or $sp(\hat{R}) < 1 = sp(R)$ if $\boldsymbol{\alpha}A_0\mathbf{1} \neq \boldsymbol{\alpha}A_2\mathbf{1}$. This implies that the matrix

$$\begin{bmatrix} I & R^M \\ \hat{R}^M & I \end{bmatrix}$$

is nonsingular and therefore that

$$\begin{bmatrix} A_0 & 0 \\ 0 & A_2 \end{bmatrix} \begin{bmatrix} N_{1j}^{(M)} \\ N_{M-1j}^{(M)} \end{bmatrix} = \begin{bmatrix} I & R^M \\ \hat{R}^M & I \end{bmatrix}^{-1} \begin{bmatrix} R^j \\ \hat{R}^{M-j} \end{bmatrix}.$$

Now, turning to Theorems 5.2.3 and 5.3.1, we partition the state space into two subsets, $E = \ell(0) \cup \ell(M)$ and $E^c = \cup_{1 \le i \le M-1} \ell(i)$, and find that

$$\begin{aligned} \boldsymbol{\pi}_i &= \begin{bmatrix} \boldsymbol{\pi}_0 & \boldsymbol{\pi}_M \end{bmatrix} \begin{bmatrix} A_0 & 0 \\ 0 & A_2 \end{bmatrix} \begin{bmatrix} N_{1i}^{(M)} \\ N_{M-1i}^{(M)} \end{bmatrix} \\ &= \begin{bmatrix} \boldsymbol{\pi}_0 & \boldsymbol{\pi}_M \end{bmatrix} \begin{bmatrix} I & R^M \\ \hat{R}^M & I \end{bmatrix}^{-1} \begin{bmatrix} R^i \\ \hat{R}^{M-i} \end{bmatrix}, \end{aligned}$$

which proves (10.19) if we define

$$(\boldsymbol{x}_0, \boldsymbol{x}_M) = (\boldsymbol{\pi}_0, \boldsymbol{\pi}_M) \begin{bmatrix} I & R^M \\ \hat{R}^M & I \end{bmatrix}^{-1}.$$

Now that we have proved that the stationary distribution has the form (10.19), we may determine $(\boldsymbol{x}_0, \boldsymbol{x}_M)$ by requiring that $\boldsymbol{\pi}(P - I) = \mathbf{0}$ and $\boldsymbol{\pi}\mathbf{1} = 1$, from which (10.20), (10.21) result. □

Although we only prove the theorem in the case where $\boldsymbol{\alpha}A_0\mathbf{1} \neq \boldsymbol{\alpha}A_2\mathbf{1}$, Hajek [32] uses a generalized inverse matrix to determine the matrices $N_{1j}^{(M)}$ and $N_{M-1j}^{(M)}$ when $\boldsymbol{\alpha}A_0\mathbf{1} = \boldsymbol{\alpha}A_2\mathbf{1}$, at least for M large enough.

The elegance of (10.19) is somewhat marred by the lack of clear probabilistic significance of the coefficient matrix in (10.20). Furthermore, since that matrix is singular when $\boldsymbol{\alpha}A_0\mathbf{1} = \boldsymbol{\alpha}A_2\mathbf{1}$, one may expect to encounter numerical instability problems in solving (10.20) when $\boldsymbol{\alpha}A_0\mathbf{1}$ and $\boldsymbol{\alpha}A_2\mathbf{1}$ are close, despite the fact that the system $\boldsymbol{\pi}P = \boldsymbol{\pi}$, $\boldsymbol{\pi}\mathbf{1} = 1$ is well posed. This is in marked contrast to the other methods described in this chapter. We are not aware of reported numerical experimentation with this solution.

Remark 10.3.3 If $P_{MM} = A_1 + A_0 G$, then (10.19) simplifies to $\boldsymbol{\pi}_i = \boldsymbol{x}_0 R^i$. In order to show this, we merely observe that the matrix (10.2) is then the transition matrix of the infinite QBD observed in the local time of $\cup_{0\le n\le M}\ell(n)$, and we use Theorem 5.3.1.

Alternatively, we may also directly use the fact that

$$\begin{aligned} R^{M+1}&(P_{MM}-I)+R^M A_0 \\ &= R^M(A_0+RA_1+RA_0G-R) \\ &= R^M(A_0+RA_1+R^2A_2-R)=0 \end{aligned}$$

and conclude that the solution of (10.20) is $(\boldsymbol{x}_0,\mathbf{0})$.

Similarly, if $P_{00}=A_1+A_2\hat{G}$, where $\hat{G}$ is the first passage probability matrix of the level-reversed process (see Lemma 8.2.1), then we find that $\boldsymbol{\pi}_i=\boldsymbol{x}_M\hat{R}^{M-i}$.

10.4 Subdiagonal Matrices of Rank 1

If the matrix A_0 or the matrix A_2 is of rank 1, then the stationary distribution has an even simpler expression. Such is the case of the M/PH/1/K and PH/M/1/K queues with a finite buffer, as observed by several authors: Carroll, Van der Liefvoort, and Lipsky [11], Marie and Pellaumail [65], and Neuts [79, 80].

We shall first assume that the matrix A_2 is of rank 1, i.e., that $A_2=\boldsymbol{c}\cdot\boldsymbol{r}$, where $\boldsymbol{c}$ is a column vector, $\boldsymbol{r}$ is a row vector, and where we assume without loss of generality that $\boldsymbol{r}\mathbf{1}=1$.

We temporarily denote by Γ_i the matrices

$$\Gamma_i=(I-U_i)^{-1}A_2, \qquad 1\le i\le M,$$

where the matrices U_i have been defined in Theorem 10.1.3. Clearly, the matrices Γ_i are also of rank 1, and we have that

$$\Gamma_i=\boldsymbol{\gamma}_i\boldsymbol{r}, \qquad i\le i\le M,$$

where $\boldsymbol{\gamma}_i$ is some column vector.

We have indicated that Γ_i records the transition probabilities from $\ell(i)$ to $\ell(i-1)$. Since the QBD is finite and irreducible, Γ_i is a stochastic matrix for all i, and we have that

$$\mathbf{1}=\Gamma_i\mathbf{1}=\boldsymbol{\gamma}_i\boldsymbol{r}\mathbf{1}=\boldsymbol{\gamma}_i,$$

so that $\Gamma_i = \mathbf{1} \cdot \boldsymbol{r}$ for all $i, 1 \leq i \leq M$.

We have thus proved the first statement in the following theorem, reminiscent of Theorem 8.5.1. The second statement is similarly proved.

Theorem 10.4.1 *If the matrix A_2 is of rank 1, and is given by $A_2 = \mathbf{c} \cdot \boldsymbol{r}$ with $\boldsymbol{r}\mathbf{1} = 1$, then the stationary distribution is given by*

$$\begin{aligned}
\boldsymbol{\pi}_i &= \boldsymbol{\pi}_0 R^i \qquad \text{for } 1 \leq i \leq M-1, \\
\boldsymbol{\pi}_M &= \boldsymbol{\pi}_0 R^{M-1} R_M, \\
\boldsymbol{\pi}\mathbf{1} &= 1,
\end{aligned}$$

and

$$\boldsymbol{\pi}_0 (I - P_{00} - A_0 \mathbf{1} \cdot \boldsymbol{r}) = \mathbf{0}, \tag{10.22}$$

where

$$\begin{aligned}
R &= A_0 (I - A_1 - A_0 \mathbf{1} \cdot \boldsymbol{r})^{-1}, \\
R_M &= A_0 (I - P_{MM})^{-1}.
\end{aligned}$$

If the matrix A_0 is of rank 1 and is given by $A_0 = \mathbf{c} \cdot \boldsymbol{r}$ with $\boldsymbol{r}\mathbf{1} = 1$, then the stationary distribution is given by

$$\begin{aligned}
\boldsymbol{\pi}_{M-i} &= \boldsymbol{\pi}_M \hat{R}^i \qquad \text{for } 1 \leq i \leq M-1, \\
\boldsymbol{\pi}_0 &= \boldsymbol{\pi}_M \hat{R}^{M-1} \hat{R}_0, \\
\boldsymbol{\pi}\mathbf{1} &= 1,
\end{aligned}$$

and

$$\boldsymbol{\pi}_M (I - P_{MM} - A_2 \mathbf{1} \cdot \boldsymbol{r}) = \mathbf{0}, \tag{10.23}$$

where

$$\begin{aligned}
\hat{R} &= A_2 (I - A_1 - A_2 \mathbf{1} \cdot \boldsymbol{r})^{-1}, \\
\hat{R}_0 &= A_2 (I - P_{00})^{-1}.
\end{aligned}$$

□

Observe that the stationary vector is obtained here much more efficiently than by using the linear level reduction or the folding approach since the only m^3 operations are the computations of the matrices R

and R_M (or $\hat{R}$ and $\hat{R}_0$) and the resolution of the system (10.22) (or the system (10.23)) *independently of the number M of levels.*

This illustrates the reduction in algorithmic complexity which may be achieved under special circumstances. See Gün and Makowski [31] and Stewart [112] for other examples.

Chapter 11

First Passage Times

We saw earlier that for a homogeneous QBD with infinitely many levels, the distribution of the first passage time from the level $\ell(n)$ to $\ell(n-1)$ is independent of n. This property was at the heart of our analysis in Chapter 7 of the stability condition. It does not hold if the QBD is finite, since the presence of the upper level $\ell(M)$ restricts in different ways the trajectories above different levels.

In the upward direction, the boundary level $\ell(0)$ always exerts its influence and the distribution of the first passage time from $\ell(n)$ to a higher level always depends on n. Without loss of generality, we may therefore consider some fixed upper level $\ell(M)$ and examine the first passage times from $\ell(n)$ to $\ell(M)$ for $n = 0, 1, \ldots, M-1$. For finite QBDs, the analysis of first passage times downward proceeds in an entirely analogous manner and is left to the reader.

We shall mostly be concerned with the first moment of the first passage time from the bottom level $\ell(0)$ to the top level $\ell(M)$. One might similarly be interested in the design of efficient algorithms for higher moments or for other first passage times. Clearly, the principles developed in the present chapter may be used in such circumstances.

The time spent in the set of levels $\mathcal{L} = \ell(0) \cup \cdots \cup \ell(M-1)$ is progressively evaluated by determining the time spent in a subset $\mathcal{S} \subset \mathcal{L}$ and letting $\mathcal{S}$ increase to $\mathcal{L}$. In section 11.2, we proceed by adding one level at a time to $\mathcal{S}$, so that at the nth step, $\mathcal{S}$ is composed of the levels $\ell(0)$ to $\ell(n)$. In section 11.3, we double the size of $\mathcal{S}$ at each iteration, so that at the nth step, $\mathcal{S}$ contains the levels $\ell(0)$ to $\ell(2^n)$.

In section 11.4, we also double the number of levels at each iteration, the difference being that the levels in $\mathcal{S}$ are evenly spread between $\ell(0)$ and $\ell(M)$ and are not bunched close to $\ell(0)$.

It is our purpose to show that computing moments of first passage times is closely connected to computing the stationary distribution. In particular, the algorithms in sections 11.2 and 11.4, respectively, quite similar to those in sections 10.1 and 10.2.

11.1 Generating Functions

We shall denote by $\tau(M)$ the first passage time to the level $\ell(M)$,

$$\tau(M) = \inf\{k > 0 : X_k \in \ell(M)\},$$

and by $\boldsymbol{f}_n(\cdot)$ the vector of first passage densities from $\ell(n)$ with elements

$$f_{n,i}(k) = \mathrm{P}[\tau(M) = k | X_0 = (n,i)]$$

for $0 \leq n \leq M-1$, $1 \leq i \leq m$.

The distribution of $\tau(M)$ remains the same if we replace the states in $\ell(M)$ and above by a single absorbing state and the transition matrix by

$$P^* = \begin{bmatrix} T & \boldsymbol{t} \\ \boldsymbol{0} & 1 \end{bmatrix},$$

where

$$T = \begin{bmatrix} P_{00} & A_0 & 0 & \cdots & 0 & 0 \\ A_2 & A_1 & A_0 & \cdots & 0 & 0 \\ 0 & A_2 & A_1 & \cdots & 0 & 0 \\ \vdots & \vdots & \vdots & \ddots & \vdots & \vdots \\ 0 & 0 & 0 & \cdots & A_1 & A_0 \\ 0 & 0 & 0 & \cdots & A_2 & A_1 \end{bmatrix} \qquad (11.1)$$

is the transition matrix on $\ell(0) \cup \cdots \cup \ell(M-1)$ and

$$\boldsymbol{t} = \boldsymbol{1} - T\boldsymbol{1} = \begin{bmatrix} \boldsymbol{0} \\ \vdots \\ \boldsymbol{0} \\ A_0\boldsymbol{1} \end{bmatrix}.$$

Observe that we are dealing with a PH distribution, so that we may directly use the material in section 2.5 and write that

$$\boldsymbol{f}(k) = T^{k-1}\boldsymbol{t},$$

where

$$\boldsymbol{f}(k) = \begin{bmatrix} \boldsymbol{f}_0(k) \\ \cdots \\ \boldsymbol{f}_{M-1}(k) \end{bmatrix}$$

is the vector of first passage probabilities to $\ell(M)$. The vector of generating functions is given by

$$\tilde{\boldsymbol{f}}(z) = \sum_{k \geq 0} z^k \boldsymbol{f}(k) = z(I - zT)^{-1}\boldsymbol{t},$$

and we obtain after taking the appropriate derivatives that

$$\boldsymbol{\theta}^{[\nu]} = \nu!(I - T)^{-\nu}T^{\nu-1}\boldsymbol{1},$$

where $\theta^{[\nu]}_{n,i}$ is the νth factorial moment of $\tau(M)$ given that the initial state is (n, i).

Clearly, the computation of the factorial moments may be tackled by various algorithmic procedures, similar to those developed in Chapter 10 for the stationary distribution of finite QBDs. We shall consider in the remainder of this chapter various methods to determine the first moment only; these methods may be generalized to higher moments, as shown in Gaver, Jacobs, and Latouche [26], Li and Sheng [59], and Remiche [105].

The matrices $(I - T)$ and $(I - zT)$ being structurally equivalent for $0 \leq z \leq 1$, the procedures which we shall describe are also applicable to the computation of the generating function $\tilde{\boldsymbol{f}}(z)$ as the solution of the linear system $(I - zT)\tilde{\boldsymbol{f}}(z) = z\boldsymbol{t}$.

11.2 Linear Level Reduction

The simplest method is similar to the one described in section 10.1 and is discussed in [26].

Denoting by $\boldsymbol{t}_{n\to k} = \mathrm{E}[\tau(k)|X_0 \in \ell(n)]$ the vector[18] of expected first passage times from the states of $\ell(n)$ to $\ell(k)$ and by $\Gamma_{n\to k} = \mathrm{P}[X_{\tau(k)}|X_0 \in \ell(n)]$ the matrix of first passage probabilities, we readily obtain the next lemma. In order to simplify the equations, we adopt the following convention: $\boldsymbol{t}_{n\to n} = 0$ and $\Gamma_{n\to n} = I$ for all n.

Lemma 11.2.1 *We have that*

$$\boldsymbol{t}_{n\to k} = \boldsymbol{t}_{n\to n+1} + \Gamma_{n\to n+1}\boldsymbol{t}_{n+1\to k} \tag{11.2}$$

and

$$\Gamma_{n\to k} = \Gamma_{n\to n+1}\Gamma_{n+1\to k} \tag{11.3}$$

for all n and k such that $0 \le n < k \le M$.

Proof. Use the fact that if $X_0 \in \ell(n)$ with $n < k$, then $\tau(n+1) \le \tau(k)$ and condition on the state at time $\tau(n+1)$. □

Clearly, we only need to determine the expected first passage times, and first passage probabilities, from one level to the next in order to determine them all. Indeed, it immediately results from the lemma above that

$$\Gamma_{n\to k} = \prod_{n\le\nu\le k-1} \Gamma_{\nu\to\nu+1} \tag{11.4}$$

and

$$\boldsymbol{t}_{n\to k} = \sum_{n\le\nu\le k-1} \Gamma_{n\to\nu}\boldsymbol{t}_{\nu\to\nu+1} \tag{11.5}$$

for $0 \le n < k$. The remaining relations are given below.

Lemma 11.2.2 *We have that*

$$\Gamma_{n\to n+1} = (I - C_n)^{-1}A_0 \tag{11.6}$$

for $0 \le n \le M-1$ and that

$$\begin{aligned} \boldsymbol{t}_{0\to 1} &= (I - C_0)^{-1}\mathbf{1}, & (11.7)\\ \boldsymbol{t}_{n\to n+1} &= (I - C_n)^{-1}\mathbf{1} + (I - C_n)^{-1}A_2\boldsymbol{t}_{n-1\to n} & (11.8) \end{aligned}$$

[18]We use the notation $n \to k$ as a mnemonic device to indicate that the level occupied at time 0 is $\ell(n)$ and that we are interested in the first passage time or first passage probability to the level $\ell(k)$.

for $1 \le n \le M-1$, *where the matrices* C_n *are defined as follows:*

$$\begin{aligned} C_0 &= P_{00}, \\ C_n &= A_1 + A_2(I - C_{n-1}^{-1})A_0 = A_1 + A_2\Gamma_{n-1\to n} \end{aligned}$$

for $1 \le n \le M-1$.

Proof. The matrices C_n were defined in Lemma 10.1.1 and record the taboo probability, starting from a state in $\ell(n)$, of returning to $\ell(n)$ before hitting $\ell(n+1)$. Equation (11.6) is identical to the last statement in that lemma.

In order to prove (11.8), we condition on the state at time 1 and find that

$$\begin{aligned} \boldsymbol{t}_{n\to n+1} &= \mathbf{1} + A_1\boldsymbol{t}_{n\to n+1} + A_2\boldsymbol{t}_{n-1\to n+1} \\ &= \mathbf{1} + (A_1 + A_2\Gamma_{n-1\to n})\boldsymbol{t}_{n\to n+1} + A_2\boldsymbol{t}_{n-1\to n} \end{aligned}$$

by (11.2), from which the proof is immediate. □

We observe that the factor $(I - C_n)^{-1}$ in the right-hand side of (11.8) records the expected number of visits to $\ell(n)$ before the first visit to $\ell(n+1)$, while $A_2\boldsymbol{t}_{n-1\to n}$ records the expected sojourn time in $\ell(0) \cup \cdots \cup \ell(n-1)$ between two such visits to $\ell(n)$.

The actual organization of the computations naturally depends on the quantities which need to be evaluated. As an example, we give in Figure 11.1 an algorithm to compute $\boldsymbol{t}_{0\to M}$.

This algorithm in effect implements the resolution of the system $(I - T)\boldsymbol{\theta}^{[1]} = \mathbf{1}$ by LU decomposition. Indeed, writing $I - T = LU$ where L is block-lower triangular and U is block-upper triangular, it is a simple matter to verify that we can choose

$$L = \begin{bmatrix} I - C_0 & 0 & \cdots & 0 & 0 \\ -A_2 & I - C_1 & \cdots & 0 & 0 \\ \vdots & \vdots & \ddots & \vdots & \vdots \\ 0 & 0 & \cdots & I - C_{M-2} & 0 \\ 0 & 0 & \cdots & -A_2 & I - C_{M-1} \end{bmatrix}$$

Linear Level Reduction Algorithm
$C_0 := P_{00}$;
Solve $(I - C_0)\Gamma_{0\to1} = A_0$;
Solve $(I - C_0)\boldsymbol{t}_{0\to1} = \mathbf{1}$;
for $i := 1$ **to** $M - 1$ **do**
  $C_i := A_1 + A_2\Gamma_{i-1\to i}$;
  Solve $(I - C_i)\Gamma_{i\to i+1} = A_0$;
  Solve $(I - C_i)\boldsymbol{t}_{i\to i+1} = \mathbf{1} + A_2\boldsymbol{t}_{i-1\to i}$
od;
for $i := M - 1$ **to** 0 **do**
  $\boldsymbol{t}_{i\to M} := \boldsymbol{t}_{i\to i+1} + \Gamma_{i\to i+1}\boldsymbol{t}_{i+1\to M}$
od

Figure 11.1: Computation of the expected first passage times from level n $(0 \le n \le M-1)$ to level M.

---

and

$$U = \begin{bmatrix} I & -\Gamma_{0\to1} & 0 & \cdots & 0 & 0 \\ 0 & I & -\Gamma_{1\to2} & \cdots & 0 & 0 \\ 0 & 0 & I & \cdots & 0 & 0 \\ \vdots & \vdots & \vdots & \ddots & \vdots & \vdots \\ 0 & 0 & 0 & \cdots & I & -\Gamma_{M-2\to M-1} \\ 0 & 0 & 0 & \cdots & 0 & I \end{bmatrix}.$$

The system $(I - T)\boldsymbol{\theta}^{[1]} = \mathbf{1}$ may be written as

$$L\boldsymbol{v} = \mathbf{1}, \tag{11.9}$$
$$U\boldsymbol{\theta}^{[1]} = \boldsymbol{v}. \tag{11.10}$$

The solution $\boldsymbol{v}$ of (11.9) is such that $\boldsymbol{v}_n = \boldsymbol{t}_{n\to n+1}$ $(0 \le n \le M-1)$ by Lemma 11.2.2 and is computed in the first loop; (11.10) is merely a reformulation of (11.2) and is solved in the second loop.

11.3 Reduction by Bisection

This approach has been described in Latouche and Ramaswami [54]; it is based on the fact that the transition matrix T in (11.1) is homogeneous if one excepts the transition probabilities in the level 0. If we define the transition matrix $P_\omega^{(n)}$ over n levels by

$$P_\omega^{(n)} = \begin{bmatrix} A_1 & A_0 & & \\ A_2 & A_1 & \ddots & \\ & \ddots & \ddots & A_0 \\ & & A_2 & A_1 \end{bmatrix}, \tag{11.11}$$

it describes not only the transitions among the states in $\ell(1) \cup \cdots \cup \ell(n)$, but also among any set $\ell(k+1) \cup \cdots \cup \ell(k+n)$ of n contiguous levels with $0 \le k \le M-1-n$.

To fix ideas, assume that the set of interest is $\cup_{1 \le k \le n} \ell(k)$, let $\boldsymbol{s}^{(n)}$ denote the vector of expected sojourn times in such a set, and denote by $G^{(n)}$ and $\Gamma^{(n)}$, respectively, the matrices of probabilities that exit occurs into the level $\ell(0)$ and the level $\ell(n+1)$. Formally, the first passage time to $\ell(0)$ or $\ell(n+1)$ is $\theta^{(n)} = \min(\tau(0), \tau(n+1))$, and we have

$$\begin{aligned} \boldsymbol{s}_i^{(n)} &= \mathrm{E}[\theta^{(n)} | X_0 \in \ell(i)], \\ G_i^{(n)} &= \mathrm{P}[\theta^{(n)} = \tau(0) < \tau(n+1) \;\&\; X_{\tau(0)} | X_0 \in \ell(i)], \end{aligned}$$

and

$$\Gamma_i^{(n)} = \mathrm{P}[\theta^{(n)} = \tau(n+1) < \tau(0) \;\&\; X_{\tau(n+1)} | X_0 \in \ell(i)]$$

for $1 \le i \le n$. Observe that the matrix G in Chapter 6 is equal to $\lim_{n\to\infty} G_1^{(n)}$. Observe also that the vector $\boldsymbol{s}_i^{(n)}$ is equal to $\sum_{1 \le j \le n} N_{ij}^{(n)} \mathbf{1}$, where the matrices $N_{ij}^{(n)}$ have been defined in section 10.3.

We show below how the first passage time $\boldsymbol{t}_{0\to M}$ may be determined, once $\boldsymbol{s}_1^{(M-1)}$ and $G_1^{(M-1)}$ are known. The practical usefulness of this property is that $\boldsymbol{s}_1^{(M-1)}$ and $G_1^{(M-1)}$ may be efficiently determined, as we shall see shortly.

Lemma 11.3.1 *The vector $\boldsymbol{t}_{0\to M}$ of expected first passage times from $\ell(0)$ to $\ell(M)$ is given by*

$$\boldsymbol{t}_{0\to M} = (I - \Gamma_{0\to 1} G_1^{(M-1)})^{-1} \{\boldsymbol{t}_{0\to 1} + \Gamma_{0\to 1} \boldsymbol{s}_1^{(M-1)}\} \tag{11.12}$$

for $M \geq 2$.

Proof. We have by (11.2) that

$$\boldsymbol{t}_{0\to M} = \boldsymbol{t}_{0\to 1} + \Gamma_{0\to 1}\boldsymbol{t}_{1\to M}.$$

Now, conditioning on the first visit to either $\ell(0)$ or $\ell(M)$, starting from the level $\ell(1)$, we find that

$$\boldsymbol{t}_{1\to M} = \boldsymbol{s}_1^{(M-1)} + G_1^{(M-1)}\boldsymbol{t}_{0\to M};$$

either the first visit is to $\ell(M)$, in which case the first passage time is equal to $\theta^{(M-1)}$, or the first visit is to $\ell(0)$, in which case we add to $\theta^{(M-1)}$ a whole first passage time from 0 to M.

Combining the two equations, we obtain that

$$\boldsymbol{t}_{0\to M} = \boldsymbol{t}_{0\to 1} + \Gamma_{0\to 1}\boldsymbol{s}_1^{(M-1)} + \Gamma_{0\to 1}G_1^{(M-1)}\boldsymbol{t}_{0\to M},$$

where the matrix $\Gamma_{0\to 1}G_1^{(M-1)}$ records the taboo probability, starting from a state in $\ell(0)$, of returning to a state in $\ell(0)$ after a visit to $\ell(1)$, but before the first visit to the level $\ell(M)$. Since the Markov chain is irreducible, the expected number of such returns is finite and the series $\sum_{\nu\geq 0}(\Gamma_{0\to 1}G_1^{(M-1)})^\nu$ converges, from which the lemma results. □

We show below that it is a simple matter, knowing the matrices $G_i^{(n)}$ and $\Gamma_i^{(n)}$, to double the number of levels and determine the matrices $G_i^{(2n)}$ and $\Gamma_i^{(2n)}$. Thus, if we had $M = 2^\alpha + 1$ for some α, we could successively determine $G_1^{(1)}, G_1^{(2)}, G_1^{(4)}, G_1^{(8)}$, etc. up to $G_1^{(2^\alpha)}$. Since $M-1$ is not always a power of two, we need a slightly more general lemma.

Lemma 11.3.2 *We have that*

$$G_i^{(n+\nu)} = G_i^{(n)} + \Gamma_i^{(n)}G_1^{(\nu)}G_n^{(n+\nu)}, \tag{11.13}$$
$$G_{n+j}^{(n+\nu)} = G_j^{(\nu)}G_n^{(n+\nu)}, \tag{11.14}$$

and that

$$\Gamma_i^{(n+\nu)} = \Gamma_i^{(n)}\Gamma_{n+1}^{(n+\nu)}, \tag{11.15}$$
$$\Gamma_{n+j}^{(n+\nu)} = \Gamma_j^{(\nu)} + G_j^{(\nu)}\Gamma_n^{(n)}\Gamma_{n+1}^{(n+\nu)} \tag{11.16}$$

for $1 \le i \le n$ *and* $1 \le j \le \nu$. *Moreover,*

$$G_n^{(n+\nu)} = (I - \Gamma_n^{(n)} G_1^{(\nu)})^{-1} G_n^{(n)}, \tag{11.17}$$

$$\Gamma_{n+1}^{(n+\nu)} = (I - G_1^{(\nu)} \Gamma_n^{(n)})^{-1} \Gamma_1^{(\nu)}. \tag{11.18}$$

Proof. We first prove (11.15). In order to reach $\ell(n+\nu+1)$ before $\ell(0)$, starting from $\ell(i)$ with $1 \le i \le n$, we need to reach $\ell(n+1)$ before $\ell(0)$ and, conditioned on the new starting state being in $\ell(n+1)$, we need to reach $\ell(n+\nu+1)$ before $\ell(0)$.

Equation (11.14) follows from the same argument: in order to reach $\ell(0)$, the process must first reach $\ell(n)$, and then $\ell(0)$, without visiting $\ell(n+\nu+1)$; we use the remark made earlier, that the probability of moving from $\ell(n+j)$ to $\ell(n)$ under taboo of $\ell(n+\nu+1)$ is the same as the probability $G_j^{(\nu)}$ of moving from $\ell(j)$ to $\ell(0)$ under taboo of $\ell(\nu)$.

We similarly obtain that

$$G_i^{(n+\nu)} = G_i^{(n)} + \Gamma_i^{(n)} G_{n+1}^{(n+\nu)}. \tag{11.19}$$

Indeed, starting from $\ell(i)$ with $i \le n$, we may either reach $\ell(0)$ before $\ell(n+1)$ or reach $\ell(n+1)$ before $\ell(0)$ and eventually reach $\ell(0)$ before $\ell(n+\nu+1)$, starting from $\ell(n+1)$.

Equation (11.13) immediately results from (11.14), (11.19). In particular, we obtain for $i = n$ that

$$(I - \Gamma_n^{(n)} G_1^{(\nu)}) G_n^{(n+\nu)} = G_n^{(n)}.$$

The matrix $\Gamma_n^{(n)} G_1^{(\nu)}$ records the probability that, starting from $\ell(n)$, the Markov chain moves to $\ell(n+1)$ at some time and then returns to $\ell(n)$, without visiting either $\ell(0)$ or $\ell(n+\nu+1)$ in between. Since the Markov chain is irreducible, this cannot happen forever and the matrix $(I - \Gamma_n^{(n)} G_1^{(\nu)})$ is nonsingular, which proves (11.17).

The remaining equations are proved in a similar manner. □

Expected sojourn times are determined in exactly the same way, and we leave the proof of the next lemma to the reader.

Lemma 11.3.3 *We have that*

$$\begin{aligned} \boldsymbol{s}_i^{(n+\nu)} &= \boldsymbol{s}_i^{(n)} + \Gamma_i^{(n)} \boldsymbol{s}_1^{(\nu)} + \Gamma_i^{(n)} G_1^{(\nu)} \boldsymbol{s}_n^{(n+\nu)}, \\ \boldsymbol{s}_n^{(n+\nu)} &= (I - \Gamma_n^{(n)} G_1^{(\nu)})^{-1} \{\boldsymbol{s}_n^{(n)} + \Gamma_n^{(n)} \boldsymbol{s}_1^{(\nu)}\}, \\ \boldsymbol{s}_{n+j}^{(n+\nu)} &= \boldsymbol{s}_j^{(\nu)} + G_j^{(\nu)} \boldsymbol{s}_n^{(n+\nu)} \end{aligned}$$

```
Reduction by Bisection
i := M − 1;
Initialize S^(1);
k := 1;
Initialize S^(0);
n := 0;
while i > 0 do
    if odd(i) then
        n' := n;
        n := n + k;
        Compute S^(n) from S^(n') and S^(k)
    fi;
    Compute S^(2k) from S^(k);
    k := 2k;
    i := ⌊i/2⌋
od
```

Figure 11.2: Computation of the expected passage time from $\ell(0)$ to $\ell(M-1)$; the algorithm follows the right-to-left binary method for exponentiation.

---

for $1 \leq i \leq n$ *and* $1 \leq j \leq \nu$. □

In order to compute $\boldsymbol{s}_1^{(M-1)}$, we may use the algorithm in Figure 11.2, which is inspired by the "right-to-left binary method for exponentiation" (see Knuth [41]). We represent by $S^{(i)}$ the set $\{G_1^{(i)}, G_i^{(i)}, \Gamma_1^{(i)}, \Gamma_i^{(i)}, \boldsymbol{s}_1^{(i)}, \boldsymbol{s}_i^{(i)}\}$, which is such that one may determine $S^{(i+j)}$, knowing $S^{(i)}$ and $S^{(j)}$, by using Lemmas 11.3.2 and 11.3.3. The variables i, k, and n are such that $n + ik$ always remains equal to $M - 1$, so that when the algorithm terminates, n is equal to $M - 1$ and the last set $S^{(n)}$ contains $G_1^{(M-1)}$ and $\boldsymbol{s}_1^{(M-1)}$, so that we may compute the expected first passage time $\boldsymbol{t}_{0 \to M}$ by (11.12).

At each iteration, one performs a limited number of matrix multiplications. Since i is systematically halved, the number of iterations is $\log_2 M$ and the overall complexity is $O(\log_2 Mm^3)$, which is better than the $O(Mm^3)$ complexity of the linear algorithm in Figure 11.1.

11.4 Odd–Even Reduction

The odd–even reduction approach of section 10.2 may also be used in the present context. This was done in Li and Sheng [59], where it is called the folding method. The presentation in [59] is mostly based on algebraic and algorithmic arguments. We shall proceed in a different manner and, in keeping with the general spirit of this work, emphasize the probabilistic significance of the procedure. Furthermore, we shall focus our attention on the evaluation of $G_1^{(M-1)}$ and $\boldsymbol{s}_1^{(M-1)}$, which, as we saw earlier, are all that one needs to compute $\boldsymbol{t}_{0\to M}$.

Reverting to Lemmas 5.1.2 and 5.2.4, we may write that

$$(I - P_\omega^{(M-1)}) \begin{bmatrix} G_1^{(M-1)} \\ G_2^{(M-1)} \\ \vdots \\ G_{M-1}^{(M-1)} \end{bmatrix} = \begin{bmatrix} A_2 \\ 0 \\ \vdots \\ 0 \end{bmatrix} \tag{11.20}$$

and that

$$(I - P_\omega^{(M-1)}) \begin{bmatrix} \boldsymbol{s}_1^{(M-1)} \\ \boldsymbol{s}_2^{(M-1)} \\ \vdots \\ \boldsymbol{s}_{M-1}^{(M-1)} \end{bmatrix} = \begin{bmatrix} \mathbf{1} \\ \mathbf{1} \\ \vdots \\ \mathbf{1} \end{bmatrix}, \tag{11.21}$$

where $P_\omega^{(n)}$ is given in (11.11). We do not need the whole solution to these systems, of course, but only those rows which correspond to initial states in the level 1.

If we knew the transition matrix $P^{\{1\}}$ of the Markov process in the local time of $\ell(1)$ before the first visit to $\ell(0)$ or $\ell(M)$, then we might simply solve the smaller system

$$(I - P^{\{1\}})G_1^{(M-1)} = A_2, \tag{11.22}$$

since we may reach a state in $\ell(0)$ only from the states in $\ell(1)$. Furthermore, if we knew the vector $\boldsymbol{v}^{\{1\}}$ of expected sojourn times in the levels 2 to $M-1$, per unit of local time of $\ell(1)$, before the first visit to $\ell(0)$ or $\ell(M)$, then we might write

$$(I - P^{\{1\}})\boldsymbol{s}_1^{(M-1)} = \mathbf{1} + \boldsymbol{v}^{\{1\}}$$

or, equivalently,

$$(I - P^{\{1\}})\boldsymbol{s}_1^{(M-1)} = \boldsymbol{b}^{\{1\}},$$

where $\boldsymbol{b}^{\{1\}}$ is the vector of expected sojourn time in the whole set $\cup_{1\le k\le M-1}\ell(k)$ per unit of the local time of $\ell(1)$.

We shall proceed one step at a time as follows. First, we determine the transition matrix $P^{\{1,3,5,\ldots\}}$ in the local time of the odd-numbered levels $\ell(1)\cup\ell(3)\cup\ell(5)\cup\cdots$ between $\ell(1)$ and $\ell(M-1)$ and the vector $\boldsymbol{b}^{\{1,3,5,\ldots\}}$ of expected sojourn times in the levels $\ell(1)$ to $\ell(M-1)$ per unit of local time in $\ell(1)\cup\ell(3)\cup\cdots$. Thus, we may replace the systems (11.20), (11.21) by the smaller systems

$$(I - P^{\{1,3,5,\cdots\}})\begin{bmatrix} G_1^{(M-1)} \\ G_3^{(M-1)} \\ G_5^{(M-1)} \\ \vdots \end{bmatrix} = \begin{bmatrix} A_2 \\ 0 \\ 0 \\ \vdots \end{bmatrix}$$

and

$$(I - P^{\{1,3,5,\cdots\}})\begin{bmatrix} \boldsymbol{s}_1^{(M-1)} \\ \boldsymbol{s}_3^{(M-1)} \\ \boldsymbol{s}_5^{(M-1)} \\ \vdots \end{bmatrix} = \begin{bmatrix} \boldsymbol{b}_1^{\{1,3,5,\cdots\}} \\ \boldsymbol{b}_3^{\{1,3,5,\cdots\}} \\ \boldsymbol{b}_5^{\{1,3,5,\cdots\}} \\ \vdots \end{bmatrix}.$$

Recursively repeating the reduction, we consider the process in the local time of $\ell(1)\cup\ell(5)\cup\ell(9)\cup\cdots$, then $\ell(1)\cup\ell(9)\cup\ell(17)\cup\cdots$, and so on, until only the level $\ell(1)$ remains. This whole procedure may be repeated because we are left at each step with the transition matrix of a QBD, as we saw in section 10.2.

We shall now describe the transformations which are required when performing one step in the reduction procedure. We assume that we are left with the n levels $\lambda(1), \lambda(2), \ldots, \lambda(n)$ with $\lambda(1) = 1$ necessarily. The systems to be solved are

$$(I - P^{\{\Lambda\}}) \begin{bmatrix} G_1^{(M-1)} \\ G_{\lambda(2)}^{(M-1)} \\ \vdots \end{bmatrix} = \begin{bmatrix} A_2 \\ 0 \\ \vdots \end{bmatrix} \tag{11.23}$$

and

$$(I - P^{\{\Lambda\}}) \begin{bmatrix} \boldsymbol{s}_1^{(M-1)} \\ \boldsymbol{s}_{\lambda(2)}^{(M-1)} \\ \vdots \end{bmatrix} = \begin{bmatrix} \boldsymbol{b}_1^{\{\Lambda\}} \\ \boldsymbol{b}_{\lambda(2)}^{\{\Lambda\}} \\ \vdots \end{bmatrix} \tag{11.24}$$

with $\Lambda = \{\lambda(1), \lambda(2), \ldots, \lambda(n)\}$. In order to simplify the equations, we shall omit the exponent $\{\Lambda\}$, use P to represent the transition matrix in the local time of Λ, and use $\boldsymbol{b}$ for the vector of expected sojourn times in $\ell(1)$ to $\ell(M-1)$ per unit of local time.

In expanded notation, we have that

$$P = \begin{bmatrix} P_{11} & C_0 & & & & \\ C_2 & C_1 & C_0 & & & \\ & C_2 & C_1 & \ddots & & \\ & & \ddots & \ddots & C_0 & \\ & & & C_2 & C_1 & C_0 \\ & & & & C_2 & P_{nn} \end{bmatrix}.$$

Observe that the blocks in the upper-left and lower-right corners are different from the other diagonal blocks; despite the fact that they are equal at the beginning, as we see in (11.11), they become different as the reduction proceeds.

Let us temporarily assume that n is odd and is equal to $2k-1$. Permuting the rows and columns of P so that $\Lambda_{\text{odd}} = \{\lambda(1), \lambda(3), \ldots, \lambda(n)\}$ is enumerated first, followed by $\Lambda_{\text{even}} = \{\lambda(2), \lambda(4), \ldots, \lambda(n-1)\}$, we find that

$$P = \left[\begin{array}{ccccc|ccccc} P_{11} & & & & & C_0 & & & & \\ & C_1 & & & & C_2 & C_0 & & & \\ & & \ddots & & & & C_2 & \ddots & & \\ & & & C_1 & & & & \ddots & C_0 \\ & & & & P_{nn} & & & & C_2 \\ \hline C_2 & C_0 & & & & C_1 & & & \\ & C_2 & \ddots & & & & C_1 & & \\ & & \ddots & C_0 & & & & \ddots & \\ & & & C_2 & C_0 & & & & C_1 \end{array}\right],$$

which we may write as

$$P = \begin{bmatrix} P_{\text{odd}} & P_{\text{oe}} \\ P_{\text{eo}} & P_{\text{even}} \end{bmatrix}.$$

The transition matrix in the local time of Λ_{odd} is

$$\begin{aligned} P^* &= P_{\text{odd}} + P_{\text{oe}}(I - P_{\text{even}})^{-1} P_{\text{eo}} \\ &= \begin{bmatrix} P_{11}^* & C_0^* & & & \\ C_2^* & C_1^* & C_0^* & & \\ & \ddots & \ddots & \ddots & \\ & & & C_1^* & C_0^* \\ & & & C_2^* & P_{kk}^* \end{bmatrix} \end{aligned}$$

with

$$\begin{aligned} P_{11}^* &= P_{11} + C_0 B_2, \\ C_0^* &= C_0 B_0, \\ C_2^* &= C_2 B_2, \\ C_1^* &= C_1 + C_0 B_2 + C_2 B_0, \end{aligned}$$

and

$$P_{kk}^* = P_{nn} + C_2 B_0, \tag{11.25}$$

where

$$B_i = (I - C_1)^{-1} C_i$$

for $i = 0$ and 2.

If n is even, $n = 2k$, say, then P^* has the same structure and we have that

$$P^*_{kk} = C_1 + C_0(I - P_{nn})^{-1}C_2 + C_2B_0,$$

instead of (11.25).

The system (11.23) has n blocks; we may find G_1^{M-1} by solving the smaller system

$$(I - P^*)\begin{bmatrix} G_1^{(M-1)} \\ G_{\lambda(3)}^{(M-1)} \\ \vdots \end{bmatrix} = \begin{bmatrix} A_2 \\ 0 \\ \vdots \end{bmatrix}$$

with $\lfloor (n+1)/2 \rfloor$ blocks. The procedure is repeated until there is only one block left, at which time the system reduces to (11.22) and may easily be solved.

After the same permutation of rows and columns, the system (11.24) may be written as

$$\begin{bmatrix} I - P_{\text{odd}} & -P_{\text{oe}} \\ -P_{\text{eo}} & I - P_{\text{even}} \end{bmatrix}\begin{bmatrix} \boldsymbol{s}_{\text{odd}} \\ \boldsymbol{s}_{\text{even}} \end{bmatrix} = \begin{bmatrix} \boldsymbol{s}_{\text{odd}} \\ \boldsymbol{b}_{\text{even}} \end{bmatrix}.$$

After reduction, this yields

$$(I - P^*)\boldsymbol{s}_{\text{odd}} = \boldsymbol{b}^*$$

with

$$\boldsymbol{b}^* = \boldsymbol{b}_{\text{odd}} + P_{\text{oe}}(I - P_{\text{even}})^{-1}\boldsymbol{b}_{\text{even}}.$$

Indeed, $\boldsymbol{b}_{\text{odd}}$ records the expected sojourn time in $\ell(1)$ to $\ell(M-1)$, starting from one of the levels in Λ_{odd}, before the next return to any of the levels in Λ; if that level happens to be in Λ_{even} (with probability P_{oe}), the second term records the expected sojourn time accumulated between all the returns to Λ_{even} until the first visit to Λ_{odd}.

One easily verifies that

$$\begin{aligned} \boldsymbol{b}_1^* &= \boldsymbol{b}_1 + C_0(I - C_1)^{-1}\boldsymbol{b}_2, \\ \boldsymbol{b}_i^* &= \boldsymbol{b}_{2i-1} + C_2(I - C_1)^{-1}\boldsymbol{b}_{2(i-1)} + C_0(I - C_1)^{-1}\boldsymbol{b}_{2i} \end{aligned}$$

for $2 \le i \le k-1$ and that

$$\boldsymbol{b}_k^* = \boldsymbol{b}_{2k-1} + C_2(I - C_1)^{-1}\boldsymbol{b}_{2(k-1)}$$

Odd–Even Reduction

Procedure odd–even $(G, \boldsymbol{s}, n, P_{11}, C_0, C_1, C_2, P_{nn}, \boldsymbol{b})$
  if $n \geq 2$ **then**
    $B_0 := (I - C_1)^{-1}C_0;$
    $B_2 := (I - C_1)^{-1}C_2;$
    $P^*_{11} := P_1 + C_0B_2;$
    $C^*_0 := C_0B_0;$
    $C^*_2 := C_2B_2;$
    $C^*_1 := C_1 + C_0B_2 + C_2B_0;$
    $k := (n+1)$ div 2;
    $\boldsymbol{b}^*_1 := \boldsymbol{b}_1 + C_0(I - C_1)^{-1}\boldsymbol{b}_2;$
    for $i := 2$ **to** $k-1$ **do**
      $\boldsymbol{b}^*_i := \boldsymbol{b}_{2i-1} + C_2(I - C_1)^{-1}\boldsymbol{b}_{2(i-1)} + C_0(I - C_1)^{-1}\boldsymbol{b}_{2i}$
    od;
    if odd(n) **then**
      $P^*_{kk} := P_{nn} + C_2B_0;$
      $\boldsymbol{b}^*_k := \boldsymbol{b}_{2k-1} + C_2(I - C_1)^{-1}\boldsymbol{b}_{2(k-1)}$
    else
      $P^*_{kk} := C_1 + C_2B_0 + C_0(I - P_{nn})^{-1}C_2;$
      $\boldsymbol{b}^*_k := \boldsymbol{b}_{2k-1} + C_2(I - C_1)^{-1}\boldsymbol{b}_{2(k-1)} + C_0(I - P_{nn})^{-1}\boldsymbol{b}_{2k}$
    fi;
    odd–even $(G, \boldsymbol{s}, k, P^*_{11}, C^*_0, C^*_1, C^*_2, P^*_{kk}, \boldsymbol{b}^*)$
  else
    solve the systems
      $(I - P_{11})G = A_2,$
      $(I - P_{11})\boldsymbol{s} = \boldsymbol{b}.$
  fi

Figure 11.3: The procedure odd–even recursively reduces the matrix $(I - P^{\{\Lambda\}})$ and the vector $b^{\{\Lambda\}}$ until only the level 1 remains, at which time one may solve the systems $(I - P^{\{1\}})G_1^{(M-1)} = A_2$ and $(I - P^{\{1\}})s_1^{(M-1)} = b^{\{1\}}$.

if n is odd and equal to $2k-1$, while

$$\boldsymbol{b}_k^* = \boldsymbol{b}_{2k-1} + C_2(I - C_1)^{-1}\boldsymbol{b}_{2(k-1)} + C_0(I - P_{nn})^{-1}\boldsymbol{b}_{2k}$$

if n is even and equal to $2k$.

We give in Figure 11.3 a recursive algorithm which implements the equations above and is invoked by the statement

$$\text{odd–even}(G_1^{(M-1)}, \boldsymbol{s}_1^{(M-1)}, M-1, A_1, A_0, A_1, A_2, A_1, \mathbf{1}).$$

This version of the algorithm is highly inefficient, and one should in practice implement an iterative version of it. Nevertheless, it clearly shows that the complexity is $O(\log_2 Mm^3)$, as in section 11.3.

Part V

Beyond Simple QBDs

Chapter 12

Nonhomogeneous QBDs

This is the first in a sequence of short chapters where we discuss various extensions to the processes which we have analyzed so far. Our intention is to show, without going into a great deal of detail, that the fundamental ideas extend beyond simple homogeneous QBDs and apply to more complex systems.

We gave in section 1.8 an example of a nonhomogeneous QBD. In general, the transition matrix has the form

$$P = \begin{bmatrix} A_1^{(0)} & A_0^{(0)} & 0 & & \\ A_2^{(1)} & A_1^{(1)} & A_0^{(1)} & 0 & \\ 0 & A_2^{(2)} & A_1^{(2)} & A_0^{(2)} & \ddots \\ & 0 & A_2^{(3)} & A_1^{(3)} & \ddots \\ & & \ddots & \ddots & \ddots \end{bmatrix}. \tag{12.1}$$

The state space is still two dimensional and partitioned into levels, and transitions are still allowed between adjacent levels only, but the transition probabilities out of the state (n, j) may depend on the level n. As a matter of fact, different levels may have different numbers of phases, so that while the blocks on the main diagonal are square matrices, those on the secondary diagonals may be rectangular matrices of appropriate dimensions. We do not specify the dimensions of the matrices which we shall introduce; the context will indicate these unambiguously. We only note that with these dimensions properly specified, all matrix operations in what follows are well defined. We give a few references below, where nonhomogeneous QBDs are treated in their full generality. This is an area of active research, and many algorithmic issues remain unresolved.

12.1 The Stationary Distribution

The stationary distribution, when the process is positive recurrent, is easily seen to have a matrix-product form, as stated in the theorem below.

Theorem 12.1.1 *Assume that the nonhomogeneous QBD is irreducible, aperiodic, and positive recurrent. Its limiting probability vector* $\boldsymbol{\pi}$ *satisfies the relation*

$$\boldsymbol{\pi}_n = \boldsymbol{\pi}_{n-1} R^{(n)} \qquad \text{for } n \geq 1, \tag{12.2}$$

where the matrix $R^{(n)}$ *records the expected number of visits to the states in the level* n *between two visits to the level* $n-1$.

Proof. This is a direct consequence of Theorem 5.2.1, where we choose $T = \ell(0) \cup \cdots \cup \ell(n-1)$ and $D = \ell(n)$. □

In the homogeneous case of Chapter 6, we noted that $R^{(n)} \equiv R$ and we introduced two related matrices G and U. The analogues in the general case are matrix sequences $\{G^{(n)} : n \geq 1\}$ and $\{U^{(n)} : n \geq 1\}$, where $G^{(n)}$ records the first passage probabilities from $\ell(n)$ to $\ell(n-1)$ and $U^{(n)}$ records the first return probabilities to $\ell(n)$ under taboo of levels $\ell(0), \ldots, \ell(n-1)$. Just as in the homogeneous case, we have the following theorem.

Theorem 12.1.2 *Any one of the sequences* $\{G^{(n)}\}$, $\{U^{(n)}\}$, *and* $\{R^{(n)}\}$ *determines the other two through the set of relations below.*

$$G^{(n)} = (I - U^{(n)})^{-1} A_2^{(n)}, \tag{12.3}$$

$$R^{(n)} = A_0^{(n-1)} (I - U^{(n)})^{-1}, \tag{12.4}$$

$$U^{(n)} = A_1^{(n)} + A_0^{(n)} G^{(n+1)}, \tag{12.5}$$

$$U^{(n)} = A_1^{(n)} + R^{(n+1)} A_2^{(n)}. \tag{12.6}$$

Proof. All the equations above are easy to establish by a direct probabilistic argument. For instance, (12.3) is obvious once we write it as $G^{(n)} = \sum_{k\geq 0}[U^{(n)}]^k A_2^{(n)}$ and interpret it as asserting that a first passage from $\ell(n)$ to $\ell(n-1)$ occurs after some $k \geq 0$ returns to $\ell(n)$ avoiding lower levels. We omit the details regarding the other equations which are similar. □

Not only can each of the three sequences of matrices be characterized in terms of the others, but also the terms in each sequence are themselves interrelated as stated below.

Theorem 12.1.3 *The matrices $\{G^{(n)}\}$, $\{R^{(n)}\}$, and $\{U^{(n)}\}$ satisfy the equations*

$$\begin{aligned} G^{(n)} &= A_2^{(n)} + A_1^{(n)} G^{(n)} + A_0^{(n)} G^{(n+1)} G^{(n)}, && (12.7) \\ R^{(n)} &= A_0^{(n-1)} + R^{(n)} A_1^{(n)} + R^{(n)} R^{(n+1)} A_2^{(n+1)}, && (12.8) \\ U^{(n)} &= A_1^{(n)} + A_0^{(n)} (I - U^{(n+1)})^{-1} A_2^{(n+1)} && (12.9) \end{aligned}$$

for $n \geq 1$ and are the minimal nonnegative solutions of their respective system of equations.

Proof. To prove (12.7), we condition on the state after the first transition. The first term corresponds to the case where the QBD directly moves from $\ell(n)$ to $\ell(n-1)$ in one transition with probabilities recorded in $A_2^{(n)}$; with probabilities recorded in $A_1^{(n)}$, the QBD remains in $\ell(n)$, from where it still has to move eventually to $\ell(n-1)$ with probabilities recorded in $G^{(n)}$; finally, with probabilities recorded in $A_0^{(n)}$, the QBD moves up to $\ell(n+1)$, from where it must eventually return to $\ell(n)$ with probabilities recorded in $G^{(n+1)}$ and then to $\ell(n-1)$, again with probabilities recorded in $G^{(n)}$.

This is the argument which was used in the proof of (6.14) and of (7.1). To prove the other two equations, we repeat the arguments given in Remark 6.2.11.

The proof of minimality, though long and arduous, is similar to that in Theorems 8.1.3 and 8.1.4 for the homogeneous case. We only outline the main steps in the case of (12.7).

The first step is to show that an iterative scheme using (12.3) and (12.5) starting with $G_0^{(n)} \equiv 0$ yields sequences $\{G_k^{(n)} : n \geq 1\}$ and

$\{U_k^{(n)} : n \geq 1\}$ for $k \geq 1$, which monotonically converge, as k tends to infinity, to the sequences $\{G^{(n)} : n \geq 1\}$ and $\{U^{(n)} : n \geq 1\}$ of interest to us.

The second step is to take any nonnegative solution $\{X^{(n)} : n \geq 1\}$ of (12.7), define $W^{(n)} = A_1^{(n)} + A_0^{(n)} X^{(n)}$ for $n \geq 1$, and show that the sequences $\{X^{(n)} : n \geq 1\}$ and $\{W^{(n)} : n \geq 1\}$, respectively, majorize the sequences $\{G_k^{(n)} : n \geq 1\}$ and $\{U_k^{(n)} : n \geq 1\}$ for all k. $\square$

The next statement is a direct consequence of Theorem 5.3.1 and completes the characterization of the steady state probability vector. Note that an empty product, such as that which occurs in (12.11) for $n = 0$, is the identity matrix.

Theorem 12.1.4 *The QBD is positive recurrent if and only if there exists a strictly positive solution of the system*

$$\boldsymbol{\pi}_0 = \boldsymbol{\pi}_0(A_1^{(0)} + A_0^{(0)} G^{(1)}) \tag{12.10}$$

normalized by

$$\boldsymbol{\pi}_0 \sum_{n \geq 0} \prod_{1 \leq k \leq n} R^{(k)} \mathbf{1} = 1. \tag{12.11}$$

The system (12.10) *is equivalent to the system*

$$\boldsymbol{\pi}_0 = \boldsymbol{\pi}_0(A_1^{(0)} + R^{(1)} A_2^{(1)}).$$

Proof. Consider the restricted process on $T = \ell(0)$. Its transition matrix is readily seen to be $P(T) = A_1^{(0)} + A_0^{(0)} G^{(1)}$, so that (12.10) is a consequence of Theorem 5.3.1. By (12.3), (12.4) we have that

$$A_1^{(0)} + A_0^{(0)} G^{(1)} = A_1^{(0)} + R^{(1)} A_2^{(0)},$$

which proves the last statement. $\square$

If the number of phases at each level is finite, then we may state the following stability condition, which is merely a reformulation of Theorem 7.1.1.

Theorem 12.1.5 *If the levels contain finitely many phases, then the QBD is recurrent if and only if $G^{(1)}$ is stochastic.* $\square$

The qualitative properties of homogeneous QBDs thus extend easily to the general case. Unfortunately, we cannot obtain in general a simple stability condition as that of Theorem 7.2.3. Also, computational algorithms are more involved, as we shall discuss in the next sections.

12.2 Algorithmic Approaches

General QBD processes with a finite number of levels and finitely many phases at each level have been analyzed in Gaver, Jacobs, and Latouche [26]. The procedure is nearly the same as the one given in section 10.1; the only difference is that the blocks in P change at each level, so that (10.8), (10.9) now become

$$\begin{aligned} U^{(M)} &= A_1^{(M)}, \\ U^{(n)} &= A_1^{(n)} + A_0^{(n)}(I - U^{(n+1)})^{-1} A_2^{(n+1)} \end{aligned}$$

for $0 \le n \le M - 1$, from which we readily obtain the matrices $R^{(n)}$ by 12.4. Unless some additional assumption is made about the matrix P, it does not seem that there exists a significantly better direct computational procedure.

If $M = \infty$, the situation is more complex. In the case of an infinite Markov chain for which no easily computable analytic expression is available, it is a standard practice to truncate the transition matrix P at some arbitrary but large level, renormalize the rows to obtain a finite stochastic matrix P^*, and solve numerically the linear system $\boldsymbol{\pi}^* P^* = \boldsymbol{\pi}^*$, $\boldsymbol{\pi}^* \mathbf{1} = 1$ as in the section above.

The renormalization step is somewhat arbitrary: one might add to the last column the missing mass on each row, or add it to the first column, or to the nonzero elements in due proportion, etc. In general, one needs to be careful: Gibson and Seneta [27] show that, with some augmentation schemes, the approximation $\boldsymbol{\pi}^*$ does not necessarily converge to the stationary vector of P, as the number of states in P^* tends to infinity; we have already commented upon this issue in Remark 8.3.2.

In the present case, assuming that one has somehow selected an upper level M, Bright and Taylor [10] suggest proceeding with an "exact" augmentation and to replace the block $A_1^{(n)}$ by the matrix $U^{(M)}$ so that

P^* is given by

$$P^* = \begin{bmatrix} A_1^{(0)} & A_0^{(0)} & & & \\ A_2^{(1)} & A_1^{(1)} & \ddots & & \\ & \ddots & \ddots & & \\ & & & A_1^{(M-1)} & A_0^{(M-1)} \\ & & & A_2^{(M)} & U^{(M)} \end{bmatrix}.$$

Then, it is a simple matter, using Theorems 12.1.1 and 12.1.4, to show that $\boldsymbol{\pi}_n^*$ satisfies the relation (12.2) for $1 \leq n \leq M$ with $\boldsymbol{\pi}_0^*$ a strictly positive solution of (12.10). In the homogeneous case, this is the same property that has been mentioned in Remark 10.3.3.

Since the solution of (12.10) is unique up to a scalar constant, this shows that for n between 0 and M, $\boldsymbol{\pi}_n^*$ and $\boldsymbol{\pi}_n$ are proportional, the only difference being due to the fact that $\boldsymbol{\pi}_0^*$ is normalized so that

$$\boldsymbol{\pi}_0^* \sum_{0 \leq n \leq M} \prod_{1 \leq k \leq n} R^{(k)} \mathbf{1} = 1,$$

while the summation in (12.11) extends to infinity. Clearly, $\boldsymbol{\pi}_n^*$ will converge to $\boldsymbol{\pi}_n$ as $M \to \infty$.

Two major questions remain: how to select M and how to compute $U^{(M)}$. Ideally, one would like M to be such that the stationary probability of the levels $M+1$ and higher is negligible. Then $\sum_{0 \leq n \leq M} \boldsymbol{\pi}_n \mathbf{1} \approx 1$ and $\boldsymbol{\pi}_n^*$ is actually close to $\boldsymbol{\pi}_n$ for $0 \leq n \leq M$. There is, of course, no general answer to this question; several approaches are discussed by Bright and Taylor [10].

With respect to the second question, instead of computing $U^{(M)}$, these authors propose directly computing $R^{(M)}$, from which one may obtain $R^{(M-1)}$, $R^{(M-2)}$, etc. by (12.4), (12.6). Their algorithm for the evaluation of $R^{(M)}$ is inspired by the logarithmic reduction algorithm of section 8.4.

In the homogeneous case, we have generally favored iteratively computing the transition probability matrix G instead of the rate matrix R. Similarly, we would recommend computing $G^{(M+1)}$ using the theorem below, from which $U^{(M)}$ is readily obtained by (12.5). The theorem is stated without proof as one merely needs to follow the argument of

Theorem 8.4.1 (a proof is given in Ramaswami and Taylor [103]). The notations are slightly more involved due to the fact that we are dealing with level-dependent transition probabilities. Recall that $\gamma(n)$ is the first passage time to the level n. We define the matrices $H_k^{[\ell]}$ and $L_k^{[\ell]}$ as follows:

$$H_k^{[\ell]} = \mathrm{P}[\gamma(k+2^\ell) < \gamma(k-2^\ell) \ \& \ X_{\gamma(k+2^\ell)} | X_0 \in \ell(k)]$$

and

$$L_k^{[\ell]} = \mathrm{P}[\gamma(k-2^\ell) < \gamma(k+2^\ell) \ \& \ X_{\gamma(k-2^\ell)} | X_0 \in \ell(k)]$$

for $k \geq 2^\ell$.

Theorem 12.2.1 *The matrix $G^{(M+1)}$ is given by*

$$G^{(M+1)} = \sum_{\ell \geq 0} \prod_{0 \leq i \leq \ell-1} H_{M+2^i}^{[i]} L_{M+2^\ell}^{[\ell]} \tag{12.12}$$

with

$$\begin{aligned} H_k^{[0]} &= (I - A_1^{(k)})^{-1} A_0^{(k)}, \\ L_k^{[0]} &= (I - A_1^{(k)})^{-1} A_2^{(k)}, \\ H_k^{[\ell+1]} &= (I - U_k^{[\ell+1]})^{-1} H_k^{[\ell]} H_{k+2^\ell}^{[\ell]}, \\ L_k^{[\ell+1]} &= (I - U_k^{[\ell+1]})^{-1} L_k^{[\ell]} L_{k-2^\ell}^{[\ell]} \end{aligned}$$

for $\ell \geq 0$ with

$$U_k^{[\ell+1]} = L_k^{[\ell]} H_{k-2^\ell}^{[\ell]} + H_k^{[\ell]} L_{k+2^\ell}^{[\ell]}.$$

□

In order to actually compute $G^{(M+1)}$, one would naturally truncate the sum in the right-hand side of (12.12) at some index L such that the resulting matrix is nearly stochastic. Some practical details are given in [10] on how to design the algorithm.

This appears to be the only practical procedure available for computing $G^{(M+1)}$. If the maximal summation index L were known somehow (e.g., by using a demonstrable upper bound through some approximation), then one might consider solving, by LU decomposition or otherwise, a finite QBD on the levels between M and 2^L, but this would be more efficient than using (12.12) for small values of L only.

In summary, the structural results for homogeneous QBDs can be extended directly to the inhomogeneous case. Computational issues are, however, much more complex. Their resolution must necessarily exploit further properties of the specific model under study.

Chapter 13

Processes, Skip-Free in One Direction

Matrix analytic methods have been extensively applied to more elaborate systems than QBDs and in particular to Markov chains on two-dimensional state spaces for which one-step transitions are allowed across several levels in one direction. Simple examples of such are queues with batch arrivals or group services.

Transition matrices of the form

$$\tilde{P} = \begin{bmatrix} D_0 & D_1 & D_2 & D_3 & \cdots \\ C_0 & C_1 & C_2 & C_3 & \cdots \\ 0 & C_0 & C_1 & C_2 & \cdots \\ 0 & 0 & C_0 & C_1 & \cdots \\ \vdots & \vdots & \vdots & \vdots & \ddots \end{bmatrix} \tag{13.1}$$

characterize Markov chains of M/G/1-type,[19] which are also called *skip-free to the left* to indicate that the chain can move down by one level at a time only, although it may skip in the upward direction several levels in one transition.

Similarly, Markov chains of GI/M/1-type,[20] also called *skip-free to*

[19]The Markov chain embedded at service completion epochs for the standard M/G/1 queue has a transition matrix of this form.

[20]This term is a reminder for the structure of the Markov chain embedded at arrival epochs for the GI/M/1 queue.

the right, have transition matrices of the form

$$\tilde{P} = \begin{bmatrix} D_0 & C_0 & 0 & 0 & \cdots \\ D_1 & C_1 & C_0 & 0 & \cdots \\ D_2 & C_2 & C_1 & C_0 & \cdots \\ D_3 & C_3 & C_2 & C_1 & \cdots \\ \vdots & \vdots & \vdots & \vdots & \ddots \end{bmatrix}. \tag{13.2}$$

Historically, matrix analytic methods were developed for these two classes along independent lines (Neuts [79, 82]) and QBDs were seen as the special class of models at the intersection of M/G/1- and GI/M/1-type Markov chains (see, in particular, [79, Chapter 3]).

Here, we shall proceed in the reverse manner and show that QBDs are general enough so that they actually encompass the two classes of Markov chains with transition matrices given by (13.1), (13.2). These can be viewed as embedded processes in a suitably defined QBD and examined with the techniques developed in previous chapters.

We do not recommend that skip-free Markov chains should only be seen as special cases of QBDs and should only be approached as described in the next sections. Indeed, we will point out that it is simpler to directly use the structure given by (13.1), (13.2). Nevertheless, we think it is interesting to show how their theory can be brought under the unifying umbrella of QBDs.

To keep the discussion simple, we limit ourselves to discrete time Markov chains. The continuous time case is left to the reader. More details may be found in Ramaswami [97].

13.1 Markov Chains of M/G/1-Type

In (13.1), a transition from (n, i) to $(n + k, j), k \geq 1$, involves moving up k levels from $\ell(n)$ and selecting the phase j. We shall pretend that this occurs through a sequence of unseen steps in which the process first selects the phase j and marks down the number k of steps it has to move up in levels and then takes a sequence of steps, one level at a time, without changing phase.

Formally, we break up the transition from (n, i) to $(n + k, j)$ into the sequence of transitions

From	To	Transition matrix	
$\ell(0,0)$	$\ell(0,0)$	D_0	
$\ell(0,0)$	$\ell(0,k)$	D_k	$k \geq 1$
$\ell(n,0)$	$\ell(n-1,0)$	C_0	$n \geq 1$
$\ell(n,0)$	$\ell(n,k)$	C_{k+1}	$n \geq 1, k \geq 0$
$\ell(n,k)$	$\ell(n+1,k-1)$	I	$n \geq 0, k \geq 1$

Table 13.1: Representation of an M/G/1-type process as a QBD. Feasible transitions among sublevels.

$$(n,i) \to (n,k,j) \to (n+1,k-1,j) \cdots \to (n+k-1,1,j) \to (n+k,j),$$

with the probabilities of these transitions equal to $(C_{k+1})_{ij}$, 1, 1, ..., 1, respectively.

In order to have homogeneous notations for the states, we shall rewrite the states (n,i) and $(n+k,j)$, respectively, as $(n,0,i)$ and $(n+k,0,j)$. We clearly see that $(C_{k+1})_{ij}$ is the probability that the next state of the form $(\cdot,0,\cdot)$ visited after $(n,0,i)$ is $(n+k,0,j)$. Therefore, our equating states of the form $(n,0,i)$ in the new model with states of the form (n,i) in (13.1) preserves the transition law given by (13.1).

We thus consider a three-dimensional process $\{(X_t, D_t, J_t) : t \geq 0\}$ on the state space $\{(n,d,j) : n, d \geq 0, 1 \leq j \leq m\}$, which we partition into levels

$$\ell(n) = \{(n,d,j) : d \geq 0, 1 \leq j \leq m\}$$

for $n \geq 0$. Each level itself is partitioned into sublevels, and we have that $\ell(n) = \cup_{k\geq 0}\ell(n,k)$ with

$$\ell(n,k) = \{(n,k,j) : 1 \leq j \leq m\}$$

for $n, k \geq 0$. The transition probabilities among sublevels are defined in Table 13.1. That three-dimensional process is a QBD, where the phase is two dimensional and its transition matrix is

$$P = \begin{bmatrix} B & A_0 & & & \\ A_2 & A_1 & A_0 & & \\ & A_2 & A_1 & \ddots & \\ & & \ddots & \ddots & \end{bmatrix}, \tag{13.3}$$

with

$$B = \begin{bmatrix} D_0 & D_1 & D_2 & \cdots \\ 0 & 0 & 0 & \cdots \\ 0 & 0 & 0 & \cdots \\ \vdots & \vdots & \vdots & \ddots \end{bmatrix}, \tag{13.4}$$

$$A_0 = \begin{bmatrix} 0 & 0 & 0 & \cdots \\ I & 0 & 0 & \cdots \\ 0 & I & 0 & \cdots \\ \vdots & \vdots & \vdots & \ddots \end{bmatrix}, \tag{13.5}$$

$$A_1 = \begin{bmatrix} C_1 & C_2 & C_3 & \cdots \\ 0 & 0 & 0 & \cdots \\ 0 & 0 & 0 & \cdots \\ \vdots & \vdots & \vdots & \ddots \end{bmatrix}, \tag{13.6}$$

and

$$A_2 = \begin{bmatrix} C_0 & 0 & 0 & \cdots \\ 0 & 0 & 0 & \cdots \\ 0 & 0 & 0 & \cdots \\ \vdots & \vdots & \vdots & \ddots \end{bmatrix}. \tag{13.7}$$

The connection between the QBD and the original process is stated as a proposition, the proof of which is straightforward and omitted.

Proposition 13.1.1 *The process* $\{(\tilde{X}_t, \tilde{J}_t) : t \geq 0\}$ *embedded in the process* $\{(X_t, D_t, J_t) : t \geq 0\}$ *at epochs of visits to sublevels of the form* $\ell(n, 0)$ *for* $n \geq 0$ *is a Markov chain skip-free to the left with transition matrix* (13.1). *The two processes are simultaneously irreducible and ergodic.* □

As a direct consequence of Theorem 5.3.1 and the proposition above, we have the following corollary.

Corollary 13.1.2 *Let* $\pi = (\pi_{00}, \pi_{01}, \pi_{02}, \ldots, \pi_{10}, \pi_{11}, \pi_{12}, \ldots)$ *denote the stationary probability vector of* $\{(X_t, D_t, J_t) : t \geq 0\}$*, partitioned by sublevels.*

The stationary probability vector $\tilde{\pi}$ *of the M/G/1-type Markov chain* (13.1) *is given by*

$$\tilde{\pi}_i = c\pi_{i,0} \qquad \text{for } i \geq 0,$$

where c is a normalizing constant. □

In principle, it suffices to compute, when it exists, the stationary probability vector of the QBD (13.3) in order to determine that of an M/G/1-type Markov chain. In practice, this would pose major problems. First, the matrices A_0, A_1, A_2, and B in (13.3) have infinite order, so that they would have to be truncated before we implement one of the algorithms from Chapter 8. This is not the only problem, for even if we assume that there exists some index K such that $C_k = 0$ and $D_k = 0$ for all $k > K$ (i.e., if we assume that the Markov chain does not jump more than K steps at a time), so that the blocks in (13.3) are of finite order mK, they still might be too large for the efficient use of general purpose algorithms.

We shall instead make use of the very special structure of the matrices A_0, A_1, and A_2. We begin by elucidating the structure of the matrix G.

Theorem 13.1.3 *The matrix G of the QBD with blocks* (13.5)–(13.7) *has the structure*

$$G = \begin{bmatrix} H & 0 & 0 & \cdots \\ H^2 & 0 & 0 & \cdots \\ H^3 & 0 & 0 & \cdots \\ \vdots & \vdots & \vdots & \ddots \end{bmatrix}, \tag{13.8}$$

where

$$H_{ij} = \mathrm{P}[\tau < \infty \,\&\, (X_\tau, D_\tau, J_\tau) = (n-1, 0, j) | \quad (X_0, D_0, J_0) = (n, 0, i)] \tag{13.9}$$

for $1 \le i, j \le m$ and τ is the first passage time from $\ell(n)$ to $\ell(n-1)$.

Proof. All transitions downward into any level occur into sublevel 0 of that level. This is reflected in (13.7), where all columns except the first are zero. Since $G = (I-U)^{-1}A_2$ by (6.10), all columns of G except for the first must be zero. Thus, we may write

$$G = \begin{bmatrix} W_0 & 0 & 0 & \cdots \\ W_1 & 0 & 0 & \cdots \\ W_2 & 0 & 0 & \cdots \\ \vdots & \vdots & \vdots & \ddots \end{bmatrix},$$

where the matrices $W_k, k \ge 0$, record the transition probability to $\ell(n-1)$, starting from the sublevel k in level n:

$$\begin{aligned}(W_k)_{ij} &= \text{P}[\tau < \infty \;\&\; (X_\tau, D_\tau, J_\tau) = (n-1, 0, j) | \\ &\qquad (X_0, D_0, J_0) = (n, k, i)].\end{aligned}$$

If $k = 0$, this is H by the definition (13.9). If $k \ge 1$, then the process starting from (n, k, i) makes the transitions

$$(n, k, i) \to (n+1, k-1, i) \to \cdots \to (n+k, 0, i)$$

in its first k steps with probability 1, so that we also have that

$$\begin{aligned}(W_k)_{ij} &= \text{P}[\tau < \infty \;\&\; (X_\tau, D_\tau, J_\tau) = (n-1, 0, j) | \\ &\qquad (X_0, D_0, J_0) = (n+k, 0, i)] \\ &= [G^k]_{(0,i)(0,j)} \\ &= (H^k)_{ij},\end{aligned}$$

which completes the proof. □

By virtue of Proposition 13.1.1 and Theorem 13.1.3, it suffices to determine the matrix H of order m in order to have G, and R, and the stationary distribution of (13.1). Therefore, it would be useful to obtain an equation for H that may be used directly. This is provided by the next theorem.

Theorem 13.1.4 *The matrix H is the minimal nonnegative solution of the equation*

$$H = \sum_{n \geq 0} C_n H^n. \tag{13.10}$$

Proof. Recall that by Theorem 8.1.4, G is the minimal solution of the equation

$$Y = A_2 + A_1 Y + A_0 Y^2 \tag{13.11}$$

in the set of nonnegative matrices. Using (13.8) and (13.5), (13.7), we see that (13.11) is equivalent to the equation

$$\begin{bmatrix} H \\ H^2 \\ H^3 \\ \vdots \end{bmatrix} = \begin{bmatrix} \sum_{n \geq 0} C_n H^n \\ H^2 \\ H^3 \\ \vdots \end{bmatrix},$$

and that leads us to the only equation $H = \sum_{n \geq 0} C_n H^n$ to be considered. Also, the matrix H defined in (13.9) is the minimal nonnegative solution of (13.10) because any other solution H^* would allow us to construct a solution of (13.11) which would not be minimal, as G must be. □

The standard approach in the literature is to obtain (13.10) by directly considering (13.1) and analyzing the first passage probabilities from any level $\ell(n)$ to $\ell(n-1)$ for the skip-free process itself.

With regard to the stationary probability vector $(\tilde{\boldsymbol{\pi}}_0, \tilde{\boldsymbol{\pi}}_1, \ldots)$ itself, we do not obtain a recursion as simple as the matrix-geometric result. Nevertheless, we do obtain a recursive scheme as given in the next theorem. Note that, for $n = 0$, the empty sum in (13.12) is taken to be zero.

Theorem 13.1.5 *We have*

$$\tilde{\boldsymbol{\pi}}_n = \left(\tilde{\boldsymbol{\pi}}_0 \bar{D}_n + \sum_{1 \leq j \leq n-1} \tilde{\boldsymbol{\pi}}_j \bar{C}_{n-j+1} \right) (I - \bar{C}_1)^{-1}, \tag{13.12}$$

where

$$\bar{C}_n = \sum_{i\geq 0} C_{n+i}H^i$$

and

$$\bar{D}_n = \sum_{i\geq 0} D_{n+i}H^i$$

for $n \geq 1$.

Proof. In view of Corollary 13.1.2, we need to prove that

$$\boldsymbol{\pi}_{n,0} = \left(\boldsymbol{\pi}_{0,0}\bar{D}_n + \sum_{1\leq j\leq n-1} \boldsymbol{\pi}_{j0}\bar{C}_{n-j+1} \right) (I - \bar{C}_1)^{-1} \qquad (13.13)$$

for $n \geq 1$. By Theorem 5.3.1, we have that $(\boldsymbol{\pi}_0, \boldsymbol{\pi}_1, \ldots, \boldsymbol{\pi}_n)$ is proportional to the stationary probability vector of the embedded QBD on the set $\cup_{0\leq k\leq n}\ell(k)$, which has the transition matrix

$$P^{[n]} = \begin{bmatrix} B & A_0 & & & \\ A_2 & A_1 & A_0 & & \\ & A_2 & A_1 & \ddots & \\ & & \ddots & \ddots & A_0 \\ & & & A_2 & A_1 + A_0G \end{bmatrix}.$$

For the last column, this yields

$$\boldsymbol{\pi}_n = \boldsymbol{\pi}_{n-1}A_0 + \boldsymbol{\pi}_n(A_1 + A_0G).$$

Using (13.5), (13.6), (13.8), we readily obtain that

$$\boldsymbol{\pi}_{n,0} = \boldsymbol{\pi}_{n-1,1} + \boldsymbol{\pi}_{n,0}C_1 + \sum_{k\geq 1} \boldsymbol{\pi}_{n,k}H^k \qquad (13.14)$$

and

$$\boldsymbol{\pi}_{n,k} = \boldsymbol{\pi}_{n-1,k+1} + \boldsymbol{\pi}_{n,0}C_{k+1} \qquad (13.15)$$

for $k \geq 1$. If we replace $\boldsymbol{\pi}_{n,k}$ in (13.14) by the right-hand side of (13.15), we find after some simple manipulations that

$$\boldsymbol{\pi}_{n,0} = \boldsymbol{\pi}_{n,0}\bar{C}_1 + \sum_{k\geq 0} \boldsymbol{\pi}_{n-1,k+1}H^k$$

or, upon noting that $I - \bar{C}_1$ is nonsingular,

$$\boldsymbol{\pi}_{n,0} = \sum_{k \geq 0} \boldsymbol{\pi}_{n-1,k+1} H^k (I - \bar{C}_1)^{-1}. \qquad (13.16)$$

Repeatedly using (13.15) in (13.16) for terms for which the second index is greater than zero, this eventually gives

$$\boldsymbol{\pi}_{n,0} = \left(\sum_{1 \leq j \leq n-1} \boldsymbol{\pi}_{n-j,0} \bar{C}_{j+1} \right) (I - \bar{C}_1)^{-1} + \sum_{k \geq 0} \boldsymbol{\pi}_{0,k+n} H^k (I - \bar{C}_1)^{-1}.$$

We know by Lemma 6.3.1 that $\boldsymbol{\pi}_0 = \boldsymbol{\pi}_0(B + A_0 G)$, so that, by (13.4), (13.8), $\boldsymbol{\pi}_{0,k+n} = \boldsymbol{\pi}_{00} D_{k+n}$ for all $n \geq 1$, which completes the proof. □

Remark 13.1.6 Equation (13.12) is proved in Ramaswami [94] by an argument based on the structure (13.1) and on the interpretation of the matrix H.

13.2 Markov Chains of GI/M/1-Type

In the case (13.2) of Markov chains which are skip-free to the right, we proceed in a manner analogous to the preceding section and we decompose a step from (n, i) to $(n - k, j)$ into a sequence of transitions from one level to the next.

We define a three-dimensional process $\{(X_t, D_t, J_t) : t \geq 0\}$ on the state space $\{(n, d, j) : n \geq 0, 0 \leq d \leq n, 1 \leq j \leq m\}$. The state $(n, 0, i)$ corresponds to the state (n, i) of the original process; the state (n, d, i) with $d > 0$ signifies that the process still has to move down by d levels. Thus, the transition from (n, i) to $(n - k, j)$ is replaced by the sequence

$$(n, 0, i) \to (n, k, j) \to (n - 1, k - 1, j) \to \cdots \to (n - k, 0, j).$$

The probability of the first transition is $(C_k)_{ij}$ for $1 \leq k \leq n - 1$ and $(D_k)_{ij}$ for $k = n$; the probabilities of the subsequent transitions are all equal to 1.

Since the number of feasible phases d varies with the level, we see that the augmented process constitutes a *nonhomogeneous* QBD of the kind studied in Chapter 12, and its transition matrix is

$$P = \begin{bmatrix} A_1^{(0)} & A_0^{(0)} & & \\ A_2^{(1)} & A_1^{(1)} & A_0^{(1)} & \\ & A_2^{(2)} & A_1^{(2)} & \ddots \\ & & \ddots & \ddots \end{bmatrix}, \tag{13.17}$$

where, for $i \geq 0$, $A_1^{(i)}$ is a square matrix of order $m(i+1)$, while $A_0^{(i)}$ and $A_2^{(i)}$ are rectangular matrices with dimensions $m(i+1) \times m(i+2)$ and $m(i+1) \times mi$, respectively, given by

$$A_1^{(0)} = D_0, \tag{13.18}$$

$$A_1^{(i)} = \begin{bmatrix} C_1 & C_2 & \cdots & C_i & D_i \\ 0 & 0 & \cdots & 0 & 0 \\ \vdots & \vdots & \ddots & \vdots & \vdots \\ 0 & 0 & \cdots & 0 & 0 \end{bmatrix}, \tag{13.19}$$

$$A_0^{(i)} = \begin{bmatrix} C_0 & 0 & \cdots & 0 \\ 0 & 0 & \cdots & 0 \\ \vdots & \vdots & \ddots & \vdots \\ 0 & 0 & \cdots & 0 \end{bmatrix}, \tag{13.20}$$

and

$$A_2^{(i)} = \begin{bmatrix} 0 & 0 & \cdots & 0 \\ I & 0 & \cdots & 0 \\ 0 & I & \cdots & 0 \\ \vdots & \vdots & \ddots & \vdots \\ 0 & 0 & \cdots & I \end{bmatrix}.$$

As in section 13.1, we may state the following proposition, where $\boldsymbol{\pi} = (\boldsymbol{\pi}_{00}, \boldsymbol{\pi}_{10}, \boldsymbol{\pi}_{11}, \boldsymbol{\pi}_{20}, \boldsymbol{\pi}_{21}, \boldsymbol{\pi}_{22}, \ldots)$ is the stationary probability vector of (13.17).

Proposition 13.2.1 *The process $\{(\tilde{X}_t, \tilde{J}_t) : t \geq 0\}$ embedded in the process $\{(X_t, D_t, J_t) : t \geq 0\}$ at epochs of visits to the sublevels of the* *form* $\ell(n,0)$ *for* $n \geq 0$ *is a Markov chain skip-free to the right with transition matrix* (13.2).

Its stationary probability distribution, when it exists, is given by

$$\tilde{\boldsymbol{\pi}}_i = c\boldsymbol{\pi}_{i0}, \tag{13.21}$$

where c *is a normalizing constant.* □

We know by Theorem 12.1.1 that

$$\boldsymbol{\pi}_n = \boldsymbol{\pi}_{n-1}R^{(n)},$$

where the matrix $R^{(n)}$ records the expected sojourn time in the level $\ell(n)$, starting from the level $\ell(n-1)$, before the first return to $\ell(n-1)$. We decompose that matrix into blocks $R^{(n)}_{dk}$, $0 \leq d \leq n-1$, $0 \leq k \leq n$, which record the sojourn time in the sublevel $\ell(n,k)$, starting from the sublevel $\ell(n-1,d)$.

Clearly, $R^{(n)}_{dk} = 0$ for $d \geq 1$ since, starting from any state of the form $(n-1,d,i)$ with $d \geq 1$, the next transition will be to the state $(n-2,d-1,i)$ with probability 1, so that no time will be spent in the level $\ell(n)$ before the first return to $\ell(n-1)$. Thus, the matrices $R^{(n)}$ have the structure

$$R^{(n)} = \begin{bmatrix} R^{(n)}_{00} & R^{(n)}_{01} & \cdots & R^{(n)}_{0n} \\ 0 & 0 & \cdots & 0 \\ \vdots & \vdots & \ddots & \vdots \\ 0 & 0 & \cdots & 0 \end{bmatrix}, \tag{13.22}$$

and (13.21) becomes

$$\boldsymbol{\pi}_{n,i} = \boldsymbol{\pi}_{n-1,0}R^{(n)}_{0,i} \qquad \text{for } 0 \leq i \leq n. \tag{13.23}$$

We observe that (13.22) may also be seen as a consequence of (12.4) and (13.20).

Lemma 13.2.2 *Let us define* $K = R^{(1)}_{00}$. *We have*

$$R^{(n)}_{00} = K \qquad \textit{for all } n \geq 1, \tag{13.24}$$

from which

$$\boldsymbol{\pi}_{n,0} = \boldsymbol{\pi}_{00}K^n \qquad \textit{for } n \geq 0. \tag{13.25}$$

Proof. Equation (13.25) immediately results from (13.23) and (13.24), so that we only need to prove (13.24).

It results from the definition of $R^{(n)}$ in Theorem 12.1.1 that $R_{00}^{(n)} = C_0 M^{(n)}$, where the matrix $M^{(n)}$ records the expected number of visits to the sublevel $\ell(n,0)$, starting from $\ell(n,0)$, before the first visit to any state in the levels $\ell(0)$ to $\ell(n-1)$.

This matrix may be obtained by calling upon Lemma 5.1.2, as we have seen in a number of circumstances (see, e.g., section 6.2 and Chapter 11). It suffices to consider the matrix $Q^{(n)}$ obtained by removing from (13.17) the rows and columns for the lower levels; i.e., let

$$Q^{(n)} = \begin{bmatrix} A_1^{(n)} & A_0^{(n)} & & \\ A_2^{(n+1)} & A_1^{(n+1)} & A_0^{(n+1)} & \\ & A_2^{(n+2)} & A_1^{(n+2)} & \ddots \\ & & \ddots & \ddots \end{bmatrix},$$

then take $M^{(n)}$ as the northwest corner of $(I - Q^{(n)})^{-1}$.

We observe, however, that we may remove from consideration all the states of the form $(n+k, d, i)$ where $d > k$ since, starting from one of these states, the chain will move down to the sublevel $\ell(n, d-k)$ and further down to one of the levels $\ell(0), \ldots, \ell(n-1)$ before returning to the sublevel $\ell(n,0)$.

Thus, $M^{(n)}$ is also the northwest corner of $(I - \tilde{Q}^{(n)})^{-1}$, where

$$\tilde{Q}^{(n)} = \begin{bmatrix} B_1^{(n)} & B_0^{(n)} & & \\ B_2^{(n+1)} & B_1^{(n+1)} & B_0^{(n+1)} & \\ & B_2^{(n+2)} & B_1^{(n+2)} & \ddots \\ & & \ddots & \ddots \end{bmatrix},$$

where the matrices $B_1^{(n+k)}$ are square matrices of order $m(k+1)$ and $B_0^{(n+k)}$ and $B_2^{(n+k)}$ are rectangular matrices with dimensions $m(k+1) \times m(k+2)$ and $m(k+1) \times mk$, respectively, with

$$B_1^{(n+k)} = \begin{bmatrix} C_1 & C_2 & \cdots & C_{k+1} \\ 0 & 0 & \cdots & 0 \\ \vdots & \vdots & \ddots & \vdots \\ 0 & 0 & \cdots & 0 \end{bmatrix},$$

$$B_0^{(n+k)} = A_0^{(k)}, \quad B_2^{(n+k)} = A_2^{(k)}.$$

Since the matrix $\tilde{Q}^{(n)}$ is independent of n, the inverse of $I - \tilde{Q}^{(n)}$, and hence $M^{(n)}$ itself, are independent of n, which completes the proof. □

The following then is the celebrated matrix-geometric result for GI/M/1-type Markov chains.

Theorem 13.2.3 *The stationary probability vector* $\tilde{\boldsymbol{\pi}}$ *of the Markov chain* (13.2) *satisfies the relation*

$$\tilde{\boldsymbol{\pi}}_n = \tilde{\boldsymbol{\pi}}_0 K^n, \tag{13.26}$$

where for all $n \geq 1$, K_{ij} *is the expected number of visits to* $(n+1, j)$ *before the first return to the level* $\ell(n-1)$, *given that the chain starts in* (n, i).

The matrix K *is the minimal nonnegative solution of the matrix equation*

$$K = C_0 + KC_1 + K^2C_2 + \cdots. \tag{13.27}$$

Proof. Equation (13.26) is an immediate consequence of (13.21), (13.25). To obtain (13.27), we first observe that K^k is the block in the northwest corner of the product $R^{(1)}R^{(2)} \cdots R^{(k)}$, as we readily conclude from the structure of $R^{(n)}$ in (13.22) and the property (13.24).

Second, we use the same argument as in Remark 6.2.8 and interpret the product $R^{(1)}R^{(2)} \cdots R^{(k)}$ as recording the expected sojourn time in the level $\ell(k)$, starting from the level $\ell(0)$, before the first return to $\ell(0)$. Therefore, we conclude that K^k records the expected sojourn time in the sublevel $\ell(k, 0)$, before the first return to $\ell(0)$, given the initial state in the sublevel $\ell(0, 0)$.

The third step consists of repeating the argument given in Remark 6.2.11 to prove (6.15): K records the expected number of visits to

$\ell(1)$; in the right-hand side of (13.27), these visits are decomposed into groups according to the level from which the visits occur.

The minimality of K as a solution of (13.27) results from the minimality of $\{R^{(n)} : n \geq 1\}$ as a solution of (12.8). □

The two classes of Markov chains discussed above have been the subject of much research. Their importance stems from the fact that they arise as embedded processes in a variety of stochastic models for queues, inventory systems, dams and storage processes, etc. In addition to the two references [79, 82] given in the introduction, we may cite the work of Bini and Meini [6, 7, 66], who develop quadratically convergent algorithms to determine the solutions of (13.10), (13.27).

Also, the similarity between (13.10) and (13.27) suggests the existence of a duality between the Markov chains of the M/G/1- and the GI/M/1-types. This has been examined in Asmussen and Ramaswami [5] and Ramaswami [95].

Chapter 14

Tree Processes

We describe a family of processes which are interesting generalizations of QBD processes. We first discuss a simple example and then consider the general case. These processes were introduced by Yeung and Sengupta [119] and Takine, Sengupta, and Yeung [114], who consider a more general class of processes than we do here. Our simplified presentation is intended to highlight the essential elements in their structure.

14.1 The M/PH/1 LIFO Queue

In order to follow more easily the discussion in this chapter, it is helpful to have in mind a simple example of a process with a tree structure. One such example is provided by the M/PH/1 queue with LIFO service discipline. Thus, we consider a single server queue with an infinite buffer where customers are piled in a stack. Each newly arriving customer joins the top of the stack, preempting the customer being served, if there is one. When an interrupted customer is again on top of the stack, its service is resumed and proceeds from where it was interrupted. Arrivals form a Poisson process, and the service time distribution is PH with representation $(\boldsymbol{\tau}, T)$ of order m.

Where customers are present, we represent the system state by k-tuples of phase indices: the state $(i_1, i_2, \ldots, i_k)$ with $1 \le i_1, \ldots, i_k \le m$ indicates that there are k customers in the system; the customer at the top of the stack is active and currently executing the phase i_k;

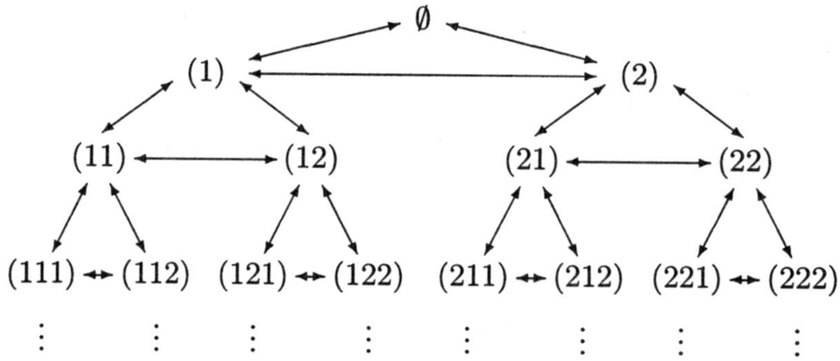

Figure 14.1: State transition diagram of the M/PH/1 queue with LIFO service discipline; the PH service time distribution has order 2.

the other customers in the stack were interrupted while executing the phases $i_1, \dots, i_{k-1}$, respectively. The empty system state is represented by $\emptyset$, an empty set of phases.

The possible transitions are as follows:

- from $(i_1, \dots, i_{k-1}, i)$ to $(i_1, \dots, i_{k-1}, j)$ with rate T_{ij} if there is a change of service phase for the active customer;
- from $(i_1, \dots, i_{k-1}, i)$ to $(i_1, \dots, i_{k-1}, i, j)$ with rate $\lambda\tau_j$ if a new customer arrives, is placed on top of the stack, and begins its service in the phase j;
- similarly, from $\emptyset$ to (j) with rate $\lambda\tau_j$;
- from $(i_1, \dots, i_{k-1}, i)$ to $(i_1, \dots, i_{k-1})$ with the rate t_i if the service of the customer at the top terminates;
- and, finally, from (i) to $\emptyset$ with the rate t_i.

In order to simplify the presentation in this section, we shall assume that $m = 2$; the description and analysis of the general case is left to the reader.

The state transition diagram is given in Figure 14.1. The states are ordered by increasing number of customers from top to bottom. We denote by $\ell(k)$ the set $\{(i_1, \ldots, i_k) : i_1, \ldots, i_k = 1 \text{ or } 2\}$ of states with k customers, and we call it the level k. Within each level, the states are ordered in lexicographical order from left to right.

One readily verifies that the infinitesimal generator is given by

$$Q = \begin{bmatrix} Q_{00} & Q_{01} & 0 & \cdots \\ Q_{10} & Q_{11} & Q_{12} & \cdots \\ 0 & Q_{21} & Q_{22} & \cdots \\ \vdots & \vdots & \vdots & \ddots \end{bmatrix},$$

where the diagonal blocks are of order 2^k,

$$Q_{00} = -\lambda,$$

$$Q_{kk} = \begin{bmatrix} T - \lambda I & 0 & \cdots & 0 \\ 0 & T - \lambda I & \cdots & 0 \\ \vdots & \vdots & \ddots & \vdots \\ 0 & 0 & \cdots & T - \lambda I \end{bmatrix},$$

$$Q_{k-1,k} = \begin{bmatrix} \lambda\boldsymbol{\tau} & \mathbf{0} & \cdots & \mathbf{0} \\ \mathbf{0} & \lambda\boldsymbol{\tau} & \cdots & \mathbf{0} \\ \vdots & \vdots & \ddots & \vdots \\ \mathbf{0} & \mathbf{0} & \cdots & \lambda\boldsymbol{\tau} \end{bmatrix},$$

and

$$Q_{k,k-1} = \begin{bmatrix} \boldsymbol{t} & \mathbf{0} & \cdots & \mathbf{0} \\ \mathbf{0} & \boldsymbol{t} & \cdots & \mathbf{0} \\ \vdots & \vdots & \ddots & \vdots \\ \mathbf{0} & \mathbf{0} & \cdots & \boldsymbol{t} \end{bmatrix}$$

for $k \geq 1$. This is the generator of a nonhomogeneous QBD of the type considered in Chapter 12, and it might be analyzed as indicated there. We shall, however, proceed differently and identify another structural property.

14.2 Tree-Like Transition Diagrams

Instead of amalgamating the states by levels as done above, we shall define new subsets, which we call *stack descriptors.* For each k-tuple of phase indices $(i_1, \ldots, i_k)$ $(k \geq 1; 1 \leq i_1, \ldots, i_k \leq m)$, we define the stack descriptor $s\{i_1, \ldots, i_k\}$ as follows:

$$s\{i_1, \ldots, i_k\} = \{(i_1, \ldots, i_k, j) : j = 1, \ldots, m\};$$

this means that there are $k+1$ customers in the system, the top customer is active and may occupy the m possible service phases, and the lower k customers in the stack are frozen at their service phases $i_1, \ldots, i_k$, respectively. We shall sometimes use the shorter notation $s\{\boldsymbol{i}\}$ to denote an arbitrary stack descriptor $s\{i_1, \ldots, i_k\}$.

We also define $s\{\cdot\} = \{1, \ldots, m\}$ to denote the possible states of the stack when there is one customer only and $\emptyset$ to denote the empty state when the server is idle.

We display the transition diagram between stack descriptors in Figure 14.2 in the case where $m = 2$; a tree structure is strikingly apparent. If we display the generator Q according to this new partitioning of the states (and only show the nonzero blocks), then we have that

$$Q = \left[\begin{array}{c|c|cc|cccc|c}
-\lambda & \lambda\boldsymbol{\tau} & & & & & & & \\
\hline
\boldsymbol{t} & C & A_1 & A_2 & & & & & \\
\hline
 & D_1 & C & & A_1 & A_2 & & & \\
 & D_2 & & C & & & A_1 & A_2 & \\
\hline
 & & D_1 & & C & & & & \\
 & & D_2 & & & C & & & \ddots \\
 & & & D_1 & & & C & & \\
 & & & D_2 & & & & C & \\
\hline
 & & & & & \ddots & & & \ddots
\end{array}\right] \qquad (14.1)$$

when $m = 2$, where

$$\begin{aligned}
A_1 &= \begin{bmatrix} \lambda\boldsymbol{\tau} \\ \mathbf{0} \end{bmatrix}, \qquad A_2 = \begin{bmatrix} \mathbf{0} \\ \lambda\boldsymbol{\tau} \end{bmatrix}, \\
D_1 &= \begin{bmatrix} \boldsymbol{t} & \mathbf{0} \end{bmatrix}, \qquad D_2 = \begin{bmatrix} \mathbf{0} & \boldsymbol{t} \end{bmatrix}, \\
C &= T - \lambda I.
\end{aligned}$$

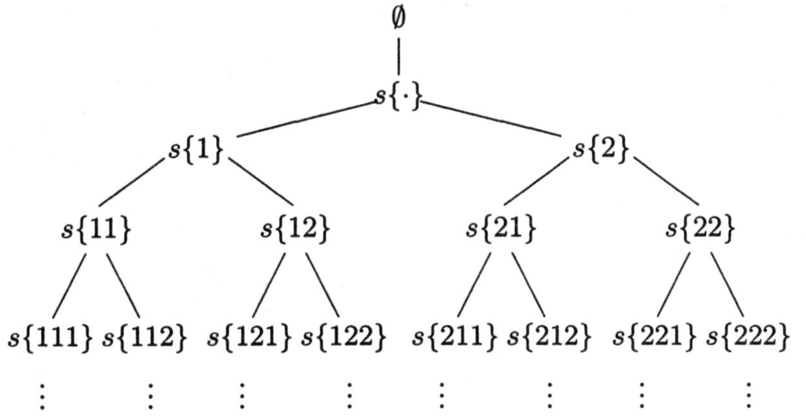

Figure 14.2: Transition diagram of the M/PH/1 queue with LIFO service discipline; the PH service time distribution has order 2; the states are grouped into subsets called stack descriptors; the graph edges indicate pairs of subsets between which individual state transitions are possible.

It is important to note that the tree in Figure 14.2 is thoroughly homogeneous: the whole tree starting at $s\{\cdot\}$ is isomorphic to the subtree starting at $s\{1\}$ or $s\{21\}$ or any other node below $\emptyset$. The only differences lie in the first part of the node labels, which reproduce the label of the root of the subtree. Moreover, the transitions within a node are governed by the same matrix C for all the nodes, the transitions from a node to its left child are governed by A_1, the transitions from a left child to the parent are governed by D_1, etc.

We shall find it more fruitful to use a different ordering of the state space and enumerate the states according to a visit of the tree diagram in preorder. This means that we recursively enumerate the states in the root, followed by the states in all the nodes of the left subtree, and then the nodes in the right subtree.[21] This gives the following sequence:

[21]The ordering which we used in order to write (14.1) corresponds to a visit in breadth first.

$$
\begin{array}{llllll}
\emptyset, & s\{\cdot\}, & & & & \\
& s\{1\}, & s\{1,1\}, & s\{1,1,1\}, & s\{1,1,1,1\}, & \cdots \\
& & & & s\{1,1,1,2\}, & \cdots \\
& & & s\{1,1,2\}, & s\{1,1,2,1\}, & \cdots \\
& & & & s\{1,1,2,2\}, & \cdots \\
& & s\{1,2\}, & s\{1,2,1\}, & s\{1,2,1,1\}, & \cdots \\
& & & & s\{1,2,1,2\}, & \cdots \\
& & & s\{1,2,2\}, & s\{1,2,2,1\}, & \cdots \\
& & & & s\{1,2,2,2\}, & \cdots \\
& s\{2\}, & s\{2,1\}, & s\{2,1,1\}, & s\{2,1,1,1\}, & \cdots \\
& & & & s\{2,1,1,2\}, & \cdots \\
& & & s\{2,1,2\}, & s\{2,1,2,1\}, & \cdots \\
& & & & s\{2,1,2,2\}, & \cdots \\
& & s\{2,2\}, & s\{2,2,1\}, & s\{2,2,1,1\}, & \cdots \\
& & & & s\{2,2,1,2\}, & \cdots \\
& & & s\{2,2,2\}, & s\{2,2,2,1\}, & \cdots \\
& & & & s\{2,2,2,2\}, & \cdots,
\end{array}
$$

where the ellipses refer to whole subtrees.

The generator is displayed in Figure 14.3 (again, we only show the nonzero blocks), and we may also write it as follows:

$$
Q = \begin{bmatrix} -\lambda & \boldsymbol{\Lambda} \\ \boldsymbol{V} & W \end{bmatrix},
$$

where

$$
\boldsymbol{\Lambda} = [\lambda \boldsymbol{\tau}, \mathbf{0}, \mathbf{0}, \ldots],
$$

$$
\boldsymbol{V} = \begin{bmatrix} \boldsymbol{t} \\ \mathbf{0} \\ \mathbf{0} \\ \vdots \end{bmatrix},
$$

and W records the transitions between the states in $s\{\cdot\}$ and its two subtrees. That matrix is given by

$$
W = \begin{bmatrix} C & \Lambda_1 & \Lambda_2 \\ V_1 & W_1 & 0 \\ V_2 & 0 & W_2 \end{bmatrix},
$$

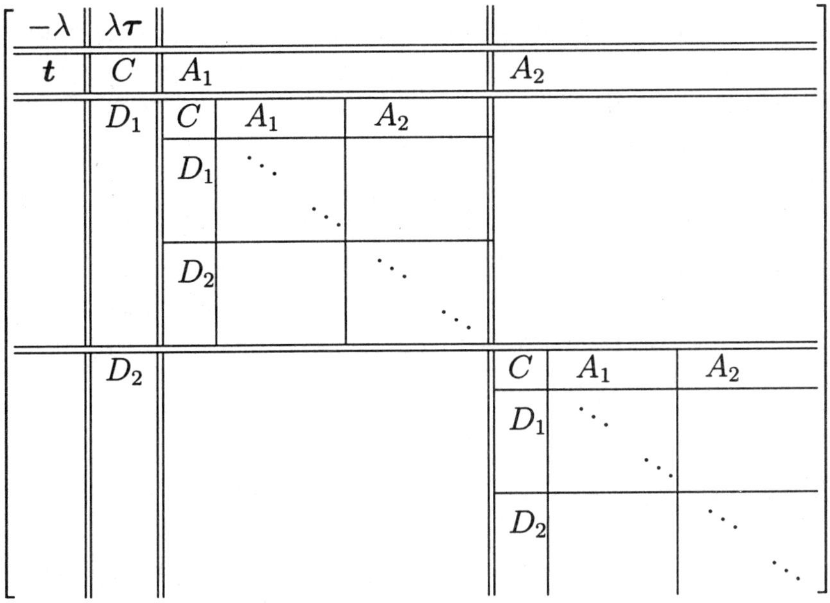

Figure 14.3: Infinitesimal generator of the M/PH/1 queue with LIFO service discipline; the PH service time distribution has order 2.

where

$$\Lambda_i = \begin{bmatrix} A_i & 0 & 0 & \cdots \end{bmatrix},$$

$$V_i = \begin{bmatrix} D_i \\ 0 \\ 0 \\ \vdots \end{bmatrix}$$

for $i = 1$ and 2 and W_i records the transitions between the states in $s\{i\}$ and its two subtrees. Since the tree with root at $s\{1\}$ is isomorphic to the tree with root at $s\{2\}$, we have that $W_1 = W_2 = W$, so that the matrix W is defined recursively by

$$W = \begin{bmatrix} C & \Lambda_1 & \Lambda_2 \\ V_1 & W & 0 \\ V_2 & 0 & W \end{bmatrix}.$$

Let us now consider a more general tree structure, where the number m of children of each node is arbitrary and the last index may change value when the process goes through a change of level. To be specific, let us assume the following:

(a) The state space is the collection

$$\emptyset \cup \{(i_1, \dots, i_k) : k \geq 1, 1 \leq i_1, \dots, i_k \leq m\}.$$

It is partitioned into stack descriptors $s\{i_1, \dots, i_k\} = \{(i_1, \dots, i_k, j) : 1 \leq j \leq m\}$ for $k \geq 0$, $1 \leq i_1, \dots, i_k \leq m$. The transition diagram between the stack descriptors form an m-ary tree, with the possible transitions being as follows:

- from $(i_1, \dots, i_{k-1}, i)$ to $(i_1, \dots, i_{k-1}, j)$ with rate C_{ij};
- from $\emptyset$ to (j) with rate a_j;
- from $(i_1, \dots, i_{k-1}, i)$ to $(i_1, \dots, i_{k-1}, j, i')$ with rate $[A_j]_{ii'}$;
- from (i) to $\emptyset$ with the rate d_i;
- from $(i_1, \dots, i_{k-1}, j, i)$ to $(i_1, \dots, i_{k-1}, i')$ with the rate $[D_j]_{ii'}$.

(b) In partitioning the infinitesimal generator, we split the states in preorder fashion in the following order: the states $\emptyset$ and those in $s\{\cdot\}$ are enumerated first, followed by all the states in $s\{1\}$ and its subtrees, then all the states in $s\{2\}$ and its subtrees, etc., and finally $s\{m\}$ and its subtrees.

(c) The infinitesimal generator is assumed to be given by

$$Q = \begin{bmatrix} \lambda & \boldsymbol{\Lambda} \\ \boldsymbol{V} & W \end{bmatrix}, \tag{14.2}$$

where

$$\boldsymbol{\Lambda} = [\boldsymbol{a}, 0, 0, \dots],$$

$$\boldsymbol{V} = \begin{bmatrix} \boldsymbol{d} \\ 0 \\ 0 \\ \vdots \end{bmatrix},$$

and

$$W = \begin{bmatrix} C & \Lambda_1 & \Lambda_2 & \cdots & \Lambda_m \\ V_1 & W & 0 & \cdots & 0 \\ V_2 & 0 & W & \cdots & 0 \\ \vdots & \vdots & \vdots & \ddots & 0 \\ V_m & 0 & 0 & \cdots & W \end{bmatrix}, \quad (14.3)$$

with

$$\Lambda_i = \begin{bmatrix} A_i & 0 & 0 & \cdots \end{bmatrix}$$

and

$$V_i = \begin{bmatrix} D_i \\ 0 \\ 0 \\ \vdots \end{bmatrix}$$

for $i = 1, \ldots, m$. The vectors $\boldsymbol{a}$ and $\boldsymbol{d}$ have dimension m and the matrices C, A_i, and D_i have order m. They must be such that Q is an irreducible generator, but they are otherwise arbitrary. This implies that $A_i \geq 0, D_i \geq 0$ for all i, that the off-diagonal elements of C are nonnegative, that the diagonal elements are strictly negative, that

$$\left(C + \sum_{1 \leq i \leq m} A_i + D_j \right) \mathbf{1} = \mathbf{0},$$

and that

$$D_j \mathbf{1} = \boldsymbol{d}$$

for all j.

14.3 Matrix-Product Form Distribution

We have observed earlier that the tree which describes the transitions between stack descriptors is homogeneous and that if one considers a subtree, its transition rates cannot be distinguished from those of any other subtree. Therefore, we may denote by N the matrix which records the expected time spent in the various states of a node different from $\emptyset$, before the first visit to the parent node, and we have that

$N_{ij}(1 \leq i, j \leq m)$ is the expected total time spent in the state $(\boldsymbol{k}, j)$ of the node $s\{\boldsymbol{k}\}$, starting from the state $(\boldsymbol{k}, i)$ at time 0, before the first visit to the parent of $s\{\boldsymbol{k}\}$ *independently* of $\boldsymbol{k}$.

This is the property which led us to the matrix-geometric distribution in Chapter 6. With ordinary QBDs, for each level $\ell(n)$, $\ell(n-1)$ is the parent and $\ell(n+1)$ is the only child. Here, we may think of the state space of being made up of an infinite number of QBDs which all share the same level 0. These are partitioned into m bunches such that all the QBDs in a given bunch share the level 1 in addition to the level 0; they are further grouped in m^2 smaller bunches which share the levels 0, 1, and 2, etc.

The tree structure leads us to the following result. We denote by π_0 the stationary probability of the empty state and by $\boldsymbol{\pi}\{\boldsymbol{n}\}$ the stationary probability subvector for the stack descriptor $s\{\boldsymbol{n}\}$.

Theorem 14.3.1 *Assume that the Markov process with generator* (14.2) *is positive recurrent. We have that*

$$\boldsymbol{\pi}\{n_1, n_2, \ldots, n_k\} = \pi_0 \boldsymbol{a} N R(n_1) R(n_2) \cdots R(n_k) \tag{14.4}$$

for all $k \geq 0$ *and* $1 \leq n_1, n_2, \ldots, n_k \leq m$, *where*

$$R(i) = A_i N \qquad \text{for } 1 \leq i \leq m.$$

Proof. Let us consider a fixed node $s\{\boldsymbol{n}, j\}$. We use Theorem 5.5.1 and Lemma 5.5.2 with E^c equal to the subtree of which $s\{\boldsymbol{n}, j\}$ is the root and E equal to the remainder of the tree. Because of the tree structure, the only states in E which directly communicate with E^c are in $s\{\boldsymbol{n}\}$, the parent of $s\{\boldsymbol{n}, j\}$. We therefore immediately obtain that

$$\boldsymbol{\pi}\{\boldsymbol{n}, j\} = \boldsymbol{\pi}\{\boldsymbol{n}\} A_j N. \tag{14.5}$$

We next consider the case where E is the root ϕ of the tree and E^c contains all the other nodes, and we obtain that

$$\boldsymbol{\pi}\{\cdot\} = \pi_0 \boldsymbol{a} N,$$

from which, together with (14.5), the theorem follows. □

As a corollary, we show below that the marginal distribution of the stack size and the active service phase has a matrix-geometric distribution. We define the vectors $\boldsymbol{p}_n$ for $n \geq 1$ such that $p_{n,j}$ is the stationary probability that the stack size is n and the phase at the top is j.

Corollary 14.3.2 *Assume that the process is positive recurrent. Then we have that*

$$\boldsymbol{p}_1 = \pi_0 \boldsymbol{a} N$$

and

$$\boldsymbol{p}_n = \pi_1 R^{n-1} \qquad \text{for } n \geq 1,$$

where

$$R = \sum_{1 \leq i \leq m} R(i) = \left(\sum_{1 \leq i \leq m} A_i \right) N.$$

Proof. Formally, we have that

$$\begin{aligned}
\boldsymbol{p}_n &= \sum_{1 \leq i_1, \ldots, i_{n-1} \leq m} \boldsymbol{\pi}\{i_1, \ldots, i_{n-1}\} \\
&= \sum_{1 \leq i_1, \ldots, i_{n-1} \leq m} \pi_0 \boldsymbol{a} N R(i_1) \cdots R(i_{n-1})
\end{aligned}$$

by (14.4), from which the result is immediate. □

Remark 14.3.3 For the M/PH/1/LIFO queue, the matrix R in the theorem above is not equal to the matrix R of the M/PH/1/FIFO queue.

As a further corollary, one proves like Theorem 6.2.6 that the marginal distribution of the stack size has a PH representation.

The matrix N records the expected time spent in the node $s\{\cdot\}$ before the first passage to the empty state. As we have seen earlier in section 6.4, it is also equal to $(-U)^{-1}$, where U is the infinitesimal generator of the Markov process in the local time of the node $s\{\cdot\}$ before the first passage to $\emptyset$. Using the equation (14.3), we obtain by a (by now) familiar argument that

$$\begin{aligned}
U &= C + [\Lambda_1 \cdots \Lambda_m] \begin{bmatrix} -W & & \\ & \ddots & \\ & & -W \end{bmatrix}^{-1} \begin{bmatrix} V_1 \\ \vdots \\ V_m \end{bmatrix} \\
&= C + \sum_{1 \leq i \leq m} \Lambda_i (-W)^{-1} V_i \\
&= C + \sum_{1 \leq i \leq m} A_i N D_i,
\end{aligned}$$

so that U and N are determined by the pair of equations

$$\begin{aligned} N &= (-U)^{-1}, \\ U &= C + \sum_{1 \le i \le m} A_i N D_i, \end{aligned} \qquad (14.6)$$

which form the basis for a computational procedure. We define the sequence $\{U_n, N_n : n \geq 1\}$ as follows:

$$U_1 = C, \qquad (14.7)$$

$$N_n = (-U_n)^{-1}, \qquad (14.8)$$

$$U_{n+1} = C + \sum_{1 \le i \le m} A_1 N_n D_i \qquad (14.9)$$

for $n \geq 1$. The matrices obtained at the nth step correspond to the case where the Markov process freely moves among the levels 1 to n and reaches the empty state before the level $n+1$, where we define the level n as comprising all the stack descriptors $\{s\{i_1, \ldots, i_n\} : 1 \leq i_1, \ldots, i_n \leq m\}$. Thus, the sequences monotonically converge to U and N, respectively; we omit the routine details (see the proof of Theorem 8.1.1).

It is difficult to find conditions under which the process is positive recurrent. No result exists that is as simple and as general as Theorem 7.2.3 for QBDs. We enumerate and briefly justify below the properties which are known.

Lemma 14.3.4 *A necessary condition for the process to be recurrent is that $N\boldsymbol{d} = \boldsymbol{1}$. If m is finite, then the process is positive recurrent if and only if $sp(R) < 1$.*

Proof. To prove the first statement, we observe that the probability, starting from the various states in $s\{\cdot\}$, of hitting the empty state in a finite amount of time is given by the first m components of the vector $(-W)^{-1}\boldsymbol{V}$, which is equal to $N\boldsymbol{d}$.

To prove the second statement, we observe that the process is positive recurrent if and only if the infinite series formed by the coefficients of π_0 in (14.4) converges. Equivalently, by Corollary 14.3.2 this happens if and only if the series of powers of R converges. □

This is not a very satisfactory result, since it requires that one computes the matrices N and R in order to determine if a stationary distribution does exist. It could be much better if one could rapidly determine if the process is ergodic or not, as in the two special cases considered in the lemmas below.

Lemma 14.3.5 *Assume that $D_i = D$ independently of i. Define $A = \sum_{1\le i\le m} A_i$. The Markov process is positive recurrent if and only if $\boldsymbol{\alpha} A\mathbf{1} < \boldsymbol{\alpha} D\mathbf{1}$, where $\boldsymbol{\alpha}$ is such that $\boldsymbol{\alpha}(A+C+D) = \mathbf{0}$, $\boldsymbol{\alpha}\mathbf{1} = 1$.*

Proof. If the matrices D_i are all equal, then (14.6) reduces to $U = C + AND$, so that the matrices U, N, R, etc. are identical to the corresponding matrices of a QBD process with $A_0 = A, A_1 = C$, and $A_2 = D$. Thus, we may apply Theorem 7.2.3. □

This lemma is, unfortunately, not very useful. The condition $D_i = D$ for all i implies, for instance, that when the state is (i_1, i_2) and a departure occurs, the new state (j) is chosen *independently* of i_1. The same is true, of course, for any reduction of the stack size. Thus, it is useless to keep track of the full stack; it is enough to retain the stack size and the active phase; in other words, the system is actually lumpable to a QBD.

The next result is useful for M/PH/1-type queues, such as the one considered in section 14.2, where we had $D_i = \boldsymbol{t}\cdot\boldsymbol{e}_i$ with $\boldsymbol{e}_i = (0,\ldots,1,\ldots,0)$ (the component 1 occupies the ith position).

Lemma 14.3.6 *Assume that the matrices D_i are of rank 1 and that*

$$D_i = \boldsymbol{d}\cdot\boldsymbol{\delta}_i, \qquad 1\le i\le m, \tag{14.10}$$

where $\boldsymbol{\delta}_i$ is a row vector such that $\boldsymbol{\delta}_i\mathbf{1} = 1$. Then we have that U is explicitly given by

$$U = C + \sum_{1\le i\le m} A_i\mathbf{1}\cdot\boldsymbol{\delta}_i$$

if the process is recurrent.

Proof. Since $D_i\mathbf{1} = \boldsymbol{d}$ for all i, we indeed may write (14.10) if D_i is of rank 1. The lemma is a direct consequence of (14.6) and Lemma 14.3.4. □

Under the assumption of this lemma, one may directly compute $R = A(-U)^{-1}$ and verify whether $sp(R) < 1$ or not.

14.4 Skip-Free Processes

We conclude this chapter by briefly describing the more general processes for which Yeung and Sengupta [119] prove that the stationary distribution has a matrix-product form.

First, they assume that the stacks are such that the top component may be different from the others: for the state $(i_1, \ldots, i_k)$, one has that $1 \leq i_1, \ldots, i_{k-1} \leq d$ and $1 \leq i_k \leq m$.

This is useful in modeling such processes as the PH/PH/1 queue with LIFO processing: one must keep track of the arrival phase and the service phase of the active customer; however, when a new arrival occurs, it is necessary to remember the service phase but not the arrival phase at the time of interruption.

Second, the authors assume in [119] that a transition from $s\{\boldsymbol{i}\}$ can occur to any node in the path between $s\{\boldsymbol{i}\}$ and the root $\emptyset$ and not just to its parent. Such processes are generalizations of those considered above just as the skip-free upward processes of Chapter 13 are generalizations of QBDs. We refer to [119] for further details.

Chapter 15

Product Form Networks

We saw in section 1.7 that a Jackson network may be viewed as a QBD, where the level is the number of customers in one of the queues and the phase is the state descriptor of the remainder of the network. Ramaswami and Taylor [102] have investigated the connection between the theory of product form networks and the theory of QBDs. We shall present here a simplified version of their results and briefly describe in section 15.3 the more general systems which are treated in [102].

Our intention here is not to advocate QBDs as the natural approach to product form networks. Instead, our view is that once we realize that the phase dimension of a QBD could literally be anything, then it clearly appears that we may represent a product form network as a QBD and it becomes quite natural to try and find out how the product form property may be discerned in the QBD formulation. This has led to the general theorems in section 15.1 and to the discovery of new classes of product form networks (see [102]).

15.1 Independence of Level and Phase

Let us consider a queueing network where each node corresponds to a single server queue with an infinite FIFO buffer; the services at node i are exponentially distributed with parameter μ_i; a customer who completes its service at node i moves to node j with probability P_{ij} and leaves the system with probability q_i; external arrivals to node i occur according to a Poisson process with rate λ_i.

The state of the system is characterized by the vector $\boldsymbol{n} = (n_1, n_2, \ldots, n_K)$, where K is the number of stations and n_i is the number of customers in node i. It is quite well known (see Lavenberg [57] or Kelly [38], for instance) that if the stationary distribution exists, then it can be expressed as a product of factors, one for each node. One may also state that the number of customers in the various nodes are independent random variables in steady state. This product form property is shown in a vast collection of processes; see [102] for a description and bibliography.

If we define the level as the number of customers in one of the queues, say queue 1, and the phase as the vector $\boldsymbol{r} = (n_2, \ldots, n_K)$ for the remainder of the network, then the Jackson network is seen as a QBD with infinite blocks and the product form property implies that the level and the phase are independent in steady state.

We shall characterize below the conditions under which this independence property holds and show in the next section how this applies to Jackson networks. Since the latter are continuous time Markov processes, this is the case which we discuss; the adaptation to discrete time QBDs is straightforward.

Theorem 15.1.1 *Consider a continuous time QBD with generator*

$$Q = \begin{bmatrix} B & A_0 & & \\ A_2 & A_1 & A_0 & \\ & A_2 & A_1 & \ddots \\ & & \ddots & \ddots \end{bmatrix}.$$

The process is positive recurrent, and the level is independent of the phase in steady state if and only if there exists a positive vector $\boldsymbol{\theta}$ with $\boldsymbol{\theta}\mathbf{1} = 1$ and a positive scalar $\eta < 1$ such that

$$\boldsymbol{\theta} R = \eta \boldsymbol{\theta} \tag{15.1}$$

and

$$\boldsymbol{\theta}(B + \eta A_2) = \mathbf{0}. \tag{15.2}$$

Then, we have that $\boldsymbol{\pi}_n = (1-\eta)\eta^n \boldsymbol{\theta}$.

Remark 15.1.2 If the number m of phases is finite, then there is always a solution to (15.1) given by the Perron–Frobenius eigenvalue of R, so that one only needs to verify if it satisfies (15.2).

Proof. The proof of sufficiency is immediate since one easily verifies that if η and $\boldsymbol{\theta}$ satisfy (15.1), (15.2), then $\pi_n = (1-\eta)\eta^n\boldsymbol{\theta}$ satisfies the steady state equations

$$\begin{aligned} \boldsymbol{\pi}_0 B + \boldsymbol{\pi}_1 A_2 &= 0, \\ \boldsymbol{\pi}_n A_0 + \boldsymbol{\pi}_{n+1} A_1 + \boldsymbol{\pi}_{n+2} A_2 &= 0. \end{aligned}$$

The proof of necessity goes as follows. If the level and the phase are independent in steady state, this implies that the stationary distribution is such that $\boldsymbol{\pi}_n = \alpha_n \boldsymbol{v}$ for some probability vector $\boldsymbol{v}$ and probability density $\{\alpha_0, \alpha_1, \ldots\}$. Since necessarily $\boldsymbol{\pi}_n = \boldsymbol{\pi}_{n-1} R$, we find that $\alpha_n \boldsymbol{v} = \alpha_{n-1} \boldsymbol{v} R$ for all n and that α_n / α_{n-1} is a constant, say, η. This proves (15.1). Since $\boldsymbol{\pi}_0 = \alpha_0 \boldsymbol{v}$ is the unique solution of $\boldsymbol{\pi}_0(B + RA_2) = \mathbf{0}$ with $\boldsymbol{\pi}_0(I-R)^{-1}\mathbf{1} = 1$, we find that (15.2) holds and that $\alpha_0 = (1-\eta)$. □

Obviously, we want a necessary and sufficient condition which does not require that we compute the matrix R. This will be obtained using Theorem 8.4.4, which gives us the following expression for R in continuous time:

$$R = \sum_{k \geq 0} C_0^{[k]} \prod_{0 \leq i \leq k-1} C_2^{[k-1-i]}$$

with

$$\begin{aligned} C_i^{[0]} &= A_i(-A_1)^{-1}, & (15.3) \\ C_i^{[k+1]} &= (C_i^{[k]})^2 [I - C_0^{[k]} C_2^{[k]} - C_2^{[k]} C_0^{[k]}]^{-1} & (15.4) \end{aligned}$$

for $i = 1$ and 2 and $k \geq 0$.

Theorem 15.1.3 *The continuous time QBD is positive recurrent, and the level and phase are independent in steady state if and only if there exists a positive vector $\boldsymbol{\theta}$ with $\boldsymbol{\theta}\mathbf{1} = 1$ and a positive scalar η with $\eta < 1$ such that*

$$\boldsymbol{\theta}(A_0 + \eta A_1 + \eta^2 A_2) = \mathbf{0} \tag{15.5}$$

and

$$\boldsymbol{\theta}(B + \eta A_2) = \mathbf{0}.$$

Then, we have that $\boldsymbol{\pi}_n = (1-\eta)\eta^n \boldsymbol{\theta}$.

Proof. We only need to prove that θ and η satisfy (15.5) if and only if they satisfy (15.1).

If $\boldsymbol{\theta}$ and η satisfy (15.1), then it suffices to premultiply the equation $A_0 + RA_1 + R^2 A_2 = 0$, of which R is a solution, by $\boldsymbol{\theta}$ and we obtain (15.5). The proof in the other direction is much more involved.

One shows that for all $k \geq 0$,

$$\boldsymbol{\theta} C_0^{[k]} + \eta^{2^{k+1}} \boldsymbol{\theta} C_2^{[k]} = \eta^{2^k} \boldsymbol{\theta}. \tag{15.6}$$

Then, it is easy to verify that

$$\begin{aligned}
\boldsymbol{\theta} R &= \boldsymbol{\theta} \lim_{N\to\infty} \sum_{0\leq k\leq N} C_0^{[k]} \prod_{0\leq i\leq k-1} C_2^{[k-1-i]} \\
&= \eta\boldsymbol{\theta} - \lim_{N\to\infty} \eta^{2^{N+1}} \prod_{0\leq i\leq N} C_2^{[N-i]} \\
&= \eta\boldsymbol{\theta}.
\end{aligned}$$

The proof of (15.6) is by induction. For $k=0$, we have that

$$\begin{aligned}
\boldsymbol{\theta} C_0^{[0]} + \eta^2 \boldsymbol{\theta}_0 C_2^{[0]} &= \boldsymbol{\theta}[A_0(-A_1)^{-1} + \eta^2 A_2(-A_1)^{-1}] \\
&= \boldsymbol{\theta}[A_0 + \eta^2 A_2](-A_1)^{-1} \\
&= \eta\boldsymbol{\theta} \qquad \text{by (15.5).}
\end{aligned}$$

Assume now that (15.6) holds for some k. By (15.4), we have that

$$\begin{aligned}
&\boldsymbol{\theta}\{C_0^{[k+1]} + \eta^{2^{k+2}} C_2^{[k+1]}\} \\
&= \boldsymbol{\theta}\{(C_0^{[k]})^2 + \eta^{2^{k+2}} (C_2^{[k]})^2\}[I - C_0^{[k]}C_2^{[k]} - C_2^{[k]}C_0^{[k]}]^{-1} \\
&= \boldsymbol{\theta}\{\eta^{2^k} C_0^{[k]} - \eta^{2^{k+1}} C_2^{[k]} C_0^{[k]} + \eta^{2^k}\eta^{2^{k+1}} C_2^{[k]} - \eta^{2^{k+1}} C_0^{[k]} C_2^{[k]}\} \\
&\quad \times [I - C_0^{[k]}C_2^{[k]} - C_2^{[k]}C_0^{[k]}]^{-1} \\
&= \eta^{2^{k+1}} \boldsymbol{\theta},
\end{aligned}$$

since $\boldsymbol{\theta} C_0^{[k]} + \eta^{2^{k+1}} \boldsymbol{\theta} C_2^{[k]} = \eta^{2^k} \boldsymbol{\theta}$ by the induction assumption. □

In the case of a finite number of phases, we might have used the same approach as in Theorem 9.3.1 to prove that if $\boldsymbol{\theta}$ and η satisfy (15.5), then they form an eigenvector/eigenvalue pair for R and satisfy (15.1). That approach fails, however, in the case of an infinite number of phases. In the proof of the theorem above, we only deal with positive vectors and matrices, so that convergence of the various inner products is absolute and the associativity of matrix products is preserved.

Observe that, as a consequence of these theorems, we may conclude that in a homogeneous QBD, if the level and phase are independent, then necessarily the marginal distribution of the level is geometric.

15.2 Network of Exponential Servers

In the case of a Jackson network, where the level is the number n_1 of customers in the first queue and the phase is the vector $\boldsymbol{r} = (n_2, \ldots, n_K)$ of number of customers in the remainder of the network, the matrices B, A_0, A_1, and A_2 are as follows:

$$\begin{aligned} A_0(\boldsymbol{r}, \boldsymbol{r}) &= \lambda_1, \\ A_0(\boldsymbol{r}, \boldsymbol{r} - \boldsymbol{e}_j) &= \mu_j P_{j1} \qquad \text{for } 2 \leq j \leq K, \end{aligned} \tag{15.7}$$

where $\boldsymbol{e}_j$ is a row vector with all components equal to 0, except for a 1 in position j, so that the difference between the states $\boldsymbol{r}$ and $\boldsymbol{r} - \boldsymbol{e}_j$ is that a customer has left the station j. Naturally, we restrict the transitions to feasible states, so that the rate (15.7) is only defined if $n_j \geq 1$.

Similarly, we have that

$$\begin{aligned} A_2(\boldsymbol{r}, \boldsymbol{r}) &= \mu_1 q_1, \\ A_2(\boldsymbol{r}, \boldsymbol{r} + \boldsymbol{e}_j) &= \mu_1 P_{1j} \qquad \text{for } 2 \leq j \leq K, \end{aligned}$$

and

$$\begin{aligned} A_1(\boldsymbol{r}, \boldsymbol{r} + \boldsymbol{e}_i) &= \lambda_i, \\ A_1(\boldsymbol{r}, \boldsymbol{r} - \boldsymbol{e}_j) &= \mu_j q_j, \\ A_1(\boldsymbol{r}, \boldsymbol{r} - \boldsymbol{e}_j + \boldsymbol{e}_i) &= \mu_j P_{ji} \end{aligned}$$

for $2 \leq i, j \leq K$. We assume that $P_{ii} = 0$ for all i. Finally, we have that $B = A_1 + \mu_1 I$.

It is tedious but simple to verify that the matrix A_1 may also be written as follows:

$$\begin{aligned} A_1(\boldsymbol{r}, \boldsymbol{r} + \boldsymbol{e}_i) &= F(\boldsymbol{r}, \boldsymbol{r} + \boldsymbol{e}_i), & (15.8) \\ A_1(\boldsymbol{r}, \boldsymbol{r} - \boldsymbol{e}_j) &= F(\boldsymbol{r}, \boldsymbol{r} - \boldsymbol{e}_j) - \mu_j P_{j1}, & (15.9) \\ A_1(\boldsymbol{r}, \boldsymbol{r} - \boldsymbol{e}_j + \boldsymbol{e}_i) &= F(\boldsymbol{r}, \boldsymbol{r} - \boldsymbol{e}_j + \boldsymbol{e}_i), & (15.10) \end{aligned}$$

and

$$A_1(\boldsymbol{r}, \boldsymbol{r}) = F(\boldsymbol{r}, \boldsymbol{r}) - \lambda_1 + \mu_1, \qquad (15.11)$$

where F is the generator of the Jackson network obtained by removing the node 1: no customer arrives there, and those customers who would be directed there, coming from another node, are redirected so that they leave the system altogether.

In order to prove that n_1 is independent from $\boldsymbol{r}$ in steady state, we need to find η and $\boldsymbol{\theta}$ which satisfy the conditions of Theorem 15.1.3, i.e., such that

$$\begin{aligned} \boldsymbol{\theta}(A_0 - \eta \mu_1 I) &= \mathbf{0}, \\ \boldsymbol{\theta}(B + \eta A_2) &= \mathbf{0}, \end{aligned}$$

or, equivalently,

$$\begin{aligned} \boldsymbol{\theta} A_0 - \eta \mu_1 \boldsymbol{\theta} &= \mathbf{0}, & (15.12) \\ \boldsymbol{\theta}(F - \lambda_1 I + \eta A_2) &= \mathbf{0}. & (15.13) \end{aligned}$$

This would be enough to prove the product form property, since by symmetry we also prove that n_2 is, in steady state, independent of $(n_1, n_3, \ldots, n_K)$, etc. However, the product form is even more obvious since we directly prove below that the components of $\boldsymbol{\theta}$ resolve into factors over the nodes 2 to K.

Theorem 15.2.1 *We denote by P the matrix of transition probabilities among the nodes and by $\boldsymbol{y}$ the solution of the traffic equations*

$$\boldsymbol{y} = \boldsymbol{\lambda} + \boldsymbol{y} P, \qquad (15.14)$$

where $\boldsymbol{\lambda} = (\lambda_1, \lambda_2, \ldots, \lambda_K)$.

If $y_i/\mu_i < 1$ *for all* i*, then the conditions of Theorem* 15.1.3 *are satisfied by the scalar*

$$\eta = y_1/\mu_1 \tag{15.15}$$

and the vector $\boldsymbol{\theta}$ *with*

$$\theta_{\boldsymbol{r}} = c \prod_{2\le j\le K} (y_j/\mu_j)^{n_j}, \tag{15.16}$$

where

$$c = \prod_{2\le j\le K} (1 - y_j/\mu_j).$$

Proof. The proof is by induction on K. The theorem clearly holds for $K = 1$ since the network then reduces to a simple M/M/1 queue. Therefore, we may assume that the theorem has been proved for networks with $K - 1$ nodes.

In expanded form, using (15.15), (15.16), the left-hand side of (15.12) is written as follows:

$$\begin{aligned}
&\lambda_1\theta_{\boldsymbol{r}} + \sum_{2\le j\le K} \mu_j P_{j1}\theta_{\boldsymbol{r}+\boldsymbol{e}_j} - \eta\mu_1\theta_{\boldsymbol{r}} \\
&\quad = (\lambda_1 + \sum_{2\le j\le K} y_j P_{j1} - y_1)\theta_{\boldsymbol{r}} \qquad \text{for all } \boldsymbol{r},
\end{aligned}$$

which is equal to zero by (15.14).

After some simple manipulations, the left-hand side of (15.13) is seen to be equal to

$$\begin{aligned}
\{\boldsymbol{\theta}(F - \lambda_1 I + \eta A_2)\}(\boldsymbol{r}) &= \theta_{\boldsymbol{r}} F'(\boldsymbol{r},\boldsymbol{r}) + \sum_{2\le j\le K} \theta_{\boldsymbol{r}-\boldsymbol{e}_j} F'(\boldsymbol{r}-\boldsymbol{e}_j,\boldsymbol{r}) \\
&+ \sum_{2\le i\le K} \theta_{\boldsymbol{r}+\boldsymbol{e}_i} F'(\boldsymbol{r}+\boldsymbol{e}_i,\boldsymbol{r}) + \sum_{2\le i\le K}\sum_{2\le j\le K} \theta_{\boldsymbol{r}-\boldsymbol{e}_j+\boldsymbol{e}_i} F'(\boldsymbol{r}-\boldsymbol{e}_j+\boldsymbol{e}_i,\boldsymbol{r}) \\
&+ \theta_{\boldsymbol{r}}(y_1 - \lambda_1) - \sum_{2\le j\le K} \theta_{\boldsymbol{r}+\boldsymbol{e}_j}\mu_j P_{j1},
\end{aligned} \tag{15.17}$$

where

$$\begin{aligned}
F'(\boldsymbol{r},\boldsymbol{r}) &= F(\boldsymbol{r},\boldsymbol{r}) - y_1 \sum_{2\le j\le N} P_{1j}, \\
F'(\boldsymbol{r}-\boldsymbol{e}_j,\boldsymbol{r}) &= F(\boldsymbol{r}-\boldsymbol{e}_j,\boldsymbol{r}) + y_1 P_{1j}, \\
F'(\boldsymbol{r}+\boldsymbol{e}_i,\boldsymbol{r}) &= F(\boldsymbol{r}+\boldsymbol{e}_i,\boldsymbol{r}),
\end{aligned}$$

and

$$F'(\boldsymbol{r} - \boldsymbol{e}_j + \boldsymbol{e}_i, \boldsymbol{r}) = F(\boldsymbol{r} - \boldsymbol{e}_j + \boldsymbol{e}_i, \boldsymbol{r}).$$

The last two terms in (15.17) add up to zero by the traffic equation; therefore, the system reduces to

$$\boldsymbol{\theta}(F - \lambda_1 I + \eta A_2) = \boldsymbol{\theta} F'. \qquad (15.18)$$

Moreover, the matrix F' is the generator of a Jackson network obtained by removing the node 1 and replacing the arrival rates to the other nodes by

$$\lambda'_j = \lambda_j + y_1 P_{1j} \qquad \text{for } 2 \leq j \leq K.$$

By the induction assumption, the right-hand side of (15.18) will be equal to zero (and the theorem will be proved) if $(y_2, \ldots, y_K)$ is a solution of the balance equations for the reduced network, i.e., if

$$\begin{aligned} y_j &= \lambda'_j + \sum_{2 \leq i \leq K} y_i P_{ij} \\ &= \lambda_j + \sum_{1 \leq i \leq K} y_i P_{ij} \qquad \text{for } 2 \leq j \leq K, \end{aligned}$$

which holds by (15.14). □

15.3 The General Case

Ramaswami and Taylor [103, 102] analyze the general class of product form networks. Since the service rates at a node may be state-dependent, they need to consider nonhomogeneous QBDs, which we presented in Chapter 12 and which have generators of the form

$$Q = \begin{bmatrix} A_1^{(0)} & A_0^{(0)} & & \\ A_2^{(1)} & A_1^{(1)} & A_0^{(1)} & \\ & A_2^{(2)} & A_1^{(2)} & \ddots \\ & & \ddots & \ddots \end{bmatrix}.$$

As we saw in Theorem 12.1.1, the stationary distribution is built upon a sequence $\{R^{(n)} : n \geq 0\}$ of rate matrices, one for each level.

The first step consists in establishing the analogues of Theorems 15.1.1 and 15.1.3. It is proved in [103] that the phase and level are independent in steady state if and only if there exists a vector $\boldsymbol{\theta}$ and a sequence of scalars $\{\eta_n : n \geq 0\}$ such that

$$\begin{aligned} \boldsymbol{\theta}(A_1^{(0)} + \eta_0 A_2^{(1)}) &= \mathbf{0}, \\ \boldsymbol{\theta}(A_0^{(k)} + \eta_k A_1^{(k+1)} + \eta_k \eta_{k+1} A_2^{(k+2)}) &= \mathbf{0} \end{aligned}$$

for $k \geq 0$. In [102], the general version of Theorem 15.2.1 is proved. A key feature which is used is that the matrices $A_1^{(k)}$ satisfy equations like (15.8)–(15.11), where F is the generator of a product form network.

Chapter 16

Nondenumerable States

So far, we have assumed that both the level and the phase coordinates are denumerable. Our analysis has been facilitated by the skip-free property in levels. The value of this property extends to certain processes with nondenumerable state spaces also. For instance, Tweedie [115] develops an operator analogue of matrix analytic methods for the case when the phase set is a continuum, and Sengupta [109] extends the matrix-geometric to a matrix-exponential result when the level is a continuum. Both authors consider the case when the underlying process is skip-free only in the upward direction in levels.

A thorough treatment of either calls for a higher level of formalism than we have used thus far. We illustrate this below by providing a simplified account of Tweedie's results, indicating the mathematical nuances necessitated by a nondenumerable phase set. In the general spirit of this book, we limit ourselves to the case where the process is skip-free in both directions.

This is a very short chapter, as it concerns an area which has largely remained untouched until recently.

16.1 General Irreducible Markov Chains

Let us consider a Markov chain $\{X_n : n \geq 0\}$ on a general measure space E and briefly review the notions of irreducibility and stationary distribution. A complete treatment may be found in Nummelin [84] and Meyn and Tweedie [67]; see Breiman [9, Chapter 7] for a brief account.

If E is denumerable, then the situation is simple enough: the Markov chain is irreducible if and only if there exists, for every pair of states i and j, a finite sequence of possible transitions which lead from i to j.

If E is not countable, then one proceeds with more caution. First, one associates to E a σ-algebra $\mathcal{E}$ of subsets of E. The transition kernel P with $P(x, S) = \mathrm{P}[X_1 \in S | X_0 = x]$ is defined such that

1. for any fixed $S \in \mathcal{E}$, the function $P(\cdot, S)$ is measurable,

2. for any fixed $x \in E$, the function $P(x, \cdot)$ is a measure on $(E, \mathcal{E})$.

Then, if we define $P^0 = I$ and

$$P^n(x, S) = \int_E P(x, dy) P^{n-1}(y, S) \qquad \text{for } n \geq 1,$$

we have that $P^n(x, S) = \mathrm{P}[X_n \in S | X_0 = x]$.

The set S is attainable from the state x, and we write $x \to S$ if and only if $P^n(x, S) > 0$ for some $n \geq 1$, or equivalently if and only if

$$\mathrm{P}[X_n \in S \text{ for some } n \geq 1 | X_0 = x] > 0.$$

Then, one defines an irreducible Markov chain as follows.

Definition 16.1.1 *A Markov chain on the general set E is irreducible if there exists a measure $\varphi(\cdot)$ on $(E, \mathcal{E})$ such that for all subsets $S \in \mathcal{E}$ with $\varphi(S) > 0$, and for all states $x \in E$, one has that $x \to S$.*

Observe that if a Markov chain is irreducible, then there is a closed set of states, i.e., a set $F \in \mathcal{E}$ such that $P(x, F^c) = 0$ for all x in F; moreover, this set is unique up to a point. More precisely, there cannot be two such disjoint sets. The definition makes it possible to have non-empty subsets Y of ephemeral states, from which F is reachable and which must have a measure $\varphi(Y)$ equal to zero.

The following proposition (Nummelin [84, Proposition 2.4]) states that if a Markov chain is irreducible, then there is a characteristic measure associated to it which identifies all the subsets which are attainable from every state.

Proposition 16.1.2 *If a Markov chain is irreducible, then there exists a maximal irreducibility measure* Φ, *i.e., a measure such that, for any measure* φ *for which Definition* 16.1.1 *applies,* $\varphi(S) > 0$ *implies* $\Phi(S) > 0$. *One writes that the Markov chain is* Φ*-irreducible.* □

Next, we consider invariant distributions for recurrent Markov chains. Recurrence is defined in terms of the potential kernel $K = \sum_{n\geq 0} P^n$. Roughly stated, an irreducible Markov chain with transition kernel P is recurrent if the series $\sum_{n\geq 0} P^n$ diverges (otherwise it is transient). If, in addition, the Markov chain has the property that for every $F \in \mathcal{E}$ with $\Phi(F) > 0$, the chain visits F infinitely often with probability 1, then the Markov chain is said to be Harris recurrent. For a more detailed account, see [84, Section 3.2] or [67, Chapters 8 and 9].

A measure π on $(E, \mathcal{E})$ is invariant for the transition kernel P if $\pi = \pi P$, i.e., if

$$\pi(S) = \int_E \pi(dy)P(y, S)$$

for all $S \in \mathcal{E}$.

The following proposition (see [67, Theorems 10.4.4 and 10.4.9]) states that irreducible recurrent Markov chains have an invariant measure which assigns a strictly positive mass to those sets which are attainable from any state x and those sets only.

Theorem 16.1.3 *Consider a* Φ*-irreducible, recurrent Markov chain. There exists a unique invariant measure* π *which is equivalent to* Φ, *i.e., for any set* $S \in \mathcal{E}$, $\pi(S) > 0$ *if and only if* $\Phi(S) > 0$. *If* $\pi(E)$ *is finite, then the Markov chain is positive recurrent and one may normalize* π *so that* $\pi(E) = 1$; *otherwise, the Markov chain is null-recurrent.* □

Clearly, the invariant measure is unique up to a multiplicative constant. For the positive recurrent case, that constant is fixed by the normalizing equation $\pi(E) = 1$.

16.2 The Operator-Geometric Property

Let us now consider a Markov process on the state space $\{(i, x) : i \in \mathbb{N}, x \in L\}$, where L is a general set such as an interval of the real line, for instance. We let $\mathcal{L}$ denote a σ-algebra on L. The level $\ell(i)$ is the set $i \times L$. A homogeneous QBD has a transition kernel with the structure 6.1, where

$$\begin{aligned} P(i,x;j,S) &= \mathrm{P}[X_1 \in j \times S | X_0 = (i,x)] \\ &= \begin{cases} A_0(x,S) & \text{for } i \geq 0, j = i+1, \\ A_1(x,S) & \text{for } i \geq 1, j = i, \\ A_2(x,S) & \text{for } i \geq 1, j = i-1, \\ B(x,S) & \text{for } i = 0, j = i, \end{cases} \end{aligned}$$

and is equal to zero otherwise. The functions A_0, A_1, A_2, and B are measures on $\mathcal{L}$ for x fixed and measurable functions on $(L, \mathcal{L})$ for S fixed. Moreover, the kernel P is stochastic, so that

$$A_0(x,L) + A_1(x,L) + A_2(x,L) = 1$$

and

$$B(x,L) + A_0(x,L) = 1$$

for all x in L.

The analysis in Tweedie [115] may be organized along the same lines as in Chapter 6. The proof needs to be adapted to the general case. Thus we shall only give the salient points, so that the reader may appreciate how unified the whole treatment of QBDs is.

First, we define the taboo probabilities of the QBD, where the level $\ell(0)$ is taboo:

$$\begin{aligned} {}_0P^1(i,x;j,S) &= P(i,x;j,S), \\ {}_0P^n(i,x;j,S) &= \sum_k \int_L {}_0P^{n-1}(i,x;k,dy) P(k,y;j,S) \end{aligned}$$

for $n \geq 2$, and the kernels

$$G(x,S) = \sum_{n \geq 1} {}_0P^n(1,x;0,S)$$

and

$$N(x,S) = \sum_{n \geq 1} {}_0P^n(1,x;1,S),$$

which have the obvious meaning $G(x, S)$ is the probability that, starting from $(1, x)$, the QBD visits the level $\ell(0)$ in a finite amount of time and does so by visiting the set $0 \times S$; $N(x, S)$ is the expected number of visits (possibly infinite) to the set $1 \times S$, starting from $(1, x)$, before the first visit to the level 0, etc.

We make the assumption that the QBD is Φ-irreducible and that $\Phi(\ell(0)) > 0$. This ensures that for every i in $\mathbb{N}$ and x in L, the level $\ell(0)$ can be reached eventually with positive probability from (i, x) (see Definition 16.1.1). In particular, we have that $G(x, L) > 0$ for all x.

The chain restricted to $\ell(0)$ has the transition kernel $P^{\{0\}}$ with

$$\begin{aligned} P^{\{0\}}(x, S) &= \sum_{n \geq 1} {}_0P^n(0, x; 0, S) \\ &= B(x, S) + \int_L A_0(x, dy) G(y, S). \end{aligned}$$

If the QBD is recurrent, then $G(y, L) = 1$ for Φ-almost every y, the kernel $P^{\{0\}}$ is stochastic, and we have the following result (see [115, Theorems 1 and 2]), which may be compared to Theorem 6.2.1 and Lemma 6.3.1.

Theorem 16.2.1 *If the QBD has a stationary probability measure, it is such that*

$$\pi(n, S) = \int_L \pi(0, dy) R^n(y, S)$$

for all $n \geq 0$ and $S \in \mathcal{L}$, where $\pi(0, \cdot)$ is a stationary measure for the kernel $P^{\{0\}}$:

$$\pi(0, S) = \int_L \pi(0, dy) P^{\{0\}}(y, S),$$

normalized so that $\pi(\mathbb{N}, L) = 1$ and where

$$R(x, S) = \int_L A_0(x, dy) N(y, S)$$

is the expected number of visits to $1 \times S$, starting from $(0, x)$, before the first return to the level 0.

The kernel R satisfies the nonlinear operator equation

$$R(x, S) = A_0(x, S) + \int_L R(x, dy) A_1(y, S) + \int_L R^2(x, dy) A_2(y, S).$$

□

In order to determine the conditions under which the QBD is positive recurrent, one defines the stochastic kernel $A = A_0 + A_1 + A_2$ and looks for properties like those in Theorem 7.2.3.

Tweedie makes the additional assumption that $B = A_2 + A_1$. With this, one clearly has that

$$P^n(i, x; \mathbb{N}, S) = A^n(x, S)$$

for all i, x, and S; in other words, the marginal phase process for the QBD has the same law as the Markov chain with kernel A. One then proves that if the QBD is Φ-irreducible, then the kernel A is ψ-irreducible with $\psi(S) = \Phi(\mathbb{N}, S), S \in \mathcal{L}$.

Under these assumptions, the following *sufficient* condition holds (see [115, Theorem 5]).

Theorem 16.2.2 *If the QBD is Φ-irreducible and if the kernel A has a stationary measure α such that*

$$\int_L \alpha(dy)A_2(y, L) > \int_L \alpha(dy)A_0(y, L), \tag{16.1}$$

then the QBD has a stationary probability measure. □

This is both stronger and weaker than Theorem 7.2.3. The new result is stronger because it applies to a general phase set; it is weaker because it assumes that $B = A_2 + A_1$, which is a very restrictive assumption; second, it does not state that the drift condition is necessary. As a matter of fact, Tweedie only shows (see [115, Theorem 6]) that if the QBD has an invariant measure π such that $\sum_{j\geq 0} j\pi(j, L)$ converges, then the mean drift condition is necessary.

16.3 Computational Issues

Clearly, we may develop for the kernel R iterative sequences similar to (8.19), (8.20) with the matrix product there replaced by composition integrals. Thus, one shows that the sequence

$$\begin{aligned} R_0(x,S) &= A_0(x,S), \\ R_{n+1}(x,S) &= A_0(x,S) + \int_L R_n(x,dy)A_1(y,S) \\ &\quad + \int_L R_n^2(x,dy)A_2(y,S) \qquad \text{for } n \geq 0 \end{aligned}$$

monotonically converges to R. The numerical evaluation of the R_n's, however, is arduous and offers a challenge in error control. Thus, it is not surprising, perhaps, that little has been reported along algorithmic lines for these models.

Recently, Nielsen and Ramaswami [83] have shown that analytical simplifications occur when the kernels are in a function space with an orthonormal basis. Denoting by $\{\psi_i(\cdot) : i \geq 1\}$ a set of base functions satisfying

$$\int_L \psi_i(x)\psi_j(x)dx = \delta_{ij},$$

where δ_{ij} is equal to 1 if $i = j$ and is equal to 0 otherwise, they assume that the kernels A_k for $k = 0, 1,$ and 2 have densities a_k which can be expressed in the form

$$a_k(x,y) = \sum_{i\geq 1}\sum_{j\geq 1}[\alpha_k]_{ij}\psi_i(x)\psi_j(y),$$

where the series in the right-hand side are absolutely convergent for all x and y.

Then, the kernels R and R_n, $n \geq 0$, respectively, have densities expressed as

$$\begin{aligned} r(x,y) &= \sum_{i\geq 1}\sum_{j\geq 1}\rho_{ij}\psi_i(x)\psi_j(y), \\ r_n(x,y) &= \sum_{i\geq 1}\sum_{j\geq 1}[\rho_n]_{ij}\psi_i(x)\psi_j(y) \qquad \text{for } n \geq 0. \end{aligned}$$

The matrices ρ_n, $n \geq 0$, satisfy the recurrence equations

$$\begin{aligned} \rho_0 &= \alpha_0, \\ \rho_{n+1} &= \alpha_0 + \rho_n\alpha_0 + \rho_n^2\alpha_0 \qquad \text{for } n \geq 0 \end{aligned}$$

and converge to ρ as n tends to infinity.

Thus, it appears possible to work with matrix multiplications only for sufficiently smooth kernels A_k, $k = 0, 1, 2$, and to avoid numerical integration altogether. We refer the reader to [83] for a detailed discussion.

Bibliography

[1] N. Akar and K. Sohraby. An invariant subspace approach in M/G/1 and G/M/1 type Markov chains. *Comm. Statist. Stochastic Models*, 13:381–416, 1997.

[2] S. C. Albright. Solving equations of the two-dimensional birth-death type. Technical report, Graduate School of Business, Indiana University, Bloomington, IN, 1981.

[3] S. Asmussen. *Applied Probability and Queues.* John Wiley, New York, 1987.

[4] S. Asmussen, O. Nerman, and M. Olsson. Fitting phase-type distributions via the EM algorithm. *Scand. J. Statist.*, 23:419–441, 1996.

[5] S. Asmussen and V. Ramaswami. Probabilistic interpretations of some duality results for the matrix paradigms in queueing theory. *Comm. Statist. Stochastic Models*, 6:715–733, 1990.

[6] D. Bini and B. Meini. On cyclic reduction applied to a class of Toeplitz-like matrices arising in queueing problems. In W. J. Stewart, editor, *Computations with Markov Chains*, pages 21–38. Kluwer Academic Publishers, Boston, MA, 1995.

[7] D. A. Bini and B. Meini. Improved cyclic reduction for solving queueing problems. *Numer. Algorithms*, 5:57–74, 1997.

[8] C. Blondia and O. Casals. Statistical multiplexing of VBR sources: A matrix analytic approach. *Performance Evaluation*, 16:5–20, 1992.

[9] L. Breiman. *Probability.* SIAM, Philadelphia, PA, 1992.

[10] L. W. Bright and P. G. Taylor. Calculating the equilibrium distribution in level dependent quasi-birth-and-death processes. *Comm. Statist. Stochastic Models*, 11:497–525, 1995.

[11] J. L. Carroll, A. Van der Liefvoort, and L. Lipsky. Solutions of M/G/1/N-type loops with extensions to M/G/1 and GI/M/1 queues. *Oper. Res.*, 30:490–514, 1982.

[12] E. Çinlar. *Introduction to Stochastic Processes.* Prentice–Hall, Englewood Cliffs, NJ, 1975.

[13] Y. Chandramouli, M. F. Neuts, and V. Ramaswami. A queueing model for meteor burst packet communication systems. *IEEE Trans. Communications*, 37:1024–1030, 1989.

[14] K. L. Chung. *Markov Chains with Stationary Transition Probabilities.* Springer-Verlag, Berlin, second edition, 1967.

[15] K. L. Chung. *A Course in Probability Theory.* Harcourt, Brace and World, New York, 1968.

[16] J. W. Cohen. *The Single Server Queue.* North–Holland, Amsterdam, 1969.

[17] R. V. Evans. Geometric distribution in some two-dimensional queueing systems. *Oper. Res.*, 15:830–846, 1967.

[18] D. K. Faddeev and V. N. Faddeeva. *Computational Methods of Linear Algebra.* W. H. Freeman, San Francisco, CA, 1963.

[19] P. Favati and B. Meini. Relaxed functional iteration techniques for the numerical solution of M/G/1 type Markov chains. *BIT*, 38:510–526, 1998.

[20] W. Feller. *An Introduction to Probability Theory and Its Applications. Volume* I. John Wiley, New York, second edition, 1964.

[21] W. Feller. *An Introduction to Probability Theory and Its Applications. Volume* II. John Wiley, New York, 1966.

[22] D. Freedman. *Approximating Countable Markov Chains.* Holden–Day, San Francisco, CA, 1972.

[23] H. R. Gail, S. L. Hantler, and B. A. Taylor. Spectral analysis of M/G/1 and GI/M/1 type Markov chains. *Adv. in Appl. Probab.*, 28:114–165, 1996.

[24] H. R. Gail, S. L. Hantler, and B. A. Taylor. Non-skip-free M/G/1 and G/M/1 type Markov chains. *Adv. in Appl. Probab.*, 29:733–758, 1997.

[25] F. Gantmacher. *The Theory of Matrices*, volume 2. Chelsea, New York, 1974.

[26] D. P. Gaver, P. A. Jacobs, and G. Latouche. Finite birth-and-death models in randomly changing environments. *Adv. in Appl. Probab.*, 16:715–731, 1984.

[27] S. Gibson and E. Seneta. Augmented truncation of infinite stochastic matrices. *J. Appl. Probab.*, 24:600–608, 1987.

[28] A. Graham. *Kronecker Products and Matrix Calculus: With Applications.* John Wiley, New York, 1981.

[29] W. K. Grassmann and D. P. Heyman. Computation of steady-state probabilities for infinite Markov chains with repeating rows. *ORSA J. Comput.*, 5:292–303, 1993.

[30] W. K. Grassmann, M. J. Taksar, and D. P. Heyman. Regenerative analysis and steady state distributions for Markov chains. *Oper. Res.*, 33:1107–1116, 1985.

[31] L. Gün and A. M. Makowski. Matrix-geometric solution for finite capacity queues with phase-type distributions. In P.-J. Courtois and G. Latouche, editors, *PERFORMANCE* '87, pages 269–282. North–Holland, Amsterdam, 1987.

[32] B. Hajek. Birth-and-death processes on the integers with phases and general boundaries. *J. Appl. Probab.*, 19:488–499, 1982.

[33] E. Isaacson and H. B. Keller. *Analysis of Numerical Methods.* John Wiley, New York, 1966.

[34] A. Jensen. Markoff chains as an aid in the study of Markoff processes. *Skand. Aktuarietidskr*, 36:87–91, 1953.

[35] E. P. Kao. Using state reduction for computing steady state probabilities of queues of GI/PH/1 types. *ORSA J. Comput.*, 3:231–240, 1991.

[36] S. Karlin and H. M. Taylor. *A First Course in Stochastic Processes.* Academic Press, New York, second edition, 1975.

[37] S. Karlin and H. M. Taylor. *A Second Course in Stochastic Processes.* Academic Press, New York, 1981.

[38] F. P. Kelly. *Reversibility and Stochastic Networks.* John Wiley, New York, 1979.

[39] G. J. Kemeny and J. L. Snell. *Finite Markov Chains.* Van Nostrand, Princeton, NJ, 1960.

[40] G. J. Kemeny, J. L. Snell, and A. W. Knapp. *Denumerable Markov Chains.* Springer-Verlag, New York, second edition, 1976.

[41] D. E. Knuth. *The Art of Computer Programming: Seminumerical Algorithms*, volume 2. Addison–Wesley, Reading, MA, 1969.

[42] G. Latouche. Algorithmic analysis of a multiprogramming multiprocessor computer system. *J. Assoc. Comput. Mach.*, 28:662–679, 1981.

[43] G. Latouche. A phase-type semi-Markov point process. *SIAM J. Algebraic Discrete Methods*, 3:77–90, 1982.

[44] G. Latouche. An exponential semi-Markov process, with applications to queueing theory. *Comm. Statist. Stochastic Models*, 1:137–169, 1985.

[45] G. Latouche. A note on two matrices occurring in the solution of quasi-birth-and-death processes. *Comm. Statist. Stochastic Models*, 3:251–257, 1987.

[46] G. Latouche. Distributions de type phase—tutorial. *Cahiers Centre Études Rech. Opér.*, 31:3–11, 1989.

[47] G. Latouche. A simple proof for the matrix-geometric theorem. *Appl. Stochastic Models Data Anal.*, 8:25–29, 1992.

[48] G. Latouche. Algorithms for infinite Markov chains with repeating columns. In C. D. Meyer and R. J. Plemmons, editors, *Linear Algebra, Markov Chains and Queueing Models*, pages 231–265. Springer-Verlag, New York, 1993.

[49] G. Latouche. Newton's iteration for nonlinear equations in Markov chains. *IMA J. Numer. Anal.*, 14:583–598, 1994.

[50] G. Latouche and M. F. Neuts. Efficient algorithmic solutions to exponential tandem queues with blocking. *SIAM J. Algebraic Discrete Methods*, 1:93–106, 1980.

[51] G. Latouche, C. Pearce, and P. Taylor. Invariant measures for quasi-birth-and-death processes. *Comm. Statist. Stochastic Models*, 14:443–460, 1998.

[52] G. Latouche and V. Ramaswami. A unified stochastic model for the arrival of packets from periodic sources. *Performance Evaluation*, 14:103–121, 1992.

[53] G. Latouche and V. Ramaswami. A logarithmic reduction algorithm for quasi-birth-and-death processes. *J. Appl. Probab.*, 30:650–674, 1993.

[54] G. Latouche and V. Ramaswami. Expected passage times in homogeneous quasi-birth-and-death processes. *Comm. Statist. Stochastic Models*, 11:103–122, 1995.

[55] G. Latouche and V. Ramaswami. The PH/PH/1 queue at epochs of queue size change. *Queueing Systems Theory Appl.*, 25:97–114, 1997.

[56] G. Latouche and P. G. Taylor. Level-phase independence in QBDs and related processes. Technical report, Université Libre de Bruxelles, Brussels, Belgium, 1997.

[57] S. S. Lavenberg. *Computer Performance Modeling Handbook.* Academic Press, New York, 1983.

[58] J.-Y. Le Boudec. An efficient solution method for Markov models of ATM links with loss priorities. *IEEE J. Select. Areas Commun.*, 9:408–417, 1991.

[59] S.-Q. Li and H.-D. Sheng. Generalized folding algorithm for transient analysis of finite QBD processes and its queueing applications. In W. J. Stewart, editor, *Computations with Markov Chains*, pages 463–481. Kluwer Academic Publishers, Boston, MA, 1995.

[60] D. M. Lucantoni. *An Algorithmic Analysis of a Communication Model with Retransmission of Flawed Messages.* Res. Notes in Math. 81. Pitman Books Limited, London, 1983.

[61] D. M. Lucantoni, K. S. Meier-Hellstern, and M. F. Neuts. A single-server queue with server vacations and a class of nonrenewal arrival processes. *Adv. in Appl. Probab.*, 22:676–705, 1990.

[62] D. M. Lucantoni and V. Ramaswami. Efficient algorithms for solving the non-linear matrix equations arising in phase type queues. *Comm. Statist. Stochastic Models*, 1:29–52, 1985.

[63] R. S. Maier. The algebraic construction of phase-type distributions. *Comm. Statist. Stochastic Models*, 7:573–602, 1991.

[64] R. S. Maier and C. A. O'Cinneide. A closure characterisation of phase-type distributions. *J. Appl. Probab.*, 29:92–103, 1992.

[65] R. A. Marie and J. M. Pellaumail. Steady state probabilities for a queue with a general service time distribution and state dependent arrivals. *IEEE Trans. Software Engineering*, 9:109–113, 1979.

[66] B. Meini. New convergence results on functional iteration techniques for the numerical solution of M/G/1 type Markov chains. *Numer. Math.*, 78:39–58, 1997.

[67] S. P. Meyn and R. L. Tweedie. *Markov Chains and Stochastic Stability.* Springer-Verlag, London, 1993.

[68] D. R. Miller. Steady-state algorithmic analysis of $m/m/c$ two-priority queues with heterogeneous rates. In *Applied Probability–Computer Science: The Interface. Vol.* II. *Proceedings of the Meeting Held at Florida Atlantic University, Boca Raton, Florida, January* 5–7, 1981, pages 207–225. Birkhäuser Boston, Cambridge, MA, 1982.

[69] D. Mitra and I. Mitrani. Analysis of a Kanban discipline for cell coordination in production lines, II: Stochastic demand. *Oper. Res.*, 39:807–823, 1991.

[70] I. Mitrani and D. Mitra. A spectral expansion method for random walks on semi-infinite strips. In R. Beauwens and P. de Groen, editors, *Iterative Methods in Linear Algebra*, pages 141–149. North–Holland, Amsterdam, 1992.

[71] C. Moler and C. Van Loan. Nineteen dubious ways to compute the exponential of a matrix. *SIAM Rev.*, 20:801–836, 1978.

[72] V. Naoumov, U. R. Krieger, and D. Wagner. Analysis of a multiserver delay-loss system with a general Markovian arrival process. In S. R. Chakravarthy and A. S. Alfa, editors, *Matrix-Analytic Methods in Stochastic Models*, pages 43–66. Marcel Dekker, New York, 1997.

[73] M. F. Neuts. Computer power or the liberation of applied probability. Technical Report 312, Department of Statistics, Purdue University, West Lafayette, IN, 1973.

[74] M. F. Neuts. Probability distributions of phase type. In *Liber Amicorum Prof. Emeritus H. Florin*, pages 173–206. University of Louvain, Belgium, 1975.

[75] M. F. Neuts. Moment formulas for the Markov renewal branching process. *Adv. in Appl. Probab.*, 8:690–711, 1976.

[76] M. F. Neuts. Markov chains with applications in queueing theory, which have a matrix-geometric invariant vector. *Adv. in Appl. Probab.*, 10:185–212, 1978.

[77] M. F. Neuts. Renewal processes of phase type. *Naval Res. Logist.*, 25:445–454, 1978.

[78] M. F. Neuts. A versatile Markovian point process. *J. Appl. Probab.*, 16:764–779, 1979.

[79] M. F. Neuts. *Matrix-Geometric Solutions in Stochastic Models. An Algorithmic Approach.* The Johns Hopkins University Press, Baltimore, MD, 1981.

[80] M. F. Neuts. Explicit steady-state solutions to some elementary queueing models. *Oper. Res.*, 30:480–489, 1982.

[81] M. F. Neuts. The caudal characteristic curve of queues. *Adv. in Appl. Probab.*, 18:221–254, 1986.

[82] M. F. Neuts. *Structured Stochastic Matrices of M/G/1 Type and Their Applications.* Marcel Dekker, New York, 1989.

[83] B. F. Nielsen and V. Ramaswami. A computational framework for a quasi birth and death process with a continuous phase variable. In V. Ramaswami and P. Wirth, editors, *Teletraffic Contributions for the Information Age. Proceedings of the 15th International Teletraffic Congress - ITC 15*, pages 477–486. Elsevier, North–Holland, Amsterdam, 1997.

[84] E. Nummelin. *General Irreducible Markov Chains and Non-negative Operators.* Cambridge University Press, Cambridge, 1984.

[85] C. A. O'Cinneide. Characterisation of phase type distributions. *Comm. Statist. Stochastic Models*, 6:1–57, 1990.

[86] C. A. O'Cinneide. Phase type distributions and invariant polytopes. *Adv. in Appl. Probab.*, 23:515–535, 1991.

[87] C. A. O'Cinneide. Relative-error bounds for the LU decomposition via the GTH algorithm. *Numer. Math.*, 73:507–519, 1996.

[88] D. P. O'Leary and Y.-J. J. Wu. A block-GTH algorithm for finding the stationary vector of a Markov chain. *SIAM J. Matrix Anal. Appl.*, 17:470–488, 1996.

[89] J. M. Ortega and W. C. Rheinboldt. *Iterative Solution of Nonlinear Equations in Several Variables.* Academic Press, New York, 1970.

[90] A. Ralston and P. Rabinowitz. *A First Course in Numerical Analysis.* McGraw–Hill, New York, 1978.

[91] G. Rama Murthy, M. Kim, and E. J. Coyle. Equilibrium analysis of skip free Markov chains: Nonlinear matrix equations. *Comm. Statist. Stochastic Models*, 7:547–571, 1991.

[92] V. Ramaswami. The N/G/1 queue and its detailed analysis. *Adv. in Appl. Probab.*, 12:222–261, 1980.

[93] V. Ramaswami. Nonlinear matrix equations in applied probability—Solution techniques and open problems. *SIAM Rev.*, 30:256–263, 1988.

[94] V. Ramaswami. Stable recursion for the steady state vector for Markov chains of M/G/1 type. *Comm. Statist. Stochastic Models*, 4:183–188, 1988.

[95] V. Ramaswami. A duality theorem for the matrix paradigms in queueing theory. *Comm. Statist. Stochastic Models*, 6:151–161, 1990.

[96] V. Ramaswami. Matrix analytic methods: A tutorial overview with some extensions and new results. In S. R. Chakravarthy and A. S. Alfa, editors, *Matrix-Analytic Methods in Stochastic Models*, pages 261–295. Marcel Dekker, New York, 1997.

[97] V. Ramaswami. The generality of quasi birth-and-death processes. In A. S. Alfa and S. R. Chakravarthy, editors, *Advances in Matrix Analytic Methods for Stochastic Models*, pages 93–113. Notable Publications Inc., Neshanic Station, NJ, 1998.

[98] V. Ramaswami and G. Latouche. A general class of Markov processes with explicit matrix-geometric solutions. *OR Spektrum*, 8:209–218, 1986.

[99] V. Ramaswami and D. M. Lucantoni. Algorithmic analysis of a dynamic priority queue. In *Applied Probability–Computer Science: The Interface. Vol.* II. *Proceedings of the Meeting Held at Florida Atlantic University, Boca Raton, Florida, January* 5–7, 1981, pages 157–204. Birkhäuser Boston, Cambridge, MA, 1982.

[100] V. Ramaswami and D. M. Lucantoni. Stationary waiting time distribution in queues with phase type service and in quasi-birth-and-death processes. *Comm. Statist. Stochastic Models*, 1:125–136, 1985.

[101] V. Ramaswami and M. F. Neuts. Some explicit formulas and computational methods for infinite server queues with phase type arrivals. *J. Appl. Probab.*, 17:498–514, 1980.

[102] V. Ramaswami and P. G. Taylor. An operator-analytic approach to product-form networks. *Comm. Statist. Stochastic Models*, 12:121–142, 1996.

[103] V. Ramaswami and P. G. Taylor. Some properties of the rate operators in level dependent quasi-birth-and-death processes with a countable number of phases. *Comm. Statist. Stochastic Models*, 12:143–164, 1996.

[104] V. Ramaswami and J. Wang. Analysis of the link error monitoring protocols in the common channel signaling network. *IEEE/ACM Trans. Networking*, 1:31–47, 1993.

[105] M.-A. Remiche. Time to congestion in homogeneous quasi-birth-and-death processes. *Opsearch*, 35, 1998, to appear.

[106] S. I. Resnick. *Adventures in Stochastic Processes.* Birkhäuser Boston, Cambridge, MA, 1992.

[107] S. M. Ross. *Stochastic Processes.* Wiley, New York, second edition, 1996.

[108] E. Seneta. *Non-Negative Matrices and Markov Chains.* Springer-Verlag, New York, second edition, 1981.

[109] B. Sengupta. Markov processes whose steady-state distribution is matrix-exponential with an application to the GI/G/1 queue. *Adv. in Appl. Probab.*, 21:159–180, 1989.

[110] B. Sengupta. Phase-type representations for matrix-geometric solutions. *Comm. Statist. Stochastic Models*, 6:163–167, 1990.

[111] G. W. Stewart. Gaussian elimination, perturbation theory, and Markov chains. In C. D. Meyer and R. J. Plemmons, editors, *Linear Algebra, Markov Chains and Queueing Models*, pages 59–69. Springer-Verlag, New York, 1993.

[112] W. J. Stewart. On the use of numerical methods for ATM models. In H. Perros, G. Pujolle, and Takahashi, editors, *Modelling and Performance Evaluation of ATM Technology*, pages 375–396. North–Holland, Amsterdam, 1993.

[113] W. J. Stewart. *Introduction to the Numerical Solution of Markov Chains.* Princeton University Press, Princeton, NJ, 1994.

[114] T. Takine, B. Sengupta, and R. W. Yeung. A generalization of the matrix M/G/1 paradigm for Markov chains with a tree structure. *Comm. Statist. Stochastic Models*, 11:411–421, 1995.

[115] R. L. Tweedie. Operator-geometric stationary distributions for Markov chains, with applications to queueing models. *Adv. in Appl. Probab.*, 14:368–391, 1982.

[116] V. Wallace. *The Solution of Quasi Birth and Death Processes Arising from Multiple Access Computer Systems.* Ph.D. thesis, Systems Engineering Laboratory, University of Michigan, Ann Arbor, MI, 1969.

[117] B. Wong, W. Giffin, and R. L. Disney. Two finite M/M/1 queues in tandem: A matrix solution for the steady state. *Opsearch*, 14:1–18, 1977.

[118] J. Ye and S. Q. Li. Folding Algorithm: A computational method for finite QBD processes with level dependent transitions. *IEEE Trans. Commun.*, 42:625–639, 1994.

[119] R. W. Yeung and B. Sengupta. Matrix product-form solutions for Markov chains with a tree structure. *Adv. in Appl. Probab.*, 26:965–987, 1994.

Index